AF323949

DYNAMICS OF MECHATRONICS SYSTEMS

Modeling, Simulation, Control, Optimization and
Experimental Investigations

DYNAMICS OF MECHATRONICS SYSTEMS

Modeling, Simulation, Control, Optimization and Experimental Investigations

Jan Awrejcewicz

Donat Lewandowski

Paweł Olejnik

Lodz University of Technology, Poland

World Scientific

NEW JERSEY · LONDON · SINGAPORE · BEIJING · SHANGHAI · HONG KONG · TAIPEI · CHENNAI · TOKYO

Published by

World Scientific Publishing Co. Pte. Ltd.

5 Toh Tuck Link, Singapore 596224

USA office: 27 Warren Street, Suite 401-402, Hackensack, NJ 07601

UK office: 57 Shelton Street, Covent Garden, London WC2H 9HE

British Library Cataloguing-in-Publication Data
A catalogue record for this book is available from the British Library.

DYNAMICS OF MECHATRONICS SYSTEMS
Modeling, Simulation, Control, Optimization and Experimental Investigations

Copyright © 2017 by World Scientific Publishing Co. Pte. Ltd.

ISBN 978-981-3146-54-9

Printed in Singapore

Preface

The progress of civilization stimulates the development of interdisciplinary science. On this basis, a modern branch of science, mechatronics, including perfectly mastered mechanics, electronics, and later developed parallel knowledge of numerical simulation, control and optimization is rapidly developing. The latter has assigned the mechanics of dynamical systems and electronics a completely new role. They have become the basis of mechatronic projects having a strong cognitive and practical meaning for engineering science. For this reason, a mechatronics, automation or electronics engineer should gain useful knowledge and acquire design skills that will enable him to meet technical challenges of the XXI century.

The monograph is directed primarily towards practicing engineers willing to support their academic knowledge with hard competences related to the ability of solving various problems of mechatronic systems extended on the essential physical and numerical modeling as well as control and optimization of dynamical systems.

We have sectioned the monograph into thirteen chapters introducing the reader to the physical modeling of magnetic, electromagnetic and piezoelectric phenomena, Maxwell's equations and atom modeling, mechanical fluid systems, electrohydraulic servomechanisms, shock response and its control, among others. Important sections are devoted to selected methods of optimization, numerical algorithms of fuzzy logic, tracking and conventional control of mechatronic systems.

An introduction to mechatronics is given in Chap. 1.

Importance of systems modeling is presented in Chap. 2. The usefulness of the theoretical, real, physical, mathematical or simulation models has been considered in this chapter on the basis of modeling in dimensional analysis and criteria of similarity. In scope of engineering, sometimes the sought analogy considered in the chapter is very important, since it denotes a similarity between different objects, phenomena or processes occurring under some conditions.

Magnetic fields affect human bodies and nature overall. They influence, even unintentionally, any technical object, and therefore, a mechatronics engineer should acquire knowledge about properties of such fields to predict the effects of their presence or a direct application. Chapter 3 pays particular attention to magnetic and electromagnetic phenomena. A wide introduction describes electric charge, induced electric fields, magnetic and electromagnetic phenomena, capacitance, resistance and laws of electrics.

Piezoelectric materials play a key role in mechatronics. Chapter 4 provides the reader with modeling of piezoelectric phenomena describing properties and laws governing piezoelectric materials. One-dimensional rod polarized along its axis has served as an exemplary actuator of which dynamics, stress-strain linear dependencies, Maxwell's equations and Hamilton's principle for mechanical continua have been analyzed.

Chapter 5 undertakes the investigation of the modeling of a mechanical-fluid system. Among others, the following aspects have been considered and numerically validated: (i) the balance of fluid flow in the case of a particular pneumomechanical system and rotary hydraulic motors; (ii) modeling of open hydromechanical system of linear displacement; (iii) modeling of a hydraulic electromechanical servomechanism of rotational motion; (iv) a proportional valve in the drive and control of a hydromechanical system; (v) physical and mathematical model of a pneumatic hydromechanical system. Each of the enumerated problems has been parametrically defined and solved by the simulation diagrams performed in Scilab.

Preceded by a simplified model of a servomechanism with proportional valve, a torque motor, a piezoelectric transducer, and a control system of load positioning are shown in Chap. 6. The positioning was realized by means of a hydraulic servo in a simulation diagram performed in LabVIEW. The numerical simulations of dynamics of an electrohydraulic servo subjected to the dynamic loading have proved that improvement of linear positioning of heavy loads carried by system's actuator can be achieved by using linearly increasing characteristics of reference current. The advantages and drawbacks of modeling of these kinds of mechatronic systems have been presented, as well as some basic types of servos were virtualized to point out areas of their applications.

A simple Newtonian and wave model of electron in quantum mechanics is shown in Chap. 7. This part of the monograph continues with an interesting modeling of an atom in a magnetic field versus a free atom, electron orbital perturbation by a moving proton, and the planar dynamics of a particle in a magnetic field, finishing at the three-dimensional dynamics of a charge.

Complete analysis of the electromagnetic phenomena, including the static and induced electric fields as well as the magnetic effects, can be carried out by the compact set of Maxwell's equations. The equations presented in Chap. 8 contribute to other important sets of equations regarding Newton's laws of classical mechanics and the three laws of thermodynamics.

Chapter 9 contains selected optimization problems being solved with the use of interesting experimental and mathematical methods. Some algorithms and simulation diagrams solving a few optimization problems are presented, i.e.: (i) a Scilab routine of polynomial fitting to a series of experimental data; (ii) linear and nonlinear programming aimed at estimation of permissible domains and optima of the maximized functions; (*iii*) dynamic programming while finding an optimal time characteristics of the displacement and the value of a corrective force which, independently of the placement of some concentrated force, eliminates any deformation of the analyzed cantilever beam; (*iv*) geometric programming while estimating dimensions of unfolded surface of a box having the maximal volume at a given boundary surface; (*v*) optimization of stiffness of a spindle system.

Chapter 10 delivers a comparative analysis of two qualitatively different approaches used for angular velocity control of a DC motor subjected to chaotic disturbances coming from a gear with a transmission belt carrying a vibrating load. The purpose was to achieve an accurate control of the speed of the DC motor for partially unknown parameters and conditions of external loading. First, the classical approach based on the PID control is considered, and then, a fuzzy logic-based alternative is proposed. Two different controllers are presented for the purpose of completion of the classical PID controller and a Takagi-Sugeno type fuzzy logic PI controller. Both control algorithms were implemented on an 8-bit AVR ATmega644PA microcontroller. On the basis of step responses of the object, an analysis as well as an interesting comparison of the controllers' performance have been presented.

Chapter 11 extends our study from Chap. 10. We present here numerical modeling of a DC motor treated as a dynamical system with stick-slip effects that appear in the transient motion, even while the direction of rotation of its rotor crosses zero velocity speed. These investigations are aimed at some future applications of the control technique serving as explanation of bifurcation phenomena existing in such kind of discontinuous systems. Putting emphasis on nonlinear effects, we apply the well-known, but a bit extended sliding-surface method allowing for compensation of nonlinear frictional effects. A limit cycle on a phase plane as well as time histories of control inputs and system outputs were obtained using numerical simulations performed in LabVIEW.

A lumped mass mechanical model shock response of a thorax subjected to a blast pressure wave is taken into consideration in Chap. 12. A thorax spring-dashpot model developed by Lobdell is implemented in numerical modeling of dynamics of the multibody system. The five-degree-of-freedom mechanical model of a chest adjacent to the elastic backrest is subjected to an impulse loading generated by the blast pressure wave released by an explosion. The so-called coupling of the pressure wave to the thorax is reconsidered. With respect to the evident existence of inherent time delays in displacements, the system of coupled bodies is described by time delay differential equations that are derived from the large-scale systems approach. Numerical solutions present interesting dynamical behavior of the bio-inspired system resulting from inherent time delays and a time of arrival of the blast pressure wave. It is even pointed out that the state time delays significantly change a dynamical response of the multibody system. Proper time of deployment of the foam-based armor plate reduces relative compression of the thorax, which is to be protected by a bullet-proof waistcoat.

Chapter 13 extends our study from Chap. 12. We focus here on application of one controlling force to minimize a relative compression of human chest cave that has been caused by some impacting action of an elastic external force. A virtual actuator controlling deformation in the analyzed rheological dynamical system of three degrees-of-freedom acts between the back of the human thorax and the back rest. Reduction of internal displacements in the thorax has been estimated solving the linear quadratic regulator (LQR) optimization problem.

Intentionally, this monograph aims at strengthening the studies in Mechatronics at the Faculty of Mechanical Engineering of Lodz University of Technology. It takes into consideration many dynamical aspects of mechatronic systems as well as provides the reader with the necessary theory that is helpful in understanding the elaborated experiments. Moreover, it is rich in numerical simulations presenting dynamical responses of the considered models, and examines a number of optimization problems seeking to improve their properties. Numerically reinforced mathematical models presented in this monograph undoubtedly gain in importance and should be for each engineer an interesting source of information on dynamical systems, numerical experiments, experimental measurements and optimization problems present in mechatronics.

We hope that for the physicists and mechanical engineers, this edition will be a popular source of knowledge in classical mechanics as well as a source of frequently practiced experiments connected with the simulation models of mechatronic systems.

Jan Awrejcewicz, Donat Lewandowski, Paweł Olejnik

Acknowledgments

This work has been supported by the Polish National Science Center grant MAE-STRO 2, No. 2012/04/A/ST8/00738. The authors greatly acknowledge the assistance of Wojciech Kunikowski, M.Sc.Eng., who has contributed to the experimental part presented in Chap. 10.

Contents

Chapter 1

Introduction

1.1 Mechatronics

The creators of curricula at technical universities have known for a long time that a good designer, apart from the skills of coding, should possess interdisciplinary knowledge. In addition to teaching mechanics, strength of materials, basics of construction and mechanical technology, many mechanical engineering faculties conducted classes in electrical engineering, electronics, automation, hydraulics and pneumatics, information technology, measuring instruments, machinery architecture, and other. The engineer educated according to this curriculum is expected to have broad knowledge and can choose the appropriate solution directly using known techniques or the knowledge of experts he had known.

In 1969, a concept name *mechatronics* was invented in Japan (and later distributed in the 70s) to determine the synergistic use of knowledge of the basic fields of technology. The word results from a combination of words *mechanics* and *electronics*. Authors [Xie (2003)] and [Lerner and Trigg (2005)] give definitions and descriptions of significance supplemented with graphs presenting the concept of mechatronics. Figure 1.1 exhibits one of the patterns showing the relationship between modern fields of technology which is presently observed in mechatronics.

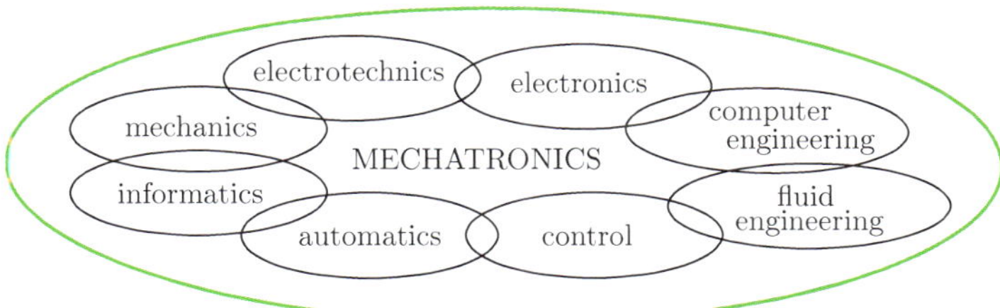

Fig. 1.1   Relationship fields in mechatronics.

One of the greatest challenges in the development of mechatronic devices and systems are, next to their variety, also progressive complexity and versatility [Scherz and Monk (2013); Budynas and Nisbett (2015); Cetinkunt (2007)]. The progress of civilization enforces researchers and engineers to search for non-coexisting solutions in various fields of science and technology [Karnopp *et al.* (2012); Di Paola and Cicirelli (2010)].

Noticeable lack of sufficiently well developed methods of dynamics analysis supporting interdisciplinary aspects of processes of the development of mechatronic devices and systems enhances the willingness to use an optimization theory and different techniques of numerical modeling. The numerical simulation and some related processing with measurement signals are highly correlated with advanced optimization methods. In this context, the device and system models presented in this monograph gain in importance and for many engineers should be an interesting source of information about mathematical modeling of dynamical systems, numerical experiments, experimental measurement, and various optimization problems of mechatronics.

1.2 Systems

System (gr. *systema* – complex object) – a physical or abstract object in which reciprocal links can be distinguished. According to one of the criteria [Schmid (2002)], systems can be divided into: abstract and physical, static and dynamic, open and closed, autonomous and nonautonomous, etc.

System – an assembly of reciprocally incorporated elements fulfilling a specific function and being treated as separated from the environment for a specific purpose, i.e., descriptive, exploratory and other. For example, a technological process defines a system. The concept of a *system* is used practically in all areas of human life and refers to either phenomena, objects or processes in nature as well as those created by people.

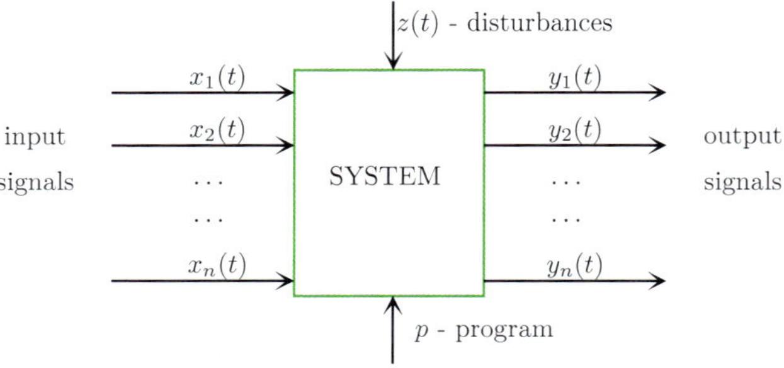

Fig. 1.2 Block diagram of system functioning.

We distinguish miscellaneous systems such as: social, political, nervous, numerical, metric, solar, radio navigation, computer, and much more. In cybernetics and exploratory systems, it is assumed that the environment influences the system through input signals, which may be targeted at interactions (control, decisions) or disturbances interferencing with the objective of the system. An important feature of real systems are their *dynamical properties*. The properties are the cause of the presence of the system in an equilibrium rest, in the steady state or in a transient state which tends to an equilibrium or not. If dynamical properties are not essential, the system is treated as *static* [Zierep (1978)]. Block diagram of a system built on the basis of that definition is shown in Fig. 1.2.

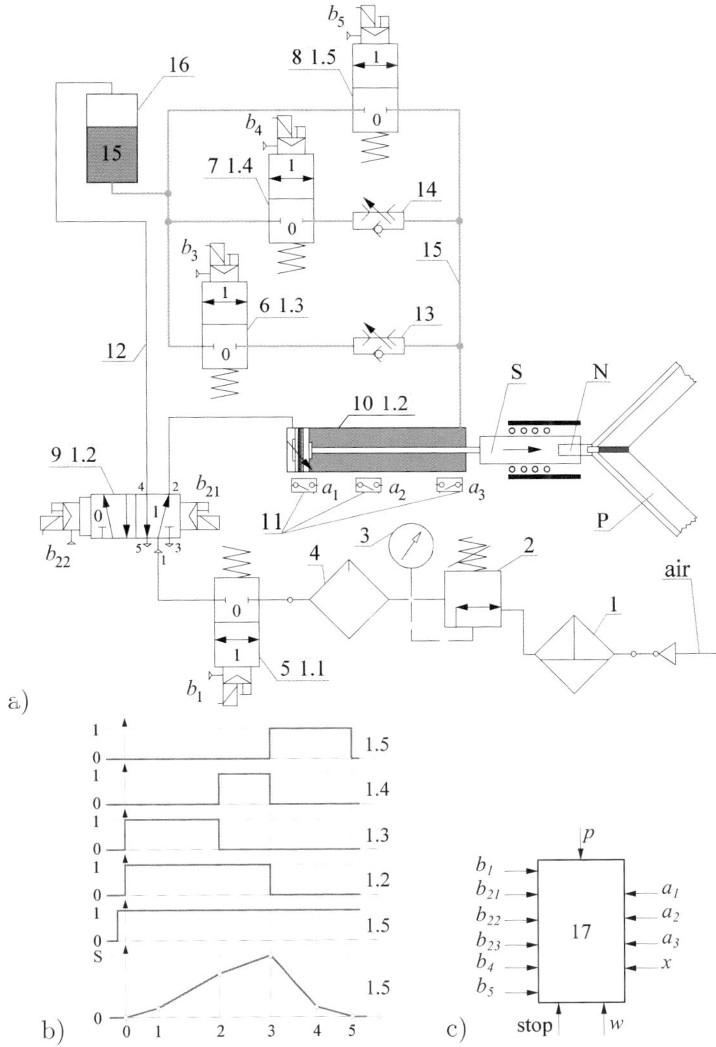

Fig. 1.3   A pneumohydraulic system of actuating control of a planer of corners of PCV windows: (a) pneumohydraulic scheme, (b) diagram of operation, (c) the controller [Lewandowski (2005)].

The pneumohydraulic system depicted in Fig. 1.3 is composed of the following parts: 1 – strainer, 2 – reducing valve, 3 – manometer, 4 – lubricator, 5-8 – monostable divide valves 2/2, 9 – bistable divide valve 5/2, 10 – pneumohydraulic actuator with mutual damping, 11 – position reed switches, 12 – air lines, 13, 14 – chock valves, 15 – oil lines, 16 – pneumohydraulic relay of pressure, 17 – PLC controller, a_i – input signals, b_{ij} – output signals, x – a signal starting the cycle, w – power "on" or "off", stop – emergency turn off.

1.3 Units of Measurement

A physical quantity **A** is defined by a value $\{\mathbf{A}\}$ and a unit of measurement $[\mathbf{A}]$:

$$\mathbf{A} = \{\mathbf{A}\}[\mathbf{A}], \quad \text{e.g.} \quad v = 30\,[\mathrm{m \cdot s^{-1}}], \ \rho = 1.29\,[\mathrm{kg \cdot m^{-3}}]. \tag{1.1}$$

The unit of measurement is a specific measure of the physical quantity which serves as a template for the quantitative determination of other measures by comparison of these measures by numbers. By convention, the numerical value of the reference measurement is equal to 1, thus:

$$[\mathbf{A}] = \frac{\mathbf{A}}{\{\mathbf{A}\}}, \quad \text{e.g.} \quad [\mathbf{A}] = \frac{45\,[\mathrm{m}]}{45} = 1\,[\mathrm{m}]. \tag{1.2}$$

In the metric system of measurement (SI), there are 7 well-defined *basic units* and 2 *supplementary units*.

Basic units include:

1) *meter* [m] – length,   2) *ampere* [A] – electric current,
3) *kilogram* [kg] – mass,   4) *mole* [mol] – amount of a substance,
5) *second* [s] – time,   6) *candela* [cd] – luminous intensity,
7) *kelvin* [K] – thermodynamic temperature.

Supplementary units include:

8) *radian* [rad] – plane angle,   9) *steradian* [sr] – solid angle.

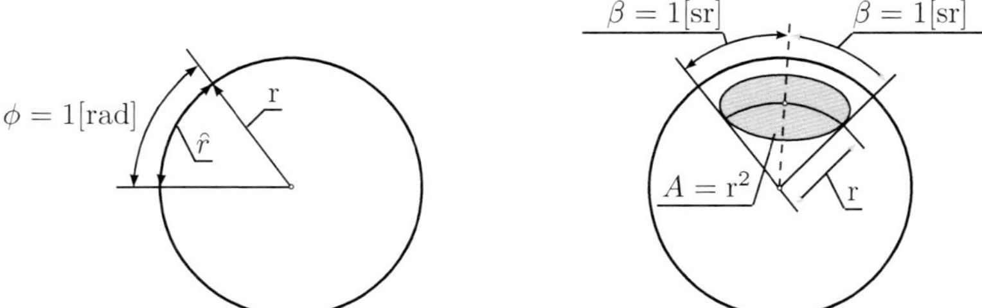

Fig. 1.4   Graphical representation of radian ϕ and steradian β.

The currently used international system of units SI was adopted in 1960 at the Ninth General Conference of Weights and Measures in Geneva. In mechanics, we use 3 basic units forming the MKS system, which are: meter [m], kilogram [kg], second

[s]. The MKS measurement system is referred to as an *absolute practical system of units*. The abovementioned set of units is also a LMT system whose name comes from the first letters of the words: length, mass, time. The earlier *absolute* unit CGS system (a part of the LMT system) used in physics consists of centimeter [cm], gram [g] and second [s] (see in Table 3.1). In the basic units one finds systems of various configurations. For example, 4-component LMTI or 6-component LMTIΘJ which distinguishes length, mass, time, electric current, thermodynamic temperature and brightness [Lerner and Trigg (2005)].

Table 1.1   Abbreviations of secondary units.

Name	Abbr.	Multiplicity	Name	Abbr.	Multiplicity
peta	P	10^{15}	deci	d	10^{-1}
tera	T	10^{12}	centi	c	10^{-2}
giga	G	$10^{\,9}$	milli	m	10^{-3}
mega	M	$10^{\,6}$	micro	μ	10^{-6}
kilo	k	$10^{\,3}$	nano	n	10^{-9}
hecto	h	10^{2}	pico	p	10^{-12}
deka	da	10^{1}	femto	f	10^{-15}

Apart from the basic units, their derivatives are also used, which are associated with basic units by some respective dependencies.

Derived units, for instance: $[N] = [kg \cdot m \cdot s^{-2}]$, $[J] = [kg \cdot m^2 \cdot s^{-2}]$, $[Pa] = [kg \cdot m^{-1} \cdot s^{-2}]$, $[W] = [kg \cdot m^2 \cdot s^{-3}]$, and others.

The basic units as well as their derivatives can act as main or secondary units.

Main unit in its value is equal to 1, and its denotation does not have any prefix, e.g. [N], [kg], [J], [s], [Pa], [m], [W].

Secondary unit is greater or less than the main unit and is distinguished by a prefix specifying an increased or decreased multiple. For example, $[kW] = [10^3 W]$, $[cm] = [10^{-2} m]$, $[ms] = [10^{-3} s]$, $[\mu m] = [10^{-6} m]$, $[MPa] = [10^6 Pa]$.

Non-SI unit is derived from the tradition of its application in a specific area. In automotive engineering, the unit of power often is horsepower $[hp] = [0.736 \, kW]$, in thermal engineering we speak about calorie $[cal] = [4.19 \, J]$, in meteorology – about tor, which is equal to the pressure exerted.

Chapter 2

Model and Modeling

Before a human had appeared on the Earth, the universe, stellar and planetary systems have already existed. Water and tectonic systems were developed, and flora and fauna were functioning on the Earth. A human developed his creative skills to eventually turn them into the ability to produce material goods. Preparation of material goods associated with abstract thinking were engaging mind and model, which was a virtual prototype of materialization. The diversity of the resulting solutions was a reason to their research and improvement for the utilitarian purposes. Therefore, the modeling of real-world objects appearing in the nature or created by other people was born. At the same time as learning about the environment was in progress, some possibility of its use for practical purposes (current and prospective) was observed. A man has created models of common cognitive systems existing in nature to satisfy human curiosity.

Model (lat. *modulus*) – measure, pattern, perfection, ideal item to follow.

Model (physical, mathematical, simulation) – a system focused on imitation of the purposes of cognitive distinguished features of another system known as the original.

Theoretical model – hypothetical and simplified mental picture of a part of reality, in which to facilitate the solution of a problem, any irrelevant elements were eliminated to achieve the specific aim. Theoretical models are introduced due to their usefulness in the creation of theory.

Real model – an object or a system of objects that meet the assumptions of the theory, sufficiently similar to the system under test, but simpler and more accessible for research.

The physical, mathematical or simulation model. To examine the object or phenomenon, one needs first to develop a physical model, which consistently with its definition, would embrace idealized phenomena, elements and parameters.

The physical model is a starting point for creating *the mathematical model* which is the formal description of an idealized object. Solving state space equations of the dynamics or even algebraic equations, which are the mathematical model for the introduced physical parameters of the object, the simulated response to the internal and external forcing (excitation) of the model is obtained. Introduction of forcing (including nonzero initial conditions) and obtaining responses of the object is called *the simulation model*. Depending on the degree of advancement of information technology, the responses resulting from the exploration of the simulation model as well as forcing can be presented in the graphical form or in a spatial or spatiotemporal monitoring.

For the purpose of elaboration of the model, we apply:

1. *Theoretical methods* based on measurement analysis, either principles of analogy or physical laws and mathematical analysis.

2. *Theoretical and experimental methods* allowing for: a) the use of a real object to provide a mathematical description to determine its selected features based on the findings of the reconstituted scale model (in accordance with the principles of similarity); b) linking equations chosen for the test – a real object of experimentally obtained parameters, giving the best approximation within the specified range of values. If there exists a graph or experimental results stored in a table, where for carrying out some mathematical operations any elementary function is needed, then an approximating function is selected in the general form as follows

$$\varphi(x) = a_0 + a_1 x + a_2 x^2 + \cdots + a_n x^n \qquad (2.1)$$

or another nonlinear function

$$\varphi(x) = A e^{rx} \sin\left(\omega t + \alpha\right) + \cdots + B e^{sx} + \cdots + K, \qquad (2.2)$$

searching for the best parameters approximating that function in the assumed range of variability $x \in [c, d]$.

3. *Experimental methods* using storage of the values of model points of objects as a function of independent variables, e.g. time, distance, velocity, and others, to control the state of the object and the response to any adverse conditions.

4. *Modal analysis* built on the methods of investigation of dynamic properties of complex mechanical objects. One of the assembly methods belonging to this field is a holistic energy model of some structural system object.

Holism (gr. *holos* – totality) is the opposite view of reductionism, according to which all phenomena create a holistic systems subject to specific regularities. Information about these systems cannot be inferred from knowledge of the regularities that govern their constituents. The totality can not be reduced to the sum of its components.

The model developed as a result of modal analysis allows us to predict the dynamic behavior of the object subjected to excitations disturbing its equilibrium.

The use of modal analysis requires a number of conditions, among which we distinguish [Ewins (2000)]:

    – linearity of the system,

– maintenance of constant coefficients of equations during modeling,
– observability and measurability of the system,
– fulfillment of Maxwell's principle of reciprocity,
– small or proportional damping.

Types of modal analysis:

* *theoretical*: aims at introduction of the theoretical description of investigated object;
* *experimental*: aims at application of the planned and controlled experiment;
* *operative*: conducted during operation at invariably located points of measurement in response of the test object to any excitation appearing in normal exploitation.

Engineers model physical phenomena and material objects, and the aim of modeling is, among others, reduction of manufacturing and exploitation costs of real objects (products).

In engineering, modeling is performed in order to:

– conduct scientific research,
– verify new concepts,
– query information that could be useful in development of control systems,
– identify existing material (tangible) objects.

Identification is about finding the relationship between a real system and its virtual model. Dynamic states of the real system are compared with the solutions generated by the model [Ewins (2000)].

Relations between the real system and its corresponding model are determined by a type of validity:

* *replicative* – if the data generated by the model match the data obtained from the real object;
* *predictive* – if compliance of the relationship between the real system and the model is known prior to obtaining data from the real system;
* *structural* – if the model not only can generate the same data as the real object, but also functions in a similar manner.

2.1 Modeling in Dimensional Analysis and Criteria of Similarity

Dimensional analysis is involved in operations on measurable values on which multiplication and exponentiation by a real exponent is applied. The analysis of dimensional equations enables for determination of mutual dependence of physical quantities being observed in the considered phenomenon. It is a tool used in physics and chemistry, but good results come from the use of dimensional analysis in mechanics, where substantial benefits result from the application of the principles of similarity. An important step in drawing up the rules of dimensional analysis was the π *theorem* formulated in 1914 by Buckingham on the basis of linear algebra [Zierep (1978)].

If we look for a relationship

$$f(Q_1, Q_2, \ldots Q_n) = 0, \tag{2.3}$$

between dimensional physical variables $Q_1, \ldots, Q_n$, our expectations will be met when the formula

$$\pi_i = Q_1^{k_1} \cdot Q_2^{k_2} \cdots Q_n^{k_n} \tag{2.4}$$

will be found, where π_i are the nondimensional parameters – the so-called *Pi groups*, the particular one of which can be equal to 1 in the special case.

Variables $Q_1, \ldots, Q_n$ can be expressed by basic measurement quantities $A_1, \ldots, A_m$ composed of $m \leq n$ terms called basis. Here, *basis* is a set of units such that none of its elements can be expressed as a result of exponentiation of the others. Consequently, in mechanics we have: meter [m], kilogram [kg], second [s] (LMT measurement system of units). A dimensional variable Q_n can be then expressed as the result of exponentiation of dimensional quantities of the basis A_m, i.e.:

$$\begin{aligned}
[Q_1] &= A_1^{a_{11}} \cdot A_2^{a_{21}} \cdots A_m^{a_{m1}}, \\
[Q_2] &= A_1^{a_{12}} \cdot A_2^{a_{22}} \cdots A_m^{a_{m2}}, \\
&\;\;\vdots \qquad\qquad \vdots \\
[Q_n] &= A_1^{a_{1n}} \cdot A_2^{a_{2n}} \cdots A_m^{a_{mn}}.
\end{aligned} \tag{2.5}$$

From Eqs. (2.4) and (2.5), the following system of algebraic equations is obtained:

$$\begin{aligned}
Q_1^{k_1} &= (A_1^{a_{11}} \cdot A_2^{a_{21}} \cdots A_m^{a_{m1}})^{k_1}, \\
Q_2^{k_2} &= (A_1^{a_{12}} \cdot A_2^{a_{22}} \cdots A_m^{a_{m2}})^{k_2}, \\
&\;\;\vdots \qquad\qquad \vdots \\
Q_n^{k_n} &= (A_1^{a_{1n}} \cdot A_2^{a_{2n}} \cdots A_m^{a_{mn}})^{k_n}.
\end{aligned} \tag{2.6}$$

Making an assumption that there exists a nondimensional parameter π, which is necessary to calculate the unknown exponents $k_1, \ldots, k_n$, the expression (2.4) can be given in the form (2.6), as it holds:

$$\begin{aligned}
\pi = {}& A_1^0 \cdot A_2^0 \cdots A_m^0 = \\
& \{A_1^{a_{11}} \cdots A_m^{a_{m1}}\}^{k_1} \cdot \{A_1^{a_{12}} \cdots A_m^{a_{m2}}\}^{k_2} \cdots \{A_1^{a_{1n}} \cdots A_m^{a_{mn}}\}^{k_n}.
\end{aligned} \tag{2.7}$$

Using basic dimensional quantities $A_1, \ldots, A_m$, one writes:

$$
\begin{aligned}
A_1^0 &= A_1^{a_{11}k_1} \cdot A_1^{a_{12}k_2} \cdots A_1^{a_{1n}k_2}, \\
A_2^0 &= A_2^{a_{21}k_1} \cdot A_1^{a_{22}k_2} \cdots A_1^{a_{2n}k_2}, \\
&\;\;\vdots \qquad\qquad \vdots \\
A_m^0 &= A_m^{a_{m1}k_1} \cdot A_m^{a_{m2}k_2} \cdots A_m^{a_{mn}k_2}.
\end{aligned}
\tag{2.8}
$$

Logarithm of relations (2.8) leads to the system of equations:

$$
\begin{aligned}
A_1 \quad &\rightarrow \quad 0 = a_{11}k_1 + a_{12}k_2 + \cdots + a_{1n}k_n, \\
A_2 \quad &\rightarrow \quad 0 = a_{21}k_1 + a_{22}k_2 + \cdots + a_{2n}k_n, \\
&\qquad\;\; \vdots \qquad\qquad \vdots \\
A_m \quad &\rightarrow \quad 0 = a_{m1}k_1 + a_{m2}k_2 + \cdots + a_{mn}k_n.
\end{aligned}
\tag{2.9}
$$

This linear system, composed of m equations and n unknowns $k_1, \ldots, k_n$, is described by the matrix $M_{(m \times n)}$ given in Tab. 2.1, allowing for determination of functions

$$
f(\pi_1, \pi_2, \ldots, \pi_{n-r}) = 0.
\tag{2.10}
$$

Table 2.1   Matrix $M_{(m \times n)}$ of a linear system.

	Q_1 k_1	Q_2 k_2	$\ldots$	Q_n k_n
A_1	a_{11}	a_{12}	$\ldots$	a_{1n}
A_2	a_{21}	a_{22}	$\ldots$	a_{2n}
$\vdots$	$\vdots$	$\vdots$		$\vdots$
A_m	a_{m1}	a_{m2}	$\ldots$	a_{mn}

In accordance to Buckingham's π theorem, *function of n dimensional variables $Q_1, \ldots, Q_n$ and m basic quantities $A_1, \ldots, A_m$ creating the dimensional matrix M of order $r \leq m$ has $n - r$ nondimensional solutions defining functions (2.10) of parameters π*, thus

$$
\pi_1 = f(\pi_2, \pi_3, \ldots, \pi_{n-r}).
\tag{2.11}
$$

The necessary and sufficient condition of similarity of two processes is fulfilled if they are qualitatively identical and all similarity parameters defining that processes are equal in pairs, i.e.: $\pi_1' = \pi_1''$, $\pi_2' = \pi_2''$, ..., $\pi_{n-r}' = \pi_{n-r}''$.

Qualitatively identical processes have the same mathematical descriptions, but differ in values of dimensional quantities.

A method of modeling is about recreation of real processes and investigation of their numerical models which are identical with those processes in terms of quality.

Results of the analysis and simulation of the numerical model can be extended to real objects if they meet the above formulated conditions.

The presented methodology implies the development of a mathematical model, according to which the following have to be determined:

a) dimensional physical quantities Q_n and their number n;

b) the number of elements of basis A_m (e.g. in mechanics $m = 3$, because there are 3 units in LMT measurement system);

c) order $r = m$ of a matrix, number of nondimensional parameters π_i for $i = n - r$ and matrix $M_{(m \times n)}$;

d) canonical form with exponents k_n of the dimensional variables Q_n.

A method for seeking of the exponents k_n is provided in the following exercise.

Find the dependence of the centripetal force F_d as a function of mass M of a body which moves in uniform motion with velocity V along the circle of radius R. The functional dependence $F_d = f(M, V, R)$ is sought.

Having 4-dimensional variables n and 3 elements of basis m, the number i of the nondimensional parameters π_i is 1. Thus, according to (2.9), one writes:

$$a_{11}k_1 + a_{12}k_2 + a_{13}k_3 + a_{14}k_4 = 0,$$
$$a_{21}k_1 + a_{22}k_2 + a_{23}k_3 + a_{24}k_4 = 0, \qquad (2.12)$$
$$a_{31}k_1 + a_{32}k_2 + a_{33}k_3 + a_{34}k_4 = 0.$$

Table 2.2   Expanded matrix $M_{(3\times4)}$ of the linear system of 3 equations and 4 unknowns, $k_1, \ldots, k_4$.

Dimensional variable		$Q_1^{k_1}$ $m^{a_{11}} kg^{a_{21}} s^{a_{31}}$ $F_d^{k_1}$	$Q_2^{k_2}$ $m^{a_{12}} kg^{a_{22}} s^{a_{32}}$ M^{k_2}	$Q_3^{k_3}$ $m^{a_{13}} kg^{a_{23}} s^{a_{33}}$ V^{k_3}	$Q_4^{k_4}$ $m^{a_{14}} kg^{a_{24}} s^{a_{34}}$ R^{k_4}
Symbol	Basic unit	$[F_d]$ $m \cdot kg \cdot s^{-2}$	$[M]$ kg	$[V]$ $m \cdot s^{-1}$	$[R]$ m
M	[m]	$a_{11}[\mathrm{M}]=1$	$a_{12}[\mathrm{M}]=0$	$a_{13}[\mathrm{M}]=1$	$a_{14}[\mathrm{M}]=1$
K	[kg]	$a_{21}[\mathrm{K}]=1$	$a_{22}[\mathrm{K}]=1$	$a_{23}[\mathrm{K}]=0$	$a_{24}[\mathrm{K}]=0$
S	[s]	$a_{31}[\mathrm{S}]=-2$	$a_{32}[\mathrm{S}]=0$	$a_{33}[\mathrm{S}]=-1$	$a_{34}[\mathrm{S}]=0$

Using Tab. 2.1 and the system of algebraic equations (2.12), we build a matrix M, expanded in Table 2.2, that facilitates finding the coefficients a_{mn} of the equations in which the unknowns are expressed by exponents k_i $(i = 1, \ldots, 4)$.

After preparing the table and entering the units next to their corresponding quantities (green font in Table 2.2), we analyze the dimension of independent quantities with respect to the dimension of the system of basic units and define the coefficients of matrix $M_{(3\times4)}$ (red font in Table 2.2). Knowing the exponent $k_1 = 1$ as the exponent of the first unknown quantity, all coefficients in the first column of the equation (2.9) are given. To calculate exponents k_2, k_3, k_4, Eq. (2.12) is to be transformed into the canonical form:

$$\begin{aligned}
a_{12}k_2 + a_{13}k_3 + a_{14}k_4 &= -a_{11}k_1, \\
a_{22}k_2 + a_{23}k_3 + a_{24}k_4 &= -a_{21}k_1, \\
a_{32}k_2 + a_{33}k_3 + a_{34}k_4 &= -a_{31}k_1.
\end{aligned} \qquad (2.13)$$

Putting known coefficients a_{mn} and $k_1 = 1$ from Table 2.2 into the equations (2.13), one gets the particular system of equations:

$$\begin{aligned}
0 \quad &+ 1 \cdot k_3 + 1 \cdot k_4 = -1, \\
1 \cdot k_2 + \quad &0 \quad + \quad 0 \quad = -1, \\
0 \quad &- 1 \cdot k_3 + \quad 0 \quad = \quad 2.
\end{aligned} \qquad (2.14)$$

The system (2.14) is solved by means of the determinants method as follows:

$$W = \begin{vmatrix} 0 & 1 & 1 \\ 1 & 0 & 0 \\ 0 & -1 & 0 \end{vmatrix}, \quad W_{k_2} = \begin{vmatrix} -1 & 1 & 1 \\ -1 & 0 & 0 \\ 2 & -1 & 0 \end{vmatrix}, \quad W_{k_3} = \begin{vmatrix} 0 & -1 & 1 \\ 1 & -1 & 0 \\ 0 & 2 & 0 \end{vmatrix}, \quad W_{k_4} = \begin{vmatrix} 0 & 1 & -1 \\ 1 & 0 & -1 \\ 0 & -1 & 2 \end{vmatrix},$$

where exponents of the dimensional quantities follow:

$$\det(W) = -1, \quad \det(W_{k_2}) = 1, \quad \det(W_{k_3}) = 2, \quad \det(W_{k_4}) = -1,$$

$$k_2 = \frac{W_{k_2}}{W} = -1, \quad k_3 = \frac{W_{k_3}}{W} = -2, \quad k_4 = \frac{W_{k_4}}{W} = 1.$$

We then use formula (2.4) calculating the nondimensional similarity parameter π

$$\pi = F_d^{k_1} \cdot M^{k_2} \cdot V^{k_3} \cdot R^{k_4}, \qquad (2.15)$$

and after a rearrangement, the centripetal force

$$F_d^{k_1} = \pi \cdot M^{-k_2} \cdot V^{-k_3} \cdot R^{-k_4}. \qquad (2.16)$$

After substitution of calculated values of the exponents $k_1, \ldots, k_4$ of dimensional quantities, and assuming $\pi = 1$, the sought formula for the centripetal force is found

$$F_d = M \cdot V^2 \cdot R^{-1}. \qquad (2.17)$$

We have obtained a mathematical model of the assumed idealized physical model of the analyzed process. For instance, if the existence of the gravitational field of the Earth is assumed, an aerodynamic force dependent on both the geometry and an environment in which the rotating body moves, then the model would be more complex, and the process could be defined by more than one parameter of similarity. Estimation of values of particular parameters of similarity $\pi_1, \ldots, \pi_n$ requires conducting some investigations on the model. It could turn out that this difficult problem is not solvable with the use of methods of dimensional analysis.

2.2　Modeling by Analogy

In engineering, *analogy* (adequacy) denotes some similarity between different objects, phenomena or processes occurring under some conditions.

The term *analogy* is used in many fields of science. Analogies are observed in law [Weinreb (2005)], philosophy [Nersessian (2002)], medicine [Pena and Andrade-Filho Jde (2010)], engineering [Sterrett (2006)]. A lot of mechanical engineering problems is solvable by means of physical analogies.

The method of analogy is about investigation of processes, being qualitatively different but having similar mathematical description, in which some differential equations and conditions of uniqueness have the same form.

Based on physical analogies, an analogue machine was developed, which during many years before implementation of digital computing machines – computers, was used for solving scientific and utilitarian problems. In the frame of analogy between electric, thermal, hydraulic and mechanical processes, experimental methods of research on various phenomena are used. The method of analogy has drawbacks which result from the influence of external conditions on propertiesof system components.

Analogies between basic hydraulic, mechanical and electrical elements existing in mechatronics are shown in Tables 2.3 and 2.4 [Guillon (1966)].

Diagrams and mathematical descriptions of the phenomena observed in the hydraulic and electrical elements indicate the possibility of using analogies to solve many practical problems as well as applying this method for the utilitarian purposes. Practical analogous models of complex systems with many system components are very worth considering.

Table 2.3　Electromechanical analogy.

Hydraulic element	Analogy	Electric element
Oscillator	$m \leftrightarrow L$ $f \leftrightarrow R$ $j \leftrightarrow \dfrac{1}{C}$ $F \leftrightarrow u$	$u = u_L + u_R + u_C$
$m\ddot{x} + f\dot{x} + jx = F$ $m\dot{v} + fv + j\int v\,dt = F$		$L\dfrac{di}{dt} + Ri + \dfrac{1}{C}\int i\,dt = u$

An example of an electrohydromechanical analogy of the actuating part of a hydraulic drive with assumption of mass, elasticity, leak and oil compressibility is shown in Fig. 2.1 [Guillon (1966)].

The discovery of analogy was the first step in the development of electronic analogue machines. The modeled quantities (actual) are mapped in real-time in the analogue machines by electrical voltages.

Machine time can serve only to map the independent variable, while other variables – the so-called *machine variables*, have to be mapped by voltage.

Table 2.4 Electrohydraulic analogies [Guillon (1966)].

Hydraulic element	Analogy	Electric element
Pressure source	$\Delta p \leftrightarrow \Delta u$ $\Delta p = \text{const}$	**Voltage source** $\Delta u = \text{const}$
Flow source	$Q \leftrightarrow i$ $Q = \text{const}$	**Source of current** $i = \text{const}$
Throttle valve with laminar flow **Piston with viscous friction**	$KS \leftrightarrow \dfrac{1}{R}$ $Q = KS\Delta p$ $\Delta p = \dfrac{1}{KS}Q$ $\dfrac{S^2}{f} \leftrightarrow \dfrac{1}{R}$ $Q = \dfrac{S^2}{f}\Delta p$ $\Delta p = \dfrac{f}{S^2}Q$	**Resistance** $i = \dfrac{1}{R}\Delta u$ $\Delta u = Ri$
Piston with a release spring	$\dfrac{S^2}{j} \leftrightarrow C$ $Q = \dfrac{S^2}{j}\dfrac{d(\Delta p)}{dt}$ $\Delta p = \dfrac{j}{S^2}\int Qdt$	**Capacitance** $i = C\dfrac{d\Delta u}{dt}$ $\Delta u = \dfrac{1}{C}\int idt$
Piston forced by an inertia	$\dfrac{S^2}{m} \leftrightarrow \dfrac{1}{L}$ $Q = \dfrac{S^2}{m}\int \Delta pdt$ $\Delta p = \dfrac{m}{S^2}\dfrac{dQ}{dt}$	**Inductance** $i = \dfrac{1}{L}\int \Delta udt$ $\Delta u = L\dfrac{di}{dt}$
Liquid or housing compressibility $Q = \dfrac{V_0}{E}\dfrac{dp}{dt}$ $Q = k_o\dfrac{dp}{dt}$	$\dfrac{V_0}{E} \leftrightarrow C$ $k \leftrightarrow C$	**Capacitance** $i = C\dfrac{du}{dt}$ $u = \dfrac{1}{C}\int idt$

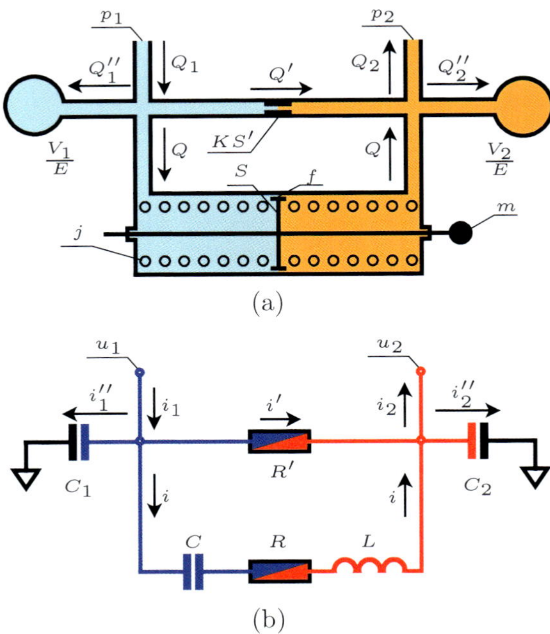

Fig. 2.1 Model of a hydromechanical system (a) and its electric counterpart elaborated using analogies (b): m – reduced mass of the driven system, j – elasticity of spring, f – viscous friction coefficient in contact between the piston and cylinder, S – piston surface, K – leakage coefficient of oil, S – cross section of the throttle (leakage), V_1, V_2 – volumes of the left and right branches of the hydraulic system, E – oil bulk modulus of elasticity, p – pressure, Q – oil flow rates, i – electric current, u – voltage, C – capacitance, R – resistance, L – inductance of coil. Analogy between variables is shown in Table 2.4.

2.3 Theory of Similarity

2.3.1 *Introduction*

In the majority of theoretical and applied sciences, we use models and apply modeling of real objects and processes [Awrejcewicz (2014); Olejnik (2013)]. It is crucial to be sure about similarity validation and the correspondence of a real object (process) to the introduced model. Once we establish a way of similarity, then similar objects (real and model) exhibit the similarity relations. In particular, there exist the so-called *similarity criteria* between the studied objects and their models.

The key role in mathematical modeling is played by *scaling*. It is known that we cannot observe (at least on a computer screen) dynamics of charges, atoms and their members, waves, etc., where we deal with distances of 10^{-10} m. On the other hand, to measure distances between planets in our universe we need years, and also we cannot directly follow phenomena with a speed being close to the light speed, etc. Therefore, we do need to introduce scaling between objects, models and secondary models associated with monitoring of solutions. In this chapter, we restrict the considerations to mathematical models described by differential equations only.

There exist three following theorems of similarity:

Theorem 2.1. *(Newton's Theorem) Similar phenomena have the same equal relations between the respective parameters.*

Theorem 2.2. *(π Theorem) Any adequate equation governing a physical process and described in a given system of units can be presented by the functional relations defined through the quantities occurred in a studied process.*

This theorem can be also formulated in its equivalent form. Any equation describing dependencies between physical quantities in a given process can be transformed to the equivalent equation with $n - k$ nondimensional quantities, where n stands for a number of all equations, and k denotes only a number of independent dimensional quantities.

Theorem 2.3. *The necessary and sufficient conditions of a similarity include proportionality of the quantities occurring in the similarity unique conditions and equality of the similarity criteria.*

2.3.2 Scaling of Equations by Similarity

Let an object be described by the parameters l_n, t_n, v_n and its counterpart model be described by l_m, t_m, v_m, where: l, t, v corresponds to a distance, time and velocity, respectively. Scales are taken as ratios between the model (l_m, t_m) and the object (l_n, t_n) parameters as follows:

$$k_l = \frac{l_m}{l_n}, \qquad k_t = \frac{t_m}{t_n}. \tag{2.18}$$

However, we cannot take an arbitrary velocity scale, since the following relation holds:

$$v_m = \frac{dl_m}{dt_m} = \frac{k_l dl_n}{k_t dt_n} = \frac{k_l}{k_t} v_n,$$
$$k_v = \frac{v_m}{v_n} = \frac{k_l}{k_t}, \tag{2.19}$$

and hence

$$\frac{k_v k_t}{k_l} = 1.$$

The obtained relation between scales is known as the *similarity indicator* (constant).

Let us take now a certain length l_0, time instant t_0, and velocity v_0 (maximal or average one). Although l_0, t_0 and v_0 are not directly coupled (they are independent), the ratio $(v_0 t_0)/l_0$ should be the same in a real object and in its model, which is manifested by the formula

$$\frac{v_0 t_0}{l_0} = \pi_0. \tag{2.20}$$

Nondimensional sets (ratios) described in similar systems by the same numbers are called the *similarity criteria* (*similarity invariants*). Any function of one or a few similarity criteria also stands for a similarity criterion. The similarity criteria are derived from the mathematical models by omitting the differential and integral operators. For example, instead of dl/dt we take l/t, or instead of $\int v dt$ we take vt.

Example 1. We consider 1-DoF nonautonomous and linear oscillator, the oscillations of which are governed by the equation

$$m\frac{d^2x}{dt^2} + c\frac{dx}{dt} + kx = F(t), \tag{2.21}$$

where m denotes the mass, c is the viscous damping coefficient, k is the stiffness, and $F(t)$ is the time-dependent excitation force. This equation may model either very small scales (atoms, electrons, neutrons, protons) or very large scales (for instance, the Earth rotation around the Sun).

In mechanics, this equation describes vibrations of a 1-DoF system of the mass m supported by the spring-damper massless system and externally excited. The parameter m and variables x and t have independent dimensions, and the variable $x(t)$ is dependent on time. Owing to the Theorem 2.2, we may construct the nondimensional similarity sets.

The integral analog of Eq. (2.21) follows

$$m\frac{x}{t^2} + c\frac{x}{t} + kx = F(t), \tag{2.22}$$

or equivalently

$$1 + c\frac{t}{m} + k\frac{t^2}{m} = \frac{Ft^2}{mx}. \tag{2.23}$$

Formula (2.23) implies three criteria of similarity

$$\pi_1 = \frac{ct}{m}, \qquad \pi_2 = \frac{kt^2}{m}, \qquad \pi_3 = \frac{Ft^2}{mx}. \tag{2.24}$$

We have

$$\frac{c_n t_n}{m_n} = \frac{c_m t_m}{m_m} = \frac{c_n k_l k_t t_n}{k_m m_n}. \tag{2.25}$$

Comparison of the first and third fraction in Eq. (2.25) yields

$$1 = \frac{k_l k_t}{k_m}. \tag{2.26}$$

On the other hand, we have

$$\frac{\gamma_m t_m}{m_m} = \frac{\gamma_n t_n}{m_n}, \tag{2.27}$$

and finally

$$k_\gamma = \frac{\gamma_m}{\gamma_n} = \frac{t_n m_m}{m_n t_m} = \frac{k_m}{k_t}. \tag{2.28}$$

Next, taking into account π_2 and π_3 and proceeding in a similar way, we obtain:

$$k_k = \frac{k_m}{k_t^2}, \qquad k_F = \frac{k_m k_l}{k_t^2}. \tag{2.29}$$

We have scaled all parameters and variables of the model equation (2.21). In order to numerically solve equation (2.21), we use the equivalent equation

$$m_m \frac{d^2 x_m}{dt_m^2} + c_m \frac{dx_m}{dt_m} + k_m x_m = F_m(t_m). \tag{2.30}$$

To return to the natural quantities $m_n, x_n, t_n, c_n, k_n, F_n$, the following relations are applied:

$$t_n = \frac{t_m}{k_t}, \quad x_n = \frac{x_m}{k_l}, \quad m_n = \frac{m_m}{k_m}, \quad k_n = \frac{k_m}{k_k}, \quad c_n = \frac{c_m}{k_c}, \quad F_n = \frac{F_m}{k_F}. \tag{2.31}$$

Example 2. We consider the movement of a cosmic object (a satellite or a comet) in a gravitational field of a planet with large mass in comparison to a movable object. The ODEs describing velocity $\vec{v}$ of the object moving in the plane OXY have the following form:

$$\frac{dv_1}{dt} = -GM \frac{x}{(x^2 + y^2)^{3/2}},$$

$$\frac{dv_2}{dt} = -GM \frac{y}{(x^2 + y^2)^{3/2}}, \tag{2.32}$$

$$v_1 = \frac{dx}{dt}, \qquad v_2 = \frac{dy}{dt},$$

where gravitational constant $G = 6.67 \cdot 10^{-11} \frac{\text{Nm}^2}{\text{kg}^2}$ and $\vec{r} = \vec{r}(x,y)$ denotes the distance between two bodies.

Let us take the Sun as a planet. The distance between the Earth and the Sun is introduced as the astronomical unit 1 [AU]$= 1.5 \cdot 10^{11}$ m. If we take year as the time unit, then the velocity unit is $[v] = \frac{1 \text{ AU}}{1 \text{ year}} = \frac{1.5 \cdot 10^{11} \text{ m}}{365 \cdot 24 \cdot 3600 \text{ s}} = 4750 \frac{\text{m}}{\text{s}}$, and the mass of the Sun is $M = 2 \cdot 10^{30}$ kg. Integral analog of the differential equations is

$$\frac{x}{t^2} = GM \frac{1}{x^2}, \tag{2.33}$$

and the nondimensional relation follows

$$1 = GM \frac{t^2}{x^3}. \tag{2.34}$$

Let us introduce the model variables and parameters in the following way:

$$t_n = t_m \cdot 10^7 \text{ s}, \ x_n = x_m \cdot 10^8 \text{ km}, \ M_n = M_m \cdot 10^{-30} \text{ kg}, \ G_n = G_m \cdot 10^{-20} \frac{\text{m}^3}{\text{s}^2 \text{kg}}.$$

Then, putting a subscript "m" in Eq. (2.32), we may use it for the computer simulations, and we have: $M_m = 2k_m$, $G_n = 6.67k_G$, $x_m = (1 - 10)k_x$, $t_m = (1 - 10)k_t$, i.e., the order of all parameters and dependent/independent variables is within the interval $(1; 10)$.

2.3.3 *Characteristic Scale Units*

In this method, the characteristic scale units regarding dependent and independent variables are introduced. This approach includes the following steps: (i) scale units are chosen; (ii) all parameters being variables are transformed to their nondimensional counterparts; (iii) a mathematical model is transformed to its corresponding nondimensional form.

This approach is widely applied in mechanics and dynamics (see [Awrejcewicz (1991); Awrejcewicz and Krysko (2008); Awrejcewicz (2012a, 2014)]). It has the following advantages: (i) obtained nondimensional equations have universal meaning, i.e., they can be directly applied to different branches of science like mechanics, mechatronics, electricity, biomechanics, etc.; (ii) usually, a number of nondimensional quantities is essentially reduced, which simplifies either numerical or analytical investigation; (iii) it allows to construct the original system in a few ways. Since there exist numerous examples of a way on nondimensionalization procedure applied to various mathematical models, including systems of linear and nonlinear algebraic equations and inequalities, linear and nonlinear ODEs and PDEs, integral and integral-differential equations as well as the combinations of the so far mentioned equations, we restrict ourselves here to only one example.

Example 3. Motion of a charge in a magnetic field is governed by the following vector equation

$$m\frac{d\vec{v}}{dt} = q\left[\vec{v} \times \vec{B}\right],\tag{2.35}$$

where B is the magnetic induction, m denotes the mass of the charge q, and $\vec{v}$ is the charge velocity. If we consider the charge moving perpendicularly to the magnetic field $\vec{B}$, then the charge movement takes place in the OXY plane. Equation (2.35) is cast to the following form:

$$\frac{d\vec{v}_\perp}{dt} = \frac{qB_0}{mB_0}\left[\vec{v}_\perp \times \vec{B}\left(\vec{r}\right)\right],$$
$$\frac{d\vec{r}}{dt} = \vec{v}_\perp,\tag{2.36}$$

where B_0 is a certain value of induction. The following initial conditions are introduced

$$x(0) = r_0, \quad y(0) = 0, \quad v_1(0) = 0, \quad v_2(0) = v_0.\tag{2.37}$$

The action of B_0 in the initial instant implies the circular motion of the charge $\vec{v}_0 \perp \vec{B}_0$ with velocity v_0, and Larmor's radius is

$$r_0 = \frac{mv_0}{qB_0}.\tag{2.38}$$

One may easily find the associated angular frequency of the periodic charge movement

$$\omega_0 = \frac{2\pi}{T} = \frac{v_\perp}{r_0} = \frac{qB_0}{m}.\tag{2.39}$$

Projection of Eq. (2.36) onto the axis OX and OY yields:

$$m\frac{d}{dt}\left(\vec{v}_1 + \vec{v}_2\right) = q\left[\left(\vec{i}v_1 + \vec{j}v_2\right) \times \vec{k}B_3\right] =$$
$$= q\left(\vec{i} + \vec{k}\right)v_1 B_3 + qv_2 B_3\left(\vec{j} + \vec{k}\right) = -\vec{j}v_1 B_3 + \vec{i}v_2 B_3 \tag{2.40}$$

which means:

$$\frac{dx}{dt} = v_1, \qquad \frac{dv_1}{dt} = \frac{\omega_0 v_2 B_3}{B_0},$$
$$\frac{dy}{dt} = v_2, \qquad \frac{dv_2}{dt} = -\frac{\omega_0 v_1 B_3}{B_0}, \qquad \vec{B}\cdot\vec{k} = B_3. \tag{2.41}$$

The characteristic time scale is here $s = 1/\omega_0$, whereas the characteristic dimension of the charge movement is r_0. Both characteristic scales are coupled by the formula $r_0 = v_0 s$.

We introduce the following nondimensional quantities:

$$t' = t\omega_0, \quad x' = x/r_0, \quad y' = y/r_0, \quad v_1' = v_1/v_0, \quad v_2' = v_2/v_0. \tag{2.42}$$

Therefore, dimensional equation (2.36) as well as initial conditions are transformed to the following nondimensional counterparts:

$$\frac{dx}{d\tau} = v_1, \qquad \frac{dy}{d\tau} = v_2,$$
$$\frac{dv_1}{d\tau} = v_2 B_3', \qquad \frac{dv_2}{d\tau} = -v_1 B_3', \tag{2.43}$$
$$x(0) = 1, \quad y(0) = 0,$$
$$v_1(0) = 0, \quad v_2(0) = 1,$$

where primes have been already omitted. Note that although in the beginning we had three parameters m, q and $\vec{B}$, in the nondimensional form (2.43) we have only one B_3'.

Chapter 3

Magnetic and Electromagnetic Phenomena

3.1 Electric Charge and its Quantization

All material objects have a mass, but some of them possess also the so-called electric charge. Although one may distinguish positive and negative charges, the most of natural bodies of our universe are neutral, i.e., two different charges cancel each other and hence the interconnections between them are not directly observed. A notation of a positive and negative charge was introduced by B. Franklin (1706-1790). The charge unit is coulomb (C). One coulomb is the charge amount flowing through a conductor cross section in 1 second, which generates the current intensity amount of 1 A (ampere).

We deal with two types of electric charges: positive and negative. Charges of opposite (the same) type attract (repel) each other.

On the atomic level, we deal with electrons and protons of negative and positive charges of the same amount, respectively. Since in classical electromagnetics, the charge is distributed continuously at a point of the line (1D), surface (2D) and volume (3D), we introduce the charge density ρ. When we rub together a piece of cloth and amber, the amber accumulates electrons. The electron carries the smallest electric charge $-1.6 \cdot 10^{-19}$C, its radius is $3.8 \cdot 10^{-15}$m, and its resting mass $m_e = 9.1 \cdot 10^{-31}$kg.

Although a classical approach to charge modeling relies on the theory of continuous fluid (Maxwell), the quantum theory tells us that our matter is quantized, i.e., any fluid charge is a multiple of a unit charge, which corresponds to the electron charge $e = 1.6 \cdot 10^{-19}$ C, and has been found experimentally. Therefore, any object charge $Q = ne$, $n = 0, \pm 1, \pm 2, \pm ... \,$. In other words, the unit of charge is indivisible.

There exist exceptions to the so far introduced terminology of the elementary charge. In 1960, the quantized charge called *quark* was detected and said to be equal to $1/3e$. However, quarks cannot exist independently and rather appear in stable groups. There exist also the so-called quasiparticles, which are fractionally charged (1982), but they are not treated as elementary particles.

In general, any material object can be considered as either a conductor or an insulator. Examples of conductors include metals, tap water or a human body, whereas typical insulators are represented by plastic, glass or pure water.

In conductors, there are 10^{23} electrons per cm^3, whereas in insulators, there is only one electron per cm^3. Therefore, in a conductor, outer electrons of atoms may freely wander within the conductor structural lattice, and they are called *conduction electrons*. On the contrary, in general, electrons do not move in an insulator.

In nature, there exists a *conservation law* of a charge. *The total charge of an isolated material object remains constant.*

If one considers gravitational forces between two particles of masses m_1 and m_2, then the gravitational force $F = G\frac{m_1 m_2}{r^2}$, where r stands for a distance between two particles. The electric force between two charged particles can be introduced in a similar way

$$F = k\frac{q_1 q_2}{r^2}, \qquad k = \frac{1}{4\pi\varepsilon}, \qquad \varepsilon = \varepsilon_0 \varepsilon_r. \tag{3.1}$$

In the case of point charges of the same (opposite) signs, the generated force is of repulsive (attractive) type.

Equation (3.1) presents Coulomb's law [Coulomb, 1785], and owing to the SI units, the force F is in newtons (N), the distance r is in meters (m), and the charge is in coulomb units (C). In the denominator, we have a normalization factor 4π used in order to avoid its occurrence in Maxwell's equations.

Although in the classical mechanics (dynamics) the SI units (fr. Systéme International d'unités) are widely used, there is a simple correspondence between the SI (meter-kilogram-second) and CGS (centimeter-gram-second) systems, since Newton's second law is invariant with respect to the chosen system units of the studied physical system. On the contrary, in electrodynamics, three different unit systems are commonly accepted, i.e., the SI system, the Gaussian system (CGS) and the Heaviside-Lorentz system (HL). Coulomb's law has different form depending on the chosen system of units. In the case of electrostatics, Coulomb's law (3.1) for a vacuum takes the form $F = \frac{q_1 q_2}{r^2}$ (CGS) and $\frac{1}{4\pi}\frac{q_1 q_2}{r^2}$ (HL), respectively. HL system is mainly used in physics, whereas a transition between SI and CGS systems is realized via substitution of ε_0 by $\frac{1}{4\pi}$. For dielectrics, the situation is more complicated, owing to different definitions of the electric induction, magnetic susceptibility, etc.

For instance, the Lorentz force is governed by the equation

$$\vec{F} = q\vec{E} + (\vec{v} \times \vec{B}) \tag{3.2}$$

in the SI system, and by the equation

$$\vec{F} = q\vec{E} + \left(\frac{\vec{V}}{c} \times \vec{B}\right) \tag{3.3}$$

in Gaussian CGS units, where c is the light speed. The latter one exhibits a correspondence between the theories of *magnetism* and *electricity*, putting emphasis on the unified approach called *electromagnetism*. In the beginning, both theories have been treated independently.

In 1820, Oersted observed that the electric current may change the position of a compass needle. In 1831, Faraday discovered that the moving magnet may induce electric current. The mentioned empirical observations were applied by Maxwell and Lorentz to construct the theory of electromagnetism.

The constant ε occurred in (3.1) is called the permittivity constant, whereas ε_r is a dimensionless constant called the relative permittivity of the material medium between the charged particles. In the case of free space (vacuum) $\varepsilon_0 = 8.85 \cdot 10^{-12} \left[\frac{\mathrm{C}^2}{\mathrm{Nm}^2} \right]$, hence $k = 8.99 \cdot 10^9 \left[\frac{\mathrm{Nm}^2}{\mathrm{C}^2} \right]$, and therefore

$$F = 8.99 \cdot 10^9 \frac{q_1 q_2}{r^2} [N]. \tag{3.4}$$

In all further examples, the problems are assumed to consider the free space. Comparison of electrostatic Coulomb's and Newton's gravitational forces for two electrons using the mentioned two formulas implies that the Coulomb forces can be omitted while considering electrostatic problems.

Table 3.1　Units in the SI and Gaussian system.

	QUANTITY	SYMBOL	SI UNIT	GAUSSIAN SYTEM
Basic	Length	L, l	Meter (m)	Centimeter (cm)
	Mass	M, m	Kilogram (kg)	Gram (g)
	Time	T, t	Second (s)	Second (s)
	Electric Current	I, i	Ampere (A)	$(10^{-1}\,\mathrm{c})$ Fr/s
Additional	Electric Charge	Q, q	Coulomb (C)	$(10^{-1}\,\mathrm{c})$ Fr
	Electric Potential	V, v	Volt (V)	$(10^8\,\mathrm{c}^{-1})$ statV
	Electric Field	E, e	Volt/Meter (V/m)	$(10^6\,\mathrm{c}^{-1})$ statV/cm
	Magnetic Induction	B	Tesla (T)	(10^4) Gs
	Magnetic Field Intensity	H	Ampere/Meter (A/m)	$(4\pi\,10^{-3})$ Oe
	Magnetic Flux	ϕ, ϕ_m	Weber (Wb)	(10^8) Gs·cm^2
	Resistance	R	Ohm (Ω)	$(10^9\,\mathrm{c}^{-2})$ s/cm
	Capacitance	C	Farad (F)	$(10^{-9}\,\mathrm{c}^2)$ cm
	Inductance	L	Henry (H)	$(10^9\,\mathrm{c}^{-2})$ s^2/cm
	Electric Displacement Field	D	Coulomb/Meter^2 (C/m^2)	$(10^{-5}\,\mathrm{c})$ Fr/cm^2
	Force	F	Newton (N)	Dyne (g·cm/s^2)
	Energy, Work	W, w	Joule (J)	Erg (g·cm^2/s^2)
	Power	P, p	Watt (W)	Erg per Second (g·cm^2/s^3)

In classical electrostatics/electrodynamics, only the SI and Gauss systems are used. In Table 3.1, the correspondence between the used units is presented.

If two charges have the same (opposite) sign, then a force is repulsive (attractive). The force $\vec{F}_{12}(\vec{F}_{21})$ acting on the charge $q_1(q_2)$ generated by the charge $q_2(q_1)$ is governed by formulas

$$\vec{F}_{12} = k\frac{q_1 q_2}{r_{12}^3}\vec{r}_{12}, \qquad \vec{F}_{21} = k\frac{q_1 q_2}{r_{21}^3}\vec{r}_{21}, \tag{3.5}$$

what is shown in Fig. 3.1

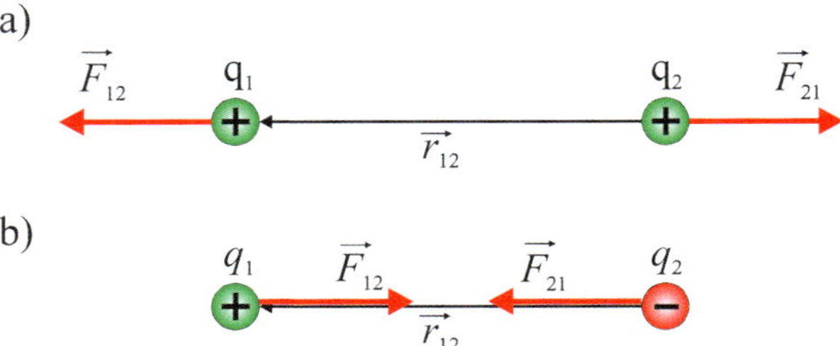

Fig. 3.1   Mutual electric forces between two charged particles of the same (a) and different (b) signs.

Figure 3.2 shows how one may find the resultant electric forces acting on each of point charges located in a plane.

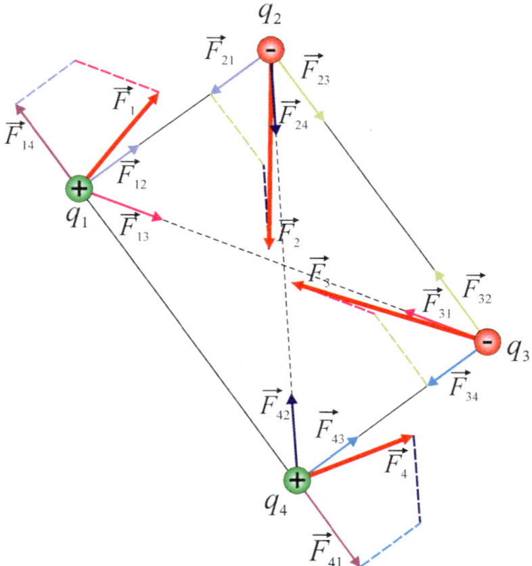

Fig. 3.2   Resultant electric forces $\vec{F}_n$ acting on each of the multiple charges.

However, it should be emphasized that this geometric simple vector superposition is valid only if a considered pair of charges is not influenced by other remaining charges. In addition, it is assumed that charges dimensions are small in comparison to their distances, and hence the charges can be treated as point charges. Furthermore, if the charges move with a speed comparable with the light speed, other modeling (by Maxwell's equations) should be taken into account.

In what follows, we introduce a concept of the electrical dipole, which is used in modeling of electromagnetic properties of material objects. The electrical dipole archetype model is mainly used in physics and chemistry, where molecules are built with negative and positive charges separated by a certain distance. Furthermore, if a molecule or an atom is embedded in either electric or magnetic field, then the mentioned objects are polarized exhibiting a dipole-like behavior.

> A dipole is a physical object consisting of two charges with equal magnitude and different sign, separated from each other by a distance d.

Since both charges are embedded in a material medium, one element charge will interact with another one through this surrounding medium, i.e., with the field produced by these two mentioned charges. Although in the dipole the charges are fixed at the distance d, they will simultaneously produce an electric field which interacts with any other charge being located in the field generated/produced by the dipole.

In general, it is assumed that there is no time delay in producing both fields, and hence also in action and reaction between two charges. However, this concept fails to model objects in our universe when the light speed is taken into account. As it has been already mentioned, the electric field intensity $\vec{E}$ is a vector the direction and magnitude of which are defined through the relationship

$$\vec{F} = \vec{E}q_0, \tag{3.6}$$

where q_0 is a positive test charge placed in the electric field, and $\vec{F}$ is the force exerted by this field on the charge q_0.

In the case of an electron orbital within the hydrogen atom $E = 10^{11}$ [N/C], while in the case of electric breakdown in the air $E = 10^6$ [N/C], and inside the wire of household circuits we have $E = 10^{-2}$ [N/C].

Owing to Coulomb's law, in order to find the electric field intensity $\vec{E}$ produced by a point charge q, we put the test charge q_0 (positive), and in the field we find

$$\vec{E} = \frac{\vec{F}}{q_0} = k\frac{q}{r^3}\vec{r}. \tag{3.7}$$

Observe that $\vec{E}$ depends on $\vec{r}$, and therefore vector $\vec{E}$ changes its direction and magnitude when location of the field point is changed.

If we have an electric multiple field produced by a few electric charges, one may use a superposition rule illustrated in Fig. 3.2. The electric dipole is shown in Fig. 3.3.

First, let us define the vector dipole moment $\vec{p} = q\vec{d}$ having its magnitude equal to the distance between point charges, and direction from $-q$ toward $+q$. We have:

$$E_1 = k\frac{q}{r_1^2}, \qquad E_2 = k\frac{q}{r_2^2}, \tag{3.8}$$

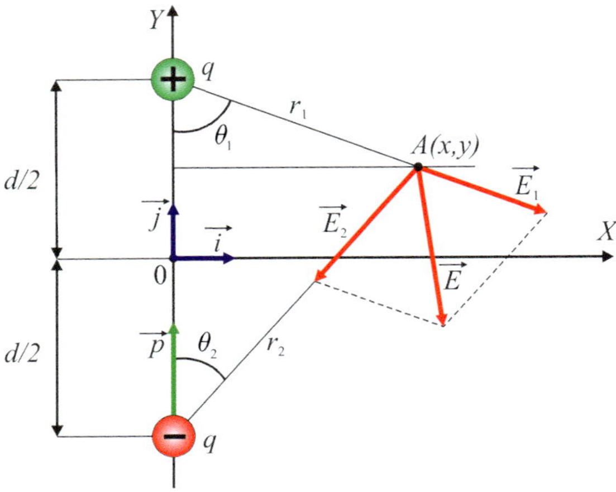

Fig. 3.3 Electric dipole and the electric field intensity $\vec{E}$ at a point A.

$$E_x = \vec{E} \cdot \vec{i} = \vec{E}_1 \cdot \vec{i} + \vec{E}_2 \cdot \vec{i} = E_1 \sin \theta_1 - E_2 \sin \theta_2,$$

$$E_y = \vec{E} \cdot \vec{j} = \vec{E}_1 \cdot \vec{j} + \vec{E}_2 \cdot \vec{j} = -E_1 \cos \theta_1 - E_2 \cos \theta_2,$$

(3.9)

where:

$$\sin \theta_1 = \frac{x}{r_1}, \qquad \cos \theta_1 = \frac{\frac{d}{2} - y}{r_1},$$

$$\sin \theta_2 = \frac{x}{r_2}, \qquad \cos \theta_2 = \frac{\frac{d}{2} + y}{r_2}.$$

(3.10)

Therefore, one finds:

$$E_x = k \frac{q}{r_1^2} \frac{x}{r_1} - k \frac{q}{r_2^2} \frac{x}{r_2} = kqx \left(\frac{1}{r_1^3} - \frac{1}{r_2^3} \right),$$

$$E_y = -k \frac{q}{r_1^2} \frac{\frac{d}{2} - y}{r_1} - k \frac{q}{r_2^2} \frac{\frac{d}{2} + y}{r_2} = -kq \left(\frac{\frac{d}{2} - y}{r_1^3} + \frac{\frac{d}{2} + y}{r_2^3} \right),$$

$$\vec{E} = E_x \vec{i} + E_y \vec{j},$$

(3.11)

where r_1, r_2 follow from Eq. (3.7).

In the case of the point lying on the axis OX $(y = 0)$, we have:

$$\theta_1 = \theta_2 = \theta, \qquad r_1 = r_2 = r = \sqrt{x^2 + \left(\frac{d}{2} \right)^2},$$

$$E_1 = E_2 = E = k \frac{q}{r^2},$$

(3.12)

and from Eq. (3.9) and (3.11) we obtain:

$$E_x = 0, \qquad E_y = -2E \cos \theta = -k \frac{p}{\left(x^2 + \frac{d^2}{4} \right)^{\frac{3}{2}}}.$$

(3.13)

In what follows, we consider an electric dipole embedded in the electric uniform field of intensity $\vec{E}$, as shown in Fig. 3.4.

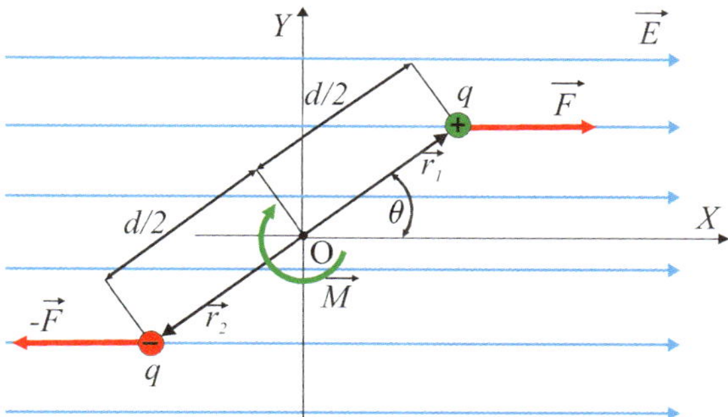

Fig. 3.4   Electric dipole lying in plane OXY embedded in the electric field $\vec{E}$.

Since magnitudes of both dipole's charges are equal and have opposite signs, we deal with a pair of force $\vec{F}$. Although the dipole does not exhibit any translation, it rotates around its center O, since it is affected by the following torque

$$\vec{M} = \vec{r}_1 \times \vec{F} + \vec{r}_2 \times (-\vec{F}) = (\vec{r}_1 - \vec{r}_2) \times q\vec{E} = \vec{p} \times \vec{E}, \tag{3.14}$$

and therefore

$$M = \frac{d}{2}F\sin\theta - \frac{d}{2}F\sin(\pi + \theta) = dF\sin\theta. \tag{3.15}$$

The dipole rotates unless $\theta = 0$, when it takes the horizontal position, i.e., $\vec{E} \parallel \vec{p}$.

However, when we remove the applied field $\vec{E}$, the distance $d \to 0$, and the charges of the dipole return to their initial superimposed position. This phenomenon is analogous to that of elastic properties of mechanical objects. Namely, the spring accumulates the potential energy generated by action of a force, and recovers it when the force is removed.

Taking into account examples presented in Fig. 3.1, 3.2, we may consider a set of charges $q_1, \ldots, q_N$ defined by their vector-positions $\vec{r}_1, \ldots, \vec{r}_N$, using the Cartesian coordinates $OXYZ$, as it is depicted in Fig. 3.5.

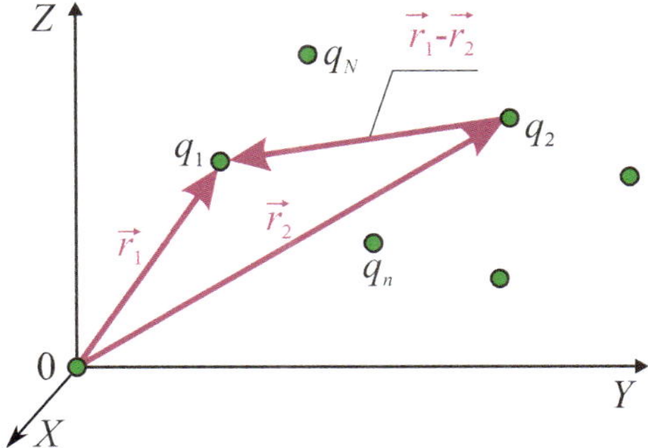

Fig. 3.5   A set of charges q_n.

It follows from Fig. 3.5 and from Coulomb's law that

$$\vec{F}_{12} = \vec{F}_{21} = \frac{q_1 q_2}{|\vec{r}_1 - \vec{r}_2|^2} \qquad (3.16)$$

and

$$F_n = \sum_{k \neq n} F_{nk} = \sum_{k \neq n} \frac{q_n q_k}{|\vec{r}_n - \vec{r}_k|^2}. \qquad (3.17)$$

Since no point charges exist in nature, we introduce a charge density $\rho(x, y, z) = \frac{dq}{d\vartheta} \left[\frac{\mathrm{C}}{\mathrm{m}^3} \right]$.

The *electric field intensity* $\vec{E}$ can be derived from (3.17), substituting the summation sign by the corresponding integral.

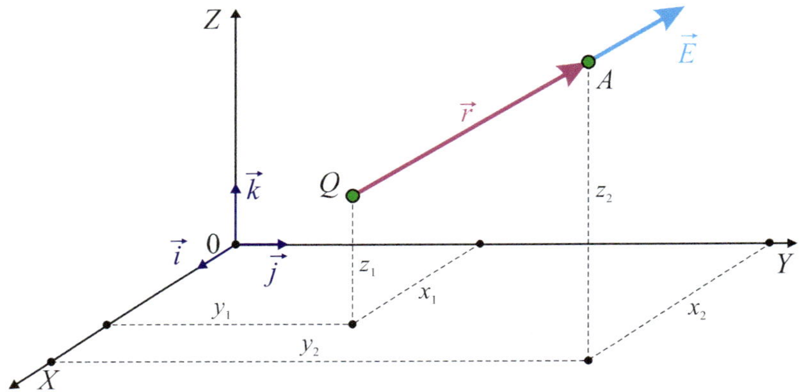

Fig. 3.6   Electric field intensity $\vec{E}$ at point A.

The electric field intensity $\vec{E}$ generated by the charge Q is the force per unit test charge q, i.e., $E = \frac{F_q}{q}$. In an arbitrarily introduced Cartesian system (see Fig. 3.6), the electric field intensity at point A is

$$\vec{E} = \frac{Q}{4\pi\varepsilon_0 r^2}\hat{\vec{r}} = \frac{Q}{4\pi\varepsilon_0 r^2} \frac{\vec{r}}{\sqrt{(x_1 - x_2)^2 + (y_1 - y_2)^2 + (z_1 - z_2)^2}}, \qquad (3.18)$$

and it is expressed in the units $\mathrm{N/C} = \mathrm{V/m}$, $\hat{\vec{r}} = \frac{\vec{r}}{|\vec{r}|}$. In the case of a volume charge, we introduce a charge density

$$\rho = \frac{dQ}{d\vartheta}, \qquad (3.19)$$

and $[\rho] = \frac{\mathrm{C}}{\mathrm{m}^3}$ (see Fig. 3.7).

Volume ϑ having only one charge Q generates total electric field $\vec{E}$ at point A obtained via integration over each differential charge dQ:

$$\vec{E} = \int_Q \frac{dQ}{4\pi\varepsilon_0 r^2}\hat{\vec{r}} = \int_\vartheta \frac{\rho(x, y, z)\hat{\vec{r}}}{4\pi\varepsilon_0 r^2} d\vartheta. \qquad (3.20)$$

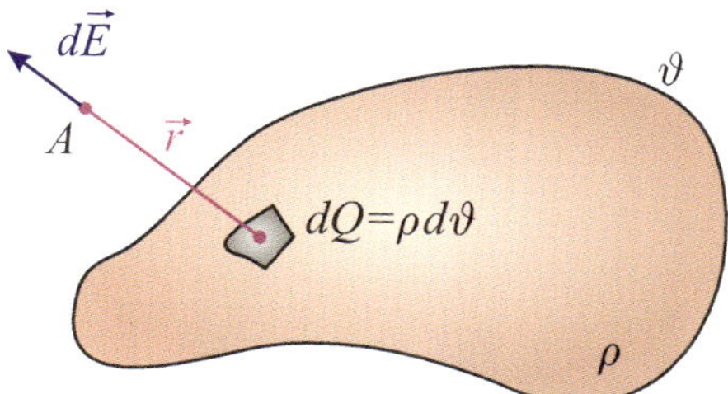

Fig. 3.7   Volume charge distribution.

In the case of a sheet charge, we have

$$\vec{E} = \int_Q \frac{dQ\hat{\vec{r}}}{4\pi\varepsilon_0 r^2} = \int_S \frac{\rho\hat{\vec{r}}}{4\pi\varepsilon_0 r^2}dS, \tag{3.21}$$

where ρ stands for the surface S charge density $[\rho] = \frac{\text{C}}{\text{m}^2}$ now.

Finally, having a total charge Q distributed over a curve, the total electric field at A is governed by the formula

$$\vec{E} = \int_L \frac{\rho\hat{\vec{r}}}{4\pi\varepsilon_0 r^2}dL, \tag{3.22}$$

where L is the curve length, ρ is the line charge density $[\rho] = \frac{\text{C}}{\text{m}}$. Observe that the unit vector $\hat{\vec{r}}$ cannot be removed from the integrand, because it changes its direction accordingly to the change of the charge coordinates. In the following cases, the use of integrand can be omitted.

(i) In spherical coordinates, the electric field of a point charge Q is

$$\vec{E} = \frac{Q}{4\pi\varepsilon_0 r^2}\hat{\vec{r}}, \tag{3.23}$$

and it is spherically symmetric.

(ii) In cylindrical coordinates, for a charge uniformly distributed along an infinite straight line, we have

$$\vec{E} = \frac{q}{2\pi\varepsilon_0 r}\hat{\vec{r}}, \tag{3.24}$$

where $[\rho] = \text{C/m}$.

(iii) For a charge uniformly distributed over an infinite plane

$$\vec{E} = \frac{q}{2\varepsilon_0}\hat{\vec{r}}, \tag{3.25}$$

where $[\rho] = \text{C/m}^2$.

3.2   Capacitance, Resistance and Electric Laws

3.2.1   *Electric Flux*

In what follows, we introduce *electric flux* ϕ (scalar) and its *density* $\vec{D}$ (vector) which cannot be directly measured. It originates on a positive charge and terminates on a negative one.

The electric flux density concept is illustrated in Fig. 3.8.

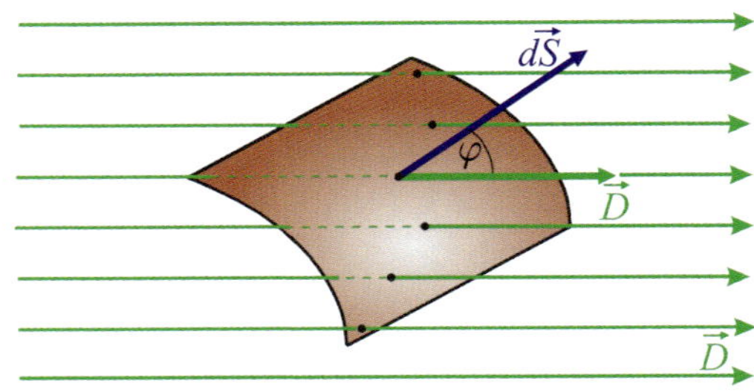

Fig. 3.8   Electric flux ($\vec{D}$ – electric flux density; $d\vec{S}$ – differential area vector).

The electric flux ϕ presents a measure of the penetration (number of the field lines) of the electric field vectors $\vec{E}$ passing through an imaginary surface represented by the surface (area) vector $\vec{S}$ (its direction is outward normal to the surface), since $\vec{D} = \varepsilon\vec{E}$. It is defined as the following dot product

$$\phi = \vec{E} \cdot \vec{S}. \tag{3.26}$$

Scalar product $\vec{E} \cdot d\vec{S}$ is proportional to a number of field lines passing through an infinitesimal surface element $d\vec{S}$. It represents a projection of $d\vec{S}$ onto direction of $\vec{E}$, and hence the surface density of the field lines is only represented by a surface perpendicular to $\vec{E}$. Gauss' law fits the total electric field flux ϕ passing through the closed surface S to the net charge Q producing the field enclosed by the surface

$$\oint \vec{D} \cdot d\vec{S} = \oint_S \varepsilon_0\vec{E} \cdot d\vec{S} = Q. \tag{3.27}$$

It is also viewed as the relation between a charge (source of the electric field) and the resultant electric field $\vec{E}$ (in a vacuum $\varepsilon = \varepsilon_0$).

We may also introduce the *electric displacement flux* ψ defined as

$$\psi = \varepsilon_0\phi, \tag{3.28}$$

being directly related to the electric field density $\vec{D}$. Electric charge of one coulomb yields an increase in the electric flux to one coulomb, since $\psi = Q$ [Q].

In general, for any electric field present in a medium, and exhibiting polarization $\vec{P}$, the vector relation is more complex:

$$\vec{D} = \varepsilon_0\vec{E} + \vec{P}, \tag{3.29}$$

as it is, for instance, exhibited by crystalline dielectrics. It means that a dipole field $\vec{P}$ is induced, where $\vec{D}$ is the electric displacement and ε_0 is vacuum's *electric permittivity*.

For $\vec{P} = 0$, both $\vec{D}$ and $\vec{E}$ have the same form and direction.

However, for an isotropic material or any linear nonconducting dielectric material, the following relation between $\vec{E}$ and $\vec{P}$ holds

$$\vec{P} = \chi_e \varepsilon_0 \vec{E}, \tag{3.30}$$

where the nondimensional constant χ_e is called the *electric susceptibility*.

Substitution (3.30) into (3.29) yields

$$\vec{D} = \varepsilon_0(1 + \chi_e)\vec{E} = \varepsilon_0 \varepsilon_r \vec{E} = \varepsilon \vec{E}, \tag{3.31}$$

where: $\varepsilon_r = 1 + \chi_e$, $\varepsilon = \varepsilon_0 \varepsilon_r$, and ε_r is called the *relative permittivity*.

It should be noted, however, that the electric field intensity $E = E(\varepsilon)$ is a function of the permittivity ε, whereas the electric flux density $\vec{D}$ does not depend on permittivity.

The choice of the closed integration surface plays a crucial role in applying Gauss' law. In order to establish the electric field $\vec{E}$ produced by the charge in a certain point, one introduces the imaginary closed (Gaussian) surface passing through this point, and then applies the formula (3.27).

In the case of only one positive charge $+Q$, flux lines are equally spaced in radial directions towards infinity (Fig. 3.9a). If we have two charges of equal magnitudes, the flux lines start at $+Q$ and terminate at $-Q$ (Fig. 3.9b). In the case of two positive charges $+Q$, direction of flux lines is as shown in Fig. 3.9c.

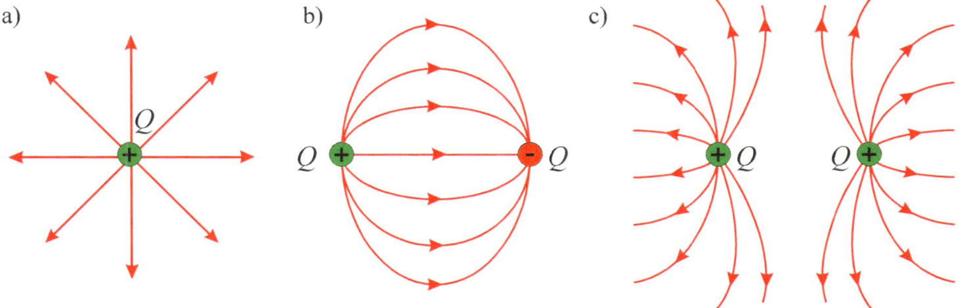

Fig. 3.9 Flux lines generated by one charge $+Q$ (a), two opposite charges $+Q, -Q$ (b), and two charges $+Q$ (c).

Diagrams presented in Fig. 3.9 require a little deeper explanation. In fact, if one charge $+Q$ is considered, each point of the field surrounding the charge location possesses its associated vector $\vec{E}(\vec{r}) \sim \frac{\vec{r}}{r^2}$. It means that $\vec{E}$ decreases with an increase in $\vec{r}$. However, instead of drawing infinitely many vectors $\vec{r}$ associated with the points lying on the radius r, we use arrows on the radial directions indicating the movement of the test charge (we connect all arrows to get the field lines). We do not loose information regarding the field strength, because the density of the field lines represents the field strength. The field lines begin on positive charges and terminate on negative charges or at infinity (they cannot intersect).

In order to illustrate the Gauss' law application, we consider an isolated point charge $+Q$, and establish the electric field E in a point located at r distance from the charge. We choose a sphere of radius r centered in the charge. Since $\vec{E} \parallel \vec{S}$ (is normal to the sphere), therefore $\vec{E} \cdot d\vec{S} = EdS$ (see Fig. 3.10).

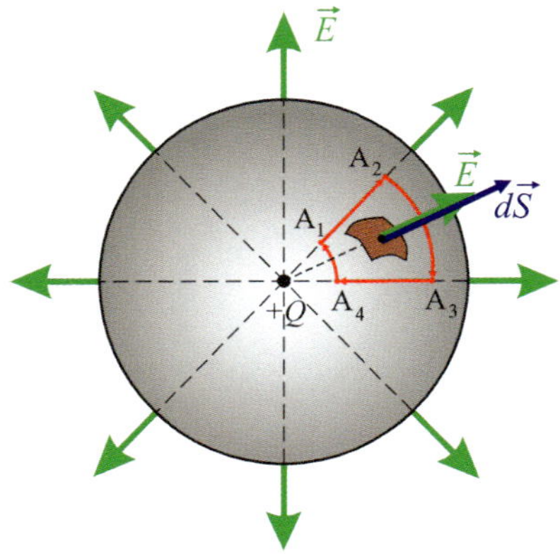

Fig. 3.10   Gaussian surface (sphere) with its center at $+Q$.

Furthermore, all points lying on the sphere have the same magnitude of E, and hence (3.27) implies

$$\varepsilon_0 \oint_S \vec{E} \cdot d\vec{S} = 4\pi\varepsilon_0 r^2 E = Q \tag{3.32}$$

which allows to find the electric field magnitude E at any point located on the sphere

$$E = \frac{Q}{4\pi\varepsilon_0 r^2}. \tag{3.33}$$

Since this result can be obtained directly from Coulomb's law, Gauss' and Coulomb's laws are totally equivalent, although the first one is more general.

If we apply spherical coordinates (r, θ, ϕ), the infinitesimal displacement of point A is (see Fig. 3.11)

$$d\vec{l} = dl_r\hat{\vec{r}} + dl_\theta\hat{\vec{\theta}} + dl_\phi\hat{\vec{\phi}} = dr\hat{\vec{r}} + rd\theta\hat{\vec{\theta}} + r\sin\theta d\phi\hat{\vec{\phi}}, \tag{3.34}$$

and $\hat{\vec{r}}, \hat{\vec{\theta}}, \hat{\vec{\phi}}$ are the unit vectors showing increase in the corresponding coordinate.

Let us conduct a direct computation of the flux generated by the field $\vec{E}$ (see

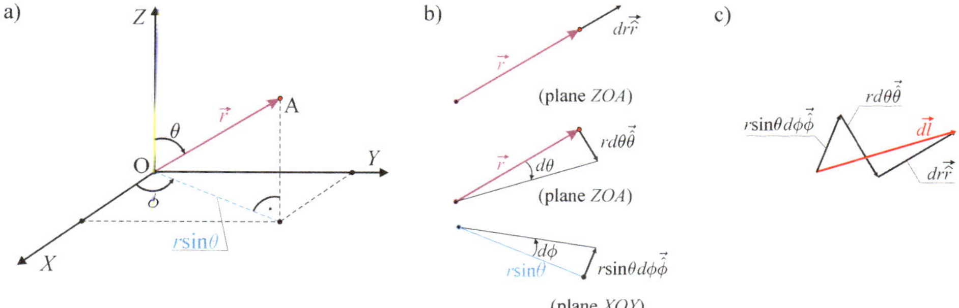

Fig. 3.11 Spherical coordinates (a), infinitesimal displacements in directions of $\vec{r}, \vec{\theta}, \vec{\phi}$ (b), and the resultant displacement $d\vec{l}$ of point A (c).

(3.26)):

$$\oint_S \vec{E} \cdot d\vec{S} = \int_\vartheta \frac{1}{4\pi\varepsilon_0}\left(\frac{Q}{r^2}\hat{r}\right) dl_\theta dl_\phi \hat{r} = \int_\vartheta \frac{1}{4\pi\varepsilon_0}\left(\frac{Q}{r^2}\right) r^2 \sin\theta d\theta d\phi =$$

$$= \frac{Q}{4\pi\varepsilon_0} \int_0^\pi \sin\theta d\theta \int_0^{2\pi} d\phi = \frac{Q}{\varepsilon_0}.$$

$$(3.35)$$

In what follows, we consider the work done in an electric field. Let us introduce another point charge q moving along the closed path from position A_1 to A_4, as the arrows indicate and as it is shown in Fig. 3.10. The work along the paths $A_1 A_2$, $A_2 A_3$ and $A_3 A_4$, and $A_4 A_1$ is as follows:

$$W_{A_1 A_2} = \int_{A_1}^{A_2} \vec{F} \cdot d\vec{l} = -q\int_{A_1}^{A_2} \vec{E} \cdot d\vec{l} = -q\frac{Q}{4\pi\varepsilon_0}\int_{r_1}^{r_2} \frac{dr}{r^2} = \frac{qQ}{4\pi\varepsilon_0}\left[\frac{1}{r}\right]_{r_1}^{r_2} = \frac{qQ}{4\pi\varepsilon_0}\left(\frac{1}{r_2} - \frac{1}{r_1}\right),$$

$$W_{A_2 A_3} = q\int_{A_2}^{A_3} \vec{E} \cdot d\vec{l} = -\frac{qQ}{4\pi\varepsilon_0}\left[\frac{1}{r}\right]_{r_2}^{r_3} = \frac{qQ}{4\pi\varepsilon_0}\left(\frac{1}{r_2} - \frac{1}{r_3}\right) = 0,$$

$$W_{A_3 A_4} = q\int_{A_3}^{A_4} \vec{E} \cdot d\vec{l} = -\frac{qQ}{4\pi\varepsilon_0}\left[\frac{1}{r}\right]_{r_3}^{r_4} = \frac{qQ}{4\pi\varepsilon_0}\left(\frac{1}{r_4} - \frac{1}{r_3}\right) = \frac{qQ}{4\pi\varepsilon_0}\left(\frac{1}{r_1} - \frac{1}{r_2}\right),$$

$$W_{A_4 A_1} = -q\int_{A_4}^{A_1} \vec{E} \cdot d\vec{l} = \frac{qQ}{4\pi\varepsilon_0}\left[\frac{1}{r}\right]_{r_4}^{r_1} = \frac{qQ}{4\pi\varepsilon_0}\left(\frac{1}{r_1} - \frac{1}{r_4}\right) = 0.$$

$$(3.36)$$

Total work done is

$$W = W_{A_1 A_2} + W_{A_2 A_3} + W_{A_3 A_4} + W_{A_4 A_1} = 0, \qquad (3.37)$$

because $r_2 = r_3$, $\quad r_1 = r_4$.

The obtained result implies an important electric field property. The work done in an electric field does not depend on the path taken in the field, or equivalently, we may say that the electrostatic force is conservative. The work done by the electric field $\vec{E}$, used for moving a charge $+q$ from point A_1 to A_2, changes its potential energy $\Delta E = E_2 - E_1$, and the following relation holds

$$\Delta E = -W_{A_1 A_2}, \tag{3.38}$$

or equivalently

$$\Delta E = -q \int_{A_1}^{A_2} \vec{E} \cdot d\vec{l}. \tag{3.39}$$

Since the electric force is conservative, the integral is independent of the path; it depends only on the initial (A_1) and final (A_2) states. It means that in a static electric field, the work done to move a point charge from location A_1 to location A_2 does not depend on the taken path.

Equations (3.36) define the following four potentials:

$$V_{A_1 A_2} = \frac{W_{A_1 A_2}}{q} = -\int_{A_1}^{A_2} \vec{E} \cdot d\vec{l}, \qquad V_{A_2 A_3} = \frac{W_{A_2 A_3}}{q} = \int_{A_2}^{A_3} \vec{E} \cdot d\vec{l},$$

$$V_{A_3 A_4} = \frac{W_{A_3 A_4}}{q} = \int_{A_3}^{A_4} \vec{E} \cdot d\vec{l}, \qquad V_{A_4 A_1} = \frac{W_{A_4 A_1}}{q} = -\int_{A_4}^{A_1} \vec{E} \cdot d\vec{l}, \tag{3.40}$$

and the potential is expressed in J/C or V.

The potential of the point A_1 with respect to the point A_2 is defined by the work needed to move a unit positive charge from A_2 to A_1. The force $\vec{F} = q\vec{E}$ acts on the charge q in each point of the moving path. In order to balance this electrostatic force, we need to apply the force $-q\vec{E}$ on the charge, which explains the occurrence of the minus sign. The potential difference between points A and B is equal to the work amount per unit charge needed to shift the charge from the point A to the point B. In this regard, it can be understood as the potential energy measured per charge unit. This definition is similar to that of the electric field intensity $\vec{E} = \frac{\vec{F}}{q}$.

Let us consider now two point charges q_1 and q_2 located at distance r and assume that the vector $\vec{E}$ points from q_1 to q_2. In the case of opposite charges $+q_1$ and $-q_2$, i.e., when the mutual electric force is attractive and q_2 moves along $\vec{E}$, we get

$$\Delta E = -(-q_2) \int_{r}^{r_f} \vec{E} \cdot d\vec{l} = \frac{1}{4\pi\varepsilon_0} q_1 q_2 \left[-\frac{1}{r} \right]_{r}^{r_f} = \frac{1}{4\pi\varepsilon_0} q_1 q_2 \left(\frac{1}{r} - \frac{1}{r_f} \right), \tag{3.41}$$

which means that the system energy increases $\left(r_f > r, \ \frac{1}{r_f} < \frac{1}{r}, \ q_2 < 0 \right)$. Here, r_f is the final state of the q_2 movement.

In this case, the external force acts against the attractive field force. On the contrary, when we release the charge from its position r_f, the field performs the work and the potential energy decreases.

Charges $+q_1$ and $+q_2$ repel each other due to the same sign. It means that moving q_2 to q_1 increases the potential energy (external force performs the work). On the contrary, releasing the charges yields the increase in the separation between them, and hence the system potential is decreased.

According to the tradition, let us take a reference point at which the potential energy is zero. Note that for $r \to \infty$, we have $1/r^2 \to 0$, which means that the associated force is zero. Thus, infinite separation of the charges implies that the potential energy is zero. Once we have a zero potential energy value reference point, it is easy to calculate the potential energy at any arbitrary point A:

$$E_A = E(\infty) - E(A) = -q \int_{\infty}^{A} \vec{E} \cdot d\vec{l} = q \int_{A}^{\infty} \vec{E} \cdot d\vec{l}. \tag{3.42}$$

If an electric field is generated by the point charge Q, the potential energy of the point A is

$$E_A = q \int_{r}^{\infty} \vec{E} \cdot d\vec{S} = \frac{1}{4\pi\varepsilon_0} \frac{qQ}{r}. \tag{3.43}$$

Formula (3.43) shows that, depending on a sign of the charges q and Q, we have $E_A < 0$ or $E_A > 0$.

When we consider the electric field generated by a system of charges, the total potential energy is computed as a sum of potential energies of each pair of charges. For three charges, we have

$$E = \frac{1}{4\pi\varepsilon_0} \left(\frac{Q_1 Q_2}{r_{12}} + \frac{Q_1 Q_3}{r_{13}} + \frac{Q_2 Q_3}{r_{23}} \right), \tag{3.44}$$

where $\vec{r}_{ij}$ points from i to j. The potential energy is associated with the work necessary to move the charges within an electric field. Therefore, a system of charges possesses the total potential energy equal to an algebraic sum of works performed by an external agent to move all charges from infinity to their current system positions. We observe that the potential energy definition requires a system of charges rather than an individual charge.

Moving a unit test charge q_0 from infinity to an arbitrary point A, we define the electric potential at this point

$$V_A = \frac{E_A}{q_0} = \int_{r}^{\infty} \vec{E} \cdot d\vec{l}. \tag{3.45}$$

The potential, being scalar, is independent of any test charge, but depends on the electric field sources, and hence it may have negative, positive or zero values. The SI unit of potential is volt, $1\ [V] = 1\ [J/C]$.

A system of N charges generates the following resultant potential

$$V = \sum_{n=1}^{N} V_n. \tag{3.46}$$

We show now that a potential and potential energy are associated with each other. The potential difference between two points A, B is

$$\Delta V_{AB} = V_B - V_A = \frac{E_B - E_A}{q_0} = \frac{\Delta E_{AB}}{q_0}. \tag{3.47}$$

It means that when an external agent moves the test charge q_0 between two points with the potential difference ΔV, it implies the change $\Delta E = \Delta V q_0$ of the potential energy of the system.

Owing to our earlier remarks, the potential difference does not depend only on their location. When we take an electron as the test charge, one may introduce another energy unit called electronvolt [eV]. It corresponds to potential energy change between two points of potential difference 1 V caused by electron movement. We note that $1[\text{eV}] = [\text{e}] \cdot [\text{V}] = 1.602 \cdot 10^{-19}[\text{J}]$.

As we have already mentioned, the electric potential V characterizes the potential energy changes (scalar), whereas the electric field $\vec{E}$ deals with electric forces (vectors). Taking any two points A, B of the electric field, one may calculate the potential difference between them

$$\Delta V_{AB} = V_B - V_A = \int_B^\infty \vec{E} \cdot \vec{dl} - \int_A^\infty \vec{E} \cdot \vec{dl} = -\int_A^B \vec{E} \cdot \vec{dl}, \tag{3.48}$$

assuming that $\vec{E}$ and $\vec{dl}$ are known. If $\vec{E}$ is generated by the positive single point charge Q, then the potential of a point located at the distance R from Q is

$$V_R = \frac{Q}{4\pi\varepsilon_0} \int_R^\infty \frac{dr}{r^2} = \frac{Q}{4\pi\varepsilon_0} \frac{1}{R}. \tag{3.49}$$

Formula (3.49) implies that $V_R \to 0$ for $R \to \infty$, and $V_R \to \infty$ for $R \to 0$.

We can treat an electrical potential in a more general way. Intensity of the electric field $\vec{E}$ has rotation $\nabla \times \vec{E} = \vec{0}$. Owing to the Stokes theorem, a curvilinear integral along an arbitrary closed curve $\oint \vec{E} \cdot \vec{dl} = 0$. We take two points A and B, and integrate from A to B using different paths of integration. The sum of integration from A to B, and then from B to A, should be zero, since both paths are arbitrarily taken. It means that the curvilinear integral does not depend on the chosen parts, but only on a distance between points.

Let us take a reference point O, and define the electric potential

$$V(\vec{r}) = -\int_0^r \vec{E} \cdot \vec{dl}. \tag{3.50}$$

The difference of potentials in points A and B is

$$V(B) - V(A) = -\int_0^B \vec{E} \cdot \vec{dl} + \int_0^A \vec{E} \cdot \vec{dl} = -\int_0^B \vec{E} \cdot \vec{dl} - \int_0^A \vec{E} \cdot \vec{dl} = -\int_A^B \vec{E} \cdot \vec{dl}. \tag{3.51}$$

From the fundamental theorem for gradients, it follows that

$$\int_A^B (\nabla V) \cdot d\vec{l} = V(B) - V(A), \tag{3.52}$$

which yields

$$\int_A^B (\nabla V) \cdot d\vec{l} = -\int_A^B \vec{E} \cdot d\vec{l}, \tag{3.53}$$

and hence

$$\vec{E} = -\nabla V. \tag{3.54}$$

It means that the electric field intensity is a gradient of a scalar potential. A potential is not uniquely defined, since a reference point O is arbitrarily taken. If we take other point O_1, then

$$V_1(\vec{r}) = -\int_{O_1}^r \vec{E} \cdot d\vec{l} = -\int_{O_1}^O \vec{E} \cdot d\vec{l} - \int_O^r \vec{E} \cdot d\vec{l} = C + V(\vec{r}), \tag{3.55}$$

where C is a constant. However, a difference between potentials of two points A and B does not depend on C, since

$$V_1(B) - V_1(A) = V(B) - V(A). \tag{3.56}$$

Equation (3.54) implies

$$\frac{\partial E_x}{\partial y} = \frac{\partial E_y}{\partial x}, \quad \frac{\partial E_z}{\partial y} = \frac{\partial E_y}{\partial z}, \quad \frac{\partial E_x}{\partial z} = \frac{\partial E_z}{\partial x}, \tag{3.57}$$

which means that E_x, E_y and E_z cannot be arbitrarily taken.

Recall the already derived equations

$$\nabla \cdot \vec{E} = \frac{\rho}{\varepsilon_0}, \qquad \nabla \times \vec{E} = \vec{0}, \tag{3.58}$$

where we apply substitution $E = -\nabla V$. We have

$$\nabla \cdot \vec{E} = \nabla \cdot (-\nabla V) = -\nabla^2 V = -\Delta V. \tag{3.59}$$

It means that the divergence of $\vec{E}$ equals the Laplacian of V, and we have

$$\Delta V = -\frac{\rho}{\varepsilon_0}. \tag{3.60}$$

We have got Poisson's equation which for $\rho = 0$ (lack of charge) takes the form of the Laplace's equation

$$\Delta V = 0. \tag{3.61}$$

The second equation implies

$$\nabla \times \vec{E} = \nabla \times (-\nabla V) = 0, \tag{3.62}$$

which means that in order to satisfy this equation, we must take

$$\vec{E} = -\nabla V. \tag{3.63}$$

Consequently, we only need Poisson's equation to find V, and then to find $\vec{E}$. However, in order to find $\vec{E}$ without introduction of the potential V, we need two equations (3.58).

We can easily define the potential energy of N point charges. Assume that in the beginning, all of them are located at infinity, and we move q_1 into the required position at first. In this case, no work is done, since there is no external field. Then we move q_2 into position $\vec{r}_2$, but in this case, there is a potential $V_1(\vec{r}_2)$, and the work done is

$$W_2 = \frac{1}{4\pi\varepsilon_0} q_2 \frac{q_1}{|\vec{r}_1 - \vec{r}_2|} = \frac{1}{4\pi\varepsilon_0} q_2 \frac{q_1}{r_{12}}. \tag{3.64}$$

In the case of the point charge q_3 located at $\vec{r}_3$, the potential $V_{1,2}$ is generated by two charges q_1 and q_2 and, owing to the superposition rule, we have

$$V_{1,2} = \frac{1}{4\pi\varepsilon_0} \left(\frac{q_1}{r_{13}} + \frac{q_2}{r_{23}} \right), \tag{3.65}$$

which means that

$$W_3 = \frac{1}{4\pi\varepsilon_0} q_3 \left(\frac{q_1}{r_{13}} + \frac{q_2}{r_{23}} \right). \tag{3.66}$$

The resultant work done in the space to locate N point charges is as follows

$$W = \frac{1}{4\pi\varepsilon_0} \left(\frac{q_1 q_2}{r_{12}} + \frac{q_1 q_3}{r_{13}} + \cdots + \frac{q_1 q_N}{r_{1N}} + \frac{q_2 q_3}{r_{23}} + \frac{q_2 q_n}{r_{2n}} + \cdots + \frac{q_2 q_N}{r_{2N}} + \frac{q_{N-1} q_N}{r_{N-1N}} \right). \tag{3.67}$$

The last formula can be presented in the following way

$$W = \frac{1}{2} \sum_{n=1}^{N} q_n \left(\sum_{\substack{m=1 \\ m \neq n}}^{N} \frac{1}{4\pi\varepsilon_0} \frac{q_m}{r_{nm}} \right) = \frac{1}{2} \sum_{n=1}^{N} q_n V(\vec{r}_n), \tag{3.68}$$

and then can be generalized to the case of a volume charge distribution

$$W = \frac{1}{2} \int \rho V \, dV. \tag{3.69}$$

We are going to eliminate ρ and V by means of introducing $\vec{E}$. Gauss' law yields $\rho = \varepsilon_0 \nabla \cdot \vec{E}$, and hence

$$W = \frac{\varepsilon_0}{2} \int (\nabla \cdot \vec{E}) \, V \, d\vartheta. \tag{3.70}$$

Now, we apply an integration by parts and the theorem on divergence. We have

$$W = \frac{\varepsilon_0}{2} \left[-\int \vec{E}(\nabla V) \, d\vartheta + \oint V\vec{E} \cdot d\vec{S} \right] =$$

$$= \frac{\varepsilon_0}{2} \left(\int_\vartheta E^2 d\vartheta + \oint_S V\vec{E} \cdot d\vec{S} \right) = \frac{\varepsilon_0}{2} \int_{\vartheta*} E^2 d\vartheta. \tag{3.71}$$

Above, we have used the formula $\nabla V = -\vec{E}$ accompanied by the observation that when we enlarge the integration space comprising the charge, a contribution coming from surface (volume) integral decreases (increases). In the case of full space $\vartheta*$, the contribution coming from the surface integral is zero.

3.2.2 *Capacitance and Capacitors*

A capacitor is a set of two conductors (plates) of an arbitrary shape, which are isolated from the surrounding environment. A capacitor is "charged" when its plates have equal opposite charges $+q, -q$. Since $+q - q = 0$, the *net* charge of the capacitor is zero. If one connects the capacitor plates to the opposite battery terminals, then equal and opposite charges are transferred to the capacitor plates. Relation between the potential difference ΔV of the plates and the carried charge is

$$\frac{q}{\Delta V} = C, \tag{3.72}$$

where the constant C is called the *electrical capacitance*. It should be noted that V is the difference between the potential of the positive conductor and the negative conductor, and since Q is the charge on the positive conductor, then always $C > 0$. The capacitance SI unit is Farad [F]=[C/V], and the following subunits are used $[\mu F] = 10^{-6}[F]$, $[nF] = 10^{-9}[F]$, $[pF] = 10^{-12}[F]$.

Consider the parallel-plate capacitor shown in Fig. 3.12 with the distance between plates d (this should not be not confused with a dipole) and a surface area S.

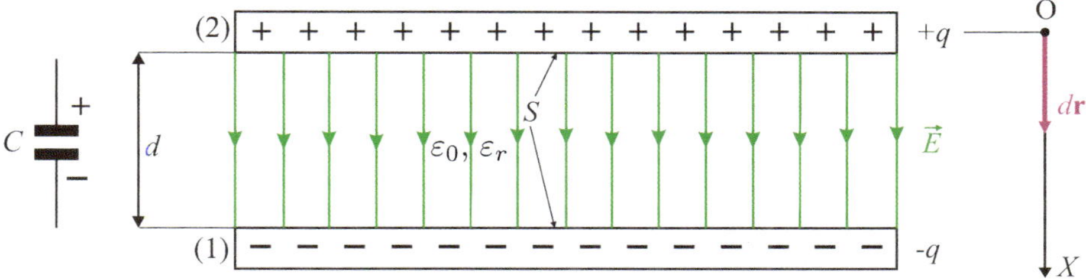

Fig. 3.12   Charged parallel-plate capacitor (cross section).

Assuming $S \gg d$, the electric field curves inside the capacitor embedded in vacuum are vertical lines, and hence the electric field is *uniform*.

We apply Gauss' law (see (3.27)) to define the electric field $\vec{E}$ between the plates, i.e., the electric field passing through S is

$$\phi = \oint_S \vec{E} \cdot d\vec{S} = \oint_S EdS = ES = \frac{q}{\varepsilon_0}, \tag{3.73}$$

and hence

$$E = \frac{qS}{\varepsilon_0}, \tag{3.74}$$

where ε_0 is the constant without a dielectric (vacuum). One may also introduce $\varepsilon = \varepsilon_0 \varepsilon_r = C/C_0$, where C_0 is the capacitance of a dielectric (isolator) put into the capacitor plates. In vacuum $\varepsilon = \varepsilon_0 = 1$, in air $\varepsilon \cong 1$, in transformer oil $\varepsilon = 4.5$, in water $\varepsilon = 80$.

The potential difference between the plates is

$$\Delta V_{12} = V_2 - V_1 = -\int_1^2 \vec{E} \cdot d\vec{r} = \vec{E}\int_2^1 dr = Ed = \frac{qSd}{\varepsilon_0}. \tag{3.75}$$

Finally, it follows from (3.72) that the capacitance

$$C = \frac{q}{\Delta V_{12}} = \varepsilon_0 \frac{S}{d}, \tag{3.76}$$

and it depends only on the capacitor geometry, i.e., S and d.

Before charging an unloaded capacitor, its potential energy E is zero. A charged capacitor stores the electric potential difference

$$\Delta V \equiv U = \frac{dE}{dq'}, \tag{3.77}$$

hence

$$dE = U dq' = \frac{q'}{C} dq', \tag{3.78}$$

and the total charge q is achieved, when

$$E = \frac{1}{C}\int_0^q q' dq' = \frac{q^2}{2C}, \tag{3.79}$$

or equivalently

$$E = \frac{qU}{2} = \frac{CU^2}{2}. \tag{3.80}$$

In what follows, we consider the parallel and serial connections of capacitors.

(i) Parallel capacitors connection is shown in Fig. 3.13.

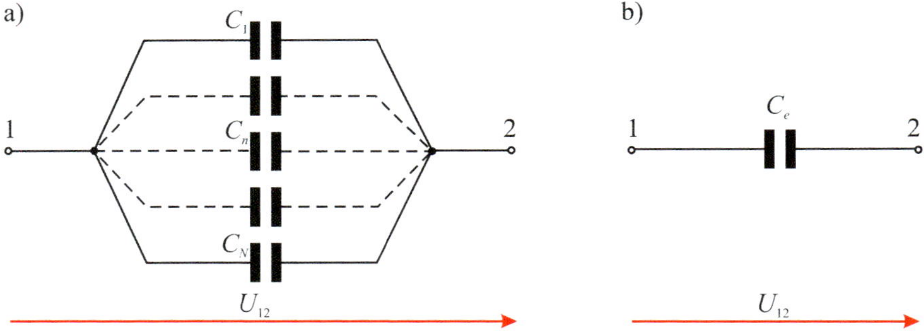

Fig. 3.13 N capacitors connected in parallel (a) and the equivalent capacitor (b).

Assuming perfect conductors, the total charge of the equivalent capacitor is

$$Q = C_e U = Q_1 + Q_2 + \cdots + Q_N = (C_1 + C_2 + \cdots + C_N)U, \tag{3.81}$$

and hence the equivalent capacitance

$$C_e = \sum_{n=1}^{N} C_n. \tag{3.82}$$

(ii) Series capacitors connection is shown in Fig. 3.14.

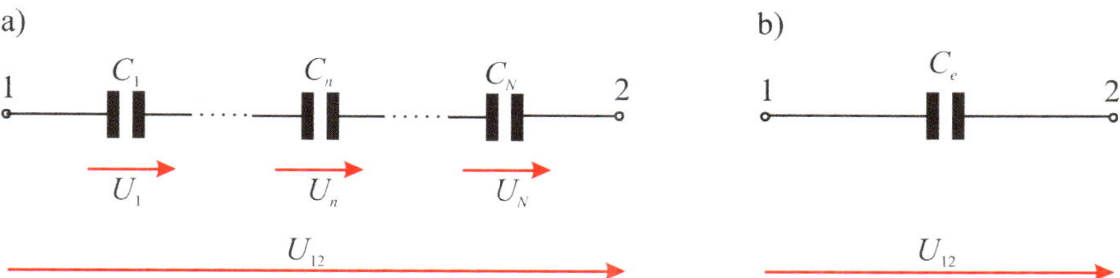

Fig. 3.14 N capacitors connected in series (a) and the equivalent capacitor (b).

The total potential difference U_{12} follows:

$$U_{12} = \frac{Q}{C_e} = U_1 + \ldots U_n + \cdots + U_N = Q\left(\frac{1}{C_1} + \cdots + \frac{1}{C_n} + \cdots + \frac{1}{C_N}\right), \quad (3.83)$$

$$Q = Q_1 = \ldots Q_n = \ldots Q_N,$$

and the equivalent capacitance

$$\frac{1}{C_e} = \sum_{n=1}^{N} \frac{1}{C_n}. \quad (3.84)$$

The energy stored in a static electric field is given by

$$W_E = \frac{1}{2} \int \vec{D} \cdot \vec{E} d\vartheta = \frac{1}{2} \int_{\vartheta} \varepsilon E^2 d\vartheta = \frac{1}{2} \int_{\vartheta} \frac{D^2}{\varepsilon} d\vartheta, \quad (3.85)$$

where $\vec{D}$ is the flux density defined by $\vec{D} = \varepsilon\vec{E}$, and its unit in the SI system is C/m^2.

In the case of a parallel-plate capacitor with constant voltage V applied across the plates, the electric field $E = V/d$ (see Fig. 3.15).

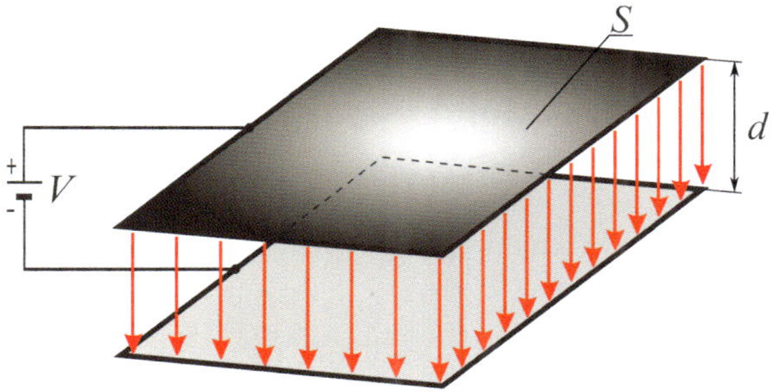

Fig. 3.15   Parallel-plate capacitor with applied constant voltage V.

Formula (3.85) implies that

$$W_E = \frac{1}{2} \int \varepsilon E^2 d\vartheta = \frac{\varepsilon}{2}\left(\frac{V}{d}\right)^2 \int d\vartheta = \frac{\varepsilon}{2}\frac{SV^2}{d} = \frac{1}{2}CV^2, \quad (3.86)$$

or alternatively

$$W_E = \frac{1}{2} \int \frac{D^2}{\varepsilon} \, d\vartheta = \frac{1}{2} \int \frac{1}{\varepsilon} \left(\frac{\varepsilon S V}{d} \right)^2 d\vartheta = \frac{1}{2} \frac{\varepsilon S^2 V^2}{d} = \frac{1}{2} C V^2. \tag{3.87}$$

3.2.3 Resistance

When positive and negative charges have different characteristics, or they are embedded in a liquid or a gas, then the current density $\vec{J}$ should be rather taken into account instead of current I. Suppose we have a particle $+Q$ in an electric field $\vec{E}$ in a vacuum. Then, the acting force $\vec{F} = +Q\vec{E}$ causes the charge to move with a constant acceleration $\vec{a}$, and velocity $\vec{v} = \vec{a}t$. However, when the same charge moves in a gas or liquid, it undergoes random impacts (collisions) with other particles of the medium, and its movement is random. In the case of a homogeneous medium, random components vanish, and as a microscopic output constant average velocity called the *drift velocity* $\vec{v}$ of the electron-gas movement can be introduced. Recall that a cubic meter of a conductor contains about 10^{28} atoms, and each atom has one or two free electrons. Therefore, the charge movement in a gas (liquid) is analogous to the movement of electrons which can be modeled with the help of the *electron-gas theory*. It means that the drift velocity of the electrons can be estimated via the following simple formula

$$\vec{v} = \mu \vec{E}, \tag{3.88}$$

where $[\mu] = \mathrm{m}^2/\mathrm{Vs}$ is the *mobility coefficient*.

Two current densities are distinguished, although both of them are governed by the same formula

$$\vec{J} = \rho \vec{v}. \tag{3.89}$$

A set of charged particles maintaining their relative positions within the volume ϑ and passing through a surface S generates a convection current density $[J] = [\mathrm{A/m}^2]$ which can be either constant or nonconstant in time.

Assume that an electric field is applied to a conductor with the same cross section S. Then, it follows from Eqs. (3.88) and (3.89) that

$$\vec{J} = \sigma \vec{E}, \qquad \sigma = \mu \rho. \tag{3.90}$$

Since σ ($[\sigma] = [\mathrm{S/m}]$ – Siemens per meter) is the material conductivity, $\vec{J}$ is known as the conduction current density. If we consider electrons as the carriers, then $\rho < 0$ and $\mu < 0$, which means that always $\sigma > 0$. However, we conventionally take $\rho > 0$, $\mu > 0$, since the current direction is defined by the direction of movement of positive carriers. For perfect conduction $\sigma = \infty$.

If we take an infinitely small surface $d\vec{S}$ and the current density $\vec{J}$ crossing this surface, then the current $dI = \vec{J} \cdot d\vec{S}$, and hence

$$I = \int_S \vec{J} \cdot d\vec{S}. \tag{3.91}$$

It is well known that the same electric potential difference acting in different conductors (made of different materials and having different geometric properties) yields different values of the current.

In general, the following proportionality condition holds

$$R = \frac{U}{I},\qquad(3.92)$$

where a constant R is called *resistance* and its SI unit is Ohm, i.e., $[\Omega]=[V/A]$ (we also deal with 1 $[k\Omega] = 10^3$ $[\Omega]$, 1 $[M\Omega] = 10^6$ $[\Omega]$). Another constant characterizing electric properties of the material, i.e. *resistivity* ρ, is defined in the following way

$$\vec{E} = \rho\vec{J},\qquad(3.93)$$

where $\vec{J}$ is the current density $[J]=[A/m^2]$, $[\rho]=[\Omega m]$ (for instance, for copper $\rho = 1.68 \cdot 10^{-8}$ Ωm, for iron $\rho = 9.61 \cdot 10^{-8}$ Ωm, for silver $\rho = 1.59 \cdot 10^{-8}$ Ωm, for silicon $\rho = 2.5 \cdot 10^3$ Ωm (semiconductor) and for glass $\rho = 1 \cdot 10^{10-14}$ Ωm (isolator), and the ρ values are given for pressure 1 atm and temperature 20°C. One may also introduce the so-called *conductivity* $\sigma = 1/\rho$. In the case of a cylindrical conductor of length l and cross section area S, the potential difference $\Delta V_{12} = U$ applied to the conductor ends generates the uniform electric field

$$E = \frac{U}{l},\qquad J = \frac{I}{S},\qquad(3.94)$$

and hence it follows from Eqs. (3.93) and (3.94) that

$$R = \frac{\rho l}{S}.\qquad(3.95)$$

Charge carriers moving in a conductor suffer from many collisions with the conductor structural lattice, and thus their kinetic energy is transformed into heat. Resistivity, in general, depends on the temperature T in an almost linear manner:

$$\rho_T = \rho_{T_0}\left[1 + \alpha(T - T_0)\right],\qquad(3.96)$$

where

$$\alpha = \frac{\Delta\rho}{\rho\Delta T},\qquad \Delta\rho = \rho_T - \rho_{T_0},\quad \Delta T = T - T_0.\qquad(3.97)$$

The coefficient α is called the *temperature coefficient of resistivity*, and $[\alpha] = [1/K]$.

A decrease in the temperature causes a decrease in the resistivity of materials. In 1911, Onnes discovered that resistivity of mercury achieves zero for the temperature around 4°K. This has opened the research directed towards superconductivity, since the established current should persist forever in such a material.

Finally, let us address the current continuity equation, which follows a concept from fluid mechanics. If one takes a closed surface S, a net current will go out if a decrease in a positive charge occurs, as described by the formula

$$I = \oint \vec{J} \cdot d\vec{S} = -\frac{dQ}{dt} = -\frac{\partial}{\partial t}\int \rho d\vartheta.\qquad(3.98)$$

From (3.98) we get

$$\lim_{\Delta\vartheta\to 0}\frac{\oint \vec{J}\cdot d\vec{S}}{\Delta\vartheta} = -\frac{\partial}{\partial t}\lim_{\Delta\vartheta\to 0}\frac{\int \rho d\vartheta}{\Delta\vartheta},\tag{3.99}$$

and thus

$$\vec{\nabla}\cdot\vec{J} = -\frac{\partial\rho}{\partial t}.\tag{3.100}$$

We show that

$$\lim_{\Delta\vartheta\to 0}\frac{\oint \vec{J}\cdot d\vec{S}}{\Delta\vartheta} = \nabla\cdot\vec{J}.\tag{3.101}$$

Let us take a cube and the vector field $\vec{J}$ in Cartesian coordinates. All six cube surfaces must be covered to obtain $\oint \vec{J}\cdot d\vec{S}$. The x component of $\vec{J}$ is $\vec{J_1}$, and we take small input face and output face of the cube:

$$I_1 = \oint \vec{J}\cdot d\vec{S} \cong -J_1(x)\Delta y\Delta z,\tag{3.102}$$

$$I_2 = \oint \vec{J}\cdot d\vec{S} \cong -J_1(x+\Delta x)\Delta y\Delta z = -\left[J_1(x)+\frac{\partial J_1}{\partial x}\Delta x\right]\Delta y\Delta z,\tag{3.103}$$

and hence

$$I_1 - I_2 = \frac{\partial J_1}{\partial x}\Delta x\Delta y\Delta z = \frac{\partial J_1}{\partial x}d\vartheta.\tag{3.104}$$

Proceeding in the similar way, the remaining two pairs of faces yield the final result

$$\oint \vec{J}\cdot d\vec{S} \cong \oint \left(\vec{J_1}+\vec{J_2}+\vec{J_3}\right)d\vec{S} \cong \left(\frac{\partial J_1}{\partial x}+\frac{\partial J_2}{\partial y}+\frac{\partial J_3}{\partial z}\right)\Delta\vartheta,\tag{3.105}$$

and consequently we get Eq. (3.101).

It should be noted that ρ in (3.101) represents a *net charge density*. Since $\dot{\rho}\neq 0$ only during a transition state, the continuity equation $\nabla\cdot\vec{J} = 0$ is equivalent to Kirchhoff's current law.

Another important remark concerns a notation of a *net charge*. In spite of that the valence electrons move due to electric field action, each conduction electron is balanced by a proton in the nucleus, and hence every $\Delta\vartheta$ exhibits *zero net charge*.

3.2.4   *Electric Laws*

In general, the function $i(U)$ is different for different materials and it is nonlinear (we denote constant current by I, and time-varying current by i). Once it is linear, then the so-called *Ohm's law* holds. It states that the resistance between two arbitrary points of a conductor does not depend on the applied potential difference, i.e., on its magnitude and polarity. It is expressed by the relationship

$$U = IR,\tag{3.106}$$

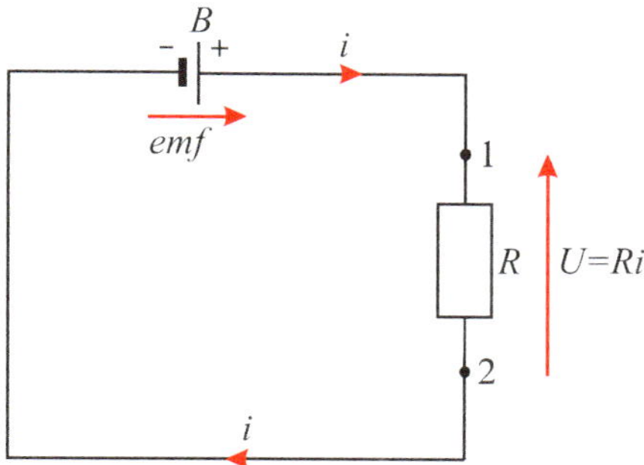

Fig. 3.16   Electric circuit with the battery B and resistor R.

and the materials obeying Ohm's law are called *Ohmic elements*. Note that relation (3.106) is called the *point form of Ohm's law*.

When the charge dq moves within an uniform electric field established between the points 1 and 2 exhibiting the potential difference $\Delta V_{12} = U$, it causes a change in the potential energy ΔE (see (3.79)). Let us consider the energy transfer in an electric circuit shown in Fig. 3.16.

In the connected circuit, positive charges move in the direction shown by arrows at i, which correspond to the used convention to indicate the current movement. When we take a positive charge inside the battery, it will move from high potential $(+)$ to low potential $(-)$, and the direction of the current inside the battery is opposite to the marked direction i in the external circuit. However, in the figure, the direction of the arrow of $\mathcal{E}$ indicates the direction of motion of positive charges from the negative terminal to the positive one. Our closed circuit loop should conserve energy of a charge carrier traveling around it. The potential difference takes place across the EMF source and the resistor.

The mentioned potential difference forces the charges to move from point 1 to 2, and hence the potential energy between these points is decreased. Since the total energy amount is conserved, the electric potential energy is transformed into another energy form. In the case shown in Fig. 3.16, the charges movement causes energy loss associated with atoms and the conductor structural lattice collisions, and thus the electric energy is transformed into heat. The energy transferred inside the resistor R follows

$$dE = dqU = IU\,dt. \tag{3.107}$$

The associated electric power, i.e., the time rate of the transferred energy is

$$P = \frac{dE}{dt} = UI = I^2 R = \frac{U^2}{R}, \tag{3.108}$$

and formula (3.108) is known as *Joule's law for electric current*. The energy transferred through the resistor in time t is governed by the Joule heating equation

$$E = Pt = UIt = I^2 Rt = \frac{U^2}{R}t, \tag{3.109}$$

where the unit of electric power in SI is $[\mathrm{P}] = [\text{volt·ampere}] = [\mathrm{W}]$, and $[\mathrm{W}]$ stands for watt.

Considering current circuits being sets of electrical devices and wires, one may distinguish *direct-current circuits* (DC) and *alternating-current circuits* (AC). The DC (AC) circuits are associated with constant (varying in time) currents and voltages.

The Ohm's law is already described by Eq. (3.89) which comes from

$$\vec{J} = \sigma(\vec{E} + \vec{v} \times \vec{B}), \tag{3.110}$$

assuming that velocity $\vec{v} \cong 0$.

Remark 1. Movement of charged particles can be generated not only by $\vec{E}$ and $\vec{B}$, but also by chemical and gravitational interactions.

Remark 2. For a perfect conductor, we have $\vec{E} = \vec{J}/\infty = 0$.

Consider a simple circuit loop consisting of a battery, two conductors and a light bulb. The first surprising observation is that the current intensity is the same in the whole circuit, and that the light appears suddenly. In the stationary state, there are only two forces: battery (source) force and the electrostatic force which transmits action of the battery force into the whole circuit. A source (battery) force can be generated by chemical reactions, piezoelectric effects, thermal effects (thermoelements), or light actions (photoelectric cells). The resultant force is

$$q\vec{F}'_r = \vec{F}'_s q + \vec{E}q, \tag{3.111}$$

and when we divide it by the charge q, we see that $\vec{F}'_r$, $\vec{F}'_s$ (source force) have the same unit as $\vec{E}$. Now we introduce the so-called *electromotive force* (EMF)

$$\mathcal{E} = \oint \vec{F}'_r \cdot d\vec{l} = \oint \vec{F}'_s \cdot d\vec{l} + \oint \vec{E} \cdot d\vec{l} = \oint \vec{F}'_s \cdot d\vec{l}, \tag{3.112}$$

since in the case of electrostatics, $\oint \vec{E} \cdot d\vec{l} = 0$.

A real battery possesses its own internal resistance R_i, and a difference of potentials between its positive B and negative A poles $V = \mathcal{E} - R_i I$. However, for an ideal battery

$$V = -\int_A^B \vec{E} \cdot d\vec{l} = \int_A^B \vec{F}'_s \cdot d\vec{l} = \oint \vec{F}'_s \cdot d\vec{l} = \mathcal{E}, \tag{3.113}$$

and the integration path has been extended to the whole circuit, since inside the battery $\vec{F}'_s = 0$.

The battery generates and maintains potential difference $V = \mathcal{E}$ which corresponds to potential energy of the battery qV, being equal to the work of the electromotive force $\mathcal{E}q$ supporting flow of the current. On the other hand, the potential energy qV supports the flow of electric charges in an opposite direction in comparison to the current movement in the circuit. Since $\vec{F}'_s$ and $\vec{E}$ are responsible for the mentioned currents movement, their direction must be opposite. It follows from Eq. (3.111) and $\vec{J} = \sigma \vec{F}'_r$, which for $\sigma = \infty$ (ideal conductor) yields $\vec{F}'_r = 0$, and hence $\vec{E} = -\vec{F}'_s$.

Let us again consider circuit loop shown in Fig. 3.16. *Kirchhoff's second law* (*voltage law*) states that the algebraic sum of changes in potential established in a complete traversal of any closed circuit loop is zero, i.e.,

$$\sum_{n=1}^{N} U_n = 0. \tag{3.114}$$

For the case shown in Fig. 3.16, changes of the potential energy are introduced by the resistor (iR) and *EMF* (battery) sources. Two rules are used while applying formula (3.114). If a resistor is traversed in the current direction (opposite), then the potential difference is $-iR$ (iR). If the battery is traversed in the *EMF* direction from its negative to positive terminal (opposite), the corresponding potential difference is $\mathcal{E}$ ($-\mathcal{E}$). For the studied case, we have

$$-IR + \mathcal{E} = 0. \tag{3.115}$$

It means that *Kirchhoff's second law* can be also formulated as follows

$$\sum_{k=1}^{K} i_k R_k + \sum_{n=1}^{N} E_n = 0, \tag{3.116}$$

which can be read in the following way: *an algebraic sum of potential differences across all resistors encountered in a complete traversal of any closed circuit loop and the algebraic sum of electromotive forces occurred in the loop is equal to zero.*

In the case of a multi-loop circuit consisting of junctions (points in the circuit where the wire segments meet) and branches (circuit paths that begin in one junction and go along the circuit to the other branches), one needs to apply the loop law to each of the distinguished branches. In addition, at any junction, the algebraic sum of currents leaving the junction ($\sum i_L$) is equal to the algebraic sum of currents entering the junction ($\sum i_E$), which is called *Kirchhoff's first law* and is governed by the following equation

$$\sum i_L = \sum i_E. \tag{3.117}$$

In other words, Kirchhoff's current law states that the net current leaving a junction of several conductors is zero.

Finally, let us discuss and illustrate *Gauss' law*.

Gauss' law states: *the total flux $\vec{D}$ out of a closed surface $\vec{S}$ is equal to the net charge within the surface.* It means that the results strongly depend on a proper choice of the integration surface, since the integral form of Gauss' law follows

$$\oint \vec{D} \cdot d\vec{S} = Q. \tag{3.118}$$

Let us consider a point charge $+Q$ and introduce Cartesian coordinates. Then, $\vec{D}$ is everywhere normal to spherical surfaces of radius r, and, Gauss' law (3.118) yields

$$Q = D \oint dS = 4\pi r^2 D.$$

(3.119)

In the vector form

$$\vec{D} = \frac{Q}{4\pi r^3}\vec{r},$$

(3.120)

and, taking into account that the electric field intensity $\vec{E}$ generated by Q is

$$\vec{E} = \frac{Q}{4\pi\varepsilon_0 r^3}\vec{r},$$

(3.121)

then

$$\vec{D} = \varepsilon_0\vec{E} \quad \left(\vec{D} = \varepsilon\vec{E}\right).$$

(3.122)

In general, Gauss' method can be directly applied for symmetric configurations of charges. Although Gaussian surface must be closed, it can be divided into a series of surface elements (*special Gaussian surfaces*) with either normal or tangential $\vec{D}$. A special Gaussian surface has the following properties: (i) It is closed; (ii) $\vec{D}$ is either normal or tangential to the surface; (iii) D is constant for the surface part where $\vec{D}$ is normal.

Gauss' law points out the difference between electric and magnetic fields. The total field flux generated by an electric field passing by a closed area is equal to the net charge (it can be either zero or nonzero)

$$\phi_E = \int_S \vec{E} \cdot d\vec{S} = \frac{1}{\varepsilon_0}Q.$$

(3.123)

If we take a bar magnet, we are able to compute the total magnetic field flux ϕ_B passing through a closed surface (Fig. 3.17).

Gauss' law applied to the magnetic field implies

$$\phi_B = \int_S \vec{B} \cdot \vec{E} = 0.$$

(3.124)

It means that magnetic field curves are closed loops which indicate the absence of isolated magnetic poles. In Fig. 3.17, three different surfaces S_1, S_2, S_3 are marked. In the case of the surface S_1, the total inward and outward fluxes are equal, and hence the total magnetic flux through S_2 is zero. The surface S_2 cuts the magnetic dipole, and this cut yields a pair of new magnets $S - N'$, and $S' - N$ having its own closed loops of field lines, thus we deal with zero magnetic field flux. The Gaussian surface S_3 possesses zero net magnetic charge, because the total inward flux equals the total outward flux.

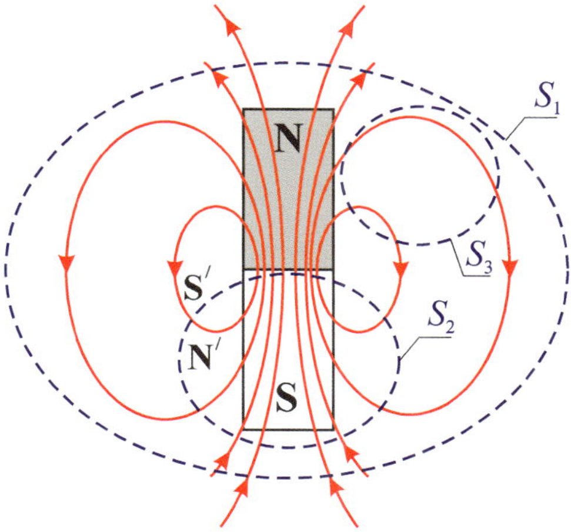

Fig. 3.17   Bar magnet and its magnetic field flux.

3.2.5 *Poisson's and Laplace's Equation*

So far, it has been shown that the fields $\vec{E}$ and $\vec{D}$ are coupled via the simple equation $\varepsilon\vec{E} = \vec{D}$, and the electric field intensity $E = -\nabla V$, where V is a potential. Therefore, for a homogeneous medium we obtain

$$\nabla \cdot \vec{D} = \varepsilon\nabla \cdot \vec{E} = -\varepsilon\nabla^2 V = -\varepsilon\nabla^2 V = \rho, \tag{3.125}$$

or equivalently

$$\Delta V \equiv \nabla^2 V = -\frac{\rho}{\varepsilon}. \tag{3.126}$$

The obtained PDE (3.126) is called *Poisson's equation*. It allows to define the potential V on the basis of known distribution of charges, and next $\vec{D}$ and $\vec{E}$ can be also defined. One may use Gauss' law to get $\vec{D}$, and next $\vec{E}$. However, in practice, this direct application of Gauss' law is questionable, since charge distributions are not a priori known. Poisson's equation is of more practical use, and the charge-free area ($\rho = 0$) is converted into the so-called Laplace's equation

$$\nabla^2 V = 0. \tag{3.127}$$

We may derive the Laplacian in Cartesian, cylindrical and spherical coordinates by finding ∇V first, and then the product $\nabla \cdot \nabla V$. Laplace's equation is directly obtained $\nabla \cdot \nabla V = \nabla^2 V = 0$.

It can be shown directly from Laplace's equation that Cartesian components of $\vec{E}$ achieve their maximum values on the boundary.

In what follows, we derive a direct solution to Laplace's equations in a few simple cases:

(i) <u>1D problem and a Cartesian solution.</u> Consider two finite parallel conducting plates with voltage $V = 0$ at $z = 0$, and $V = 50$ at $z = d$. We neglect the fringing effect, and hence the problem is reduced to one dimension, and Laplace's equation reduces to the following one

$$\frac{d^2 V}{dz^2} = 0. \tag{3.128}$$

After integration we get

$$V(z) = A_1 z + A_2. \tag{3.129}$$

The introduced boundary conditions yield $A_2 = 0$, $A_1 = 50/d$, and hence

$$V(z) = \left(\frac{50}{d}\right) z, \quad [\text{V}]. \tag{3.130}$$

Knowing V, we may find $\vec{E}$ and $\vec{D}$:

$$\vec{E} = -\nabla V = -\frac{50}{d} = -\frac{50}{d}\vec{k}, \quad [\text{V/m}], \tag{3.131}$$

and

$$\vec{D} = \varepsilon \vec{E} = -\frac{50}{d}\varepsilon \vec{k}, \quad [\text{C/m}^2]. \tag{3.132}$$

At the plate $z = 0$ we have $\rho = -50\varepsilon/d$, whereas at the plate $z = d$, we have $\rho = +50\varepsilon/d$.

(ii) **2D problem and a Cartesian product solution.**
Let $V = V(x, z)$, and let us assume the following solution form $V = X(x)Z(z)$.
From Laplace's equation we get

$$\frac{\partial^2 (XZ)}{\partial x^2} + \frac{\partial^2 (XZ)}{\partial z^2} = 0, \tag{3.133}$$

or equivalently

$$Z\frac{\partial^2 X}{\partial x^2} + X\frac{\partial^2 Z}{\partial z^2} = 0. \tag{3.134}$$

Separation of the variables yields

$$\frac{1}{X}\frac{\partial^2 X}{\partial x^2} + \frac{1}{Z}\frac{\partial^2 Z}{\partial z^2} = 0. \tag{3.135}$$

Observe that the first term of LHS is independent of $Z(X)$. Therefore, we may set them equal to a constant $\mp\alpha^2$. Equation (3.135) is split into two second order ODEs of the forms:

$$\frac{d^2 X}{\partial x^2} - \alpha^2 X = 0, \tag{3.136}$$

$$\frac{d^2 Z}{\partial z^2} + \alpha^2 Z = 0. \tag{3.137}$$

General solution to equation (3.136) is

$$X(x) = C_1^{(1)} e^{\alpha x} + C_2^{(1)} e^{-\alpha x}, \tag{3.138}$$

or equivalently

$$X(x) = C_3^{(1)} \cosh(\alpha x) + C_4^{(1)} \sinh(\alpha x). \tag{3.139}$$

General solution to equation (3.137) is

$$Z(z) = C_1^{(2)} e^{i\alpha z} + C_2^{(2)} e^{-i\alpha z}, \quad i^2 = -1, \tag{3.140}$$

or equivalently

$$Z(z) = C_3^{(2)} \cos(\alpha z) + C_4^{(2)} \sin(\alpha z). \tag{3.141}$$

Finally, we obtain

$$V(x,z) = \left(C_1^{(1)} e^{\alpha x} + C_2^{(1)} e^{-\alpha x} \right) \left(C_1^{(2)} e^{i\alpha z} + C_2^{(2)} e^{-i\alpha z} \right), \tag{3.142}$$

or equivalently

$$V(x,z) = \left(C_3^{(1)} \cosh(\alpha x) + C_4^{(1)} \sinh(\alpha x) \right) \left(C_3^{(2)} \cos(\alpha z) + C_4^{(2)} \sin(\alpha z) \right). \tag{3.143}$$

(iii) <u>3D problem and a cylindrical product solution.</u>
We assume the following form of solution $V(r, \Theta, z) = R(r)\Theta(\theta)Z(z)$ and we apply Laplace's equation in the form

$$\nabla^2 V = \frac{1}{r} \frac{\partial}{\partial r} \left(r \frac{\partial V}{\partial r} \right) + \frac{1}{r^2} \frac{\partial^2 V}{\partial \theta^2} + \frac{\partial^2 V}{\partial z^2} = 0. \tag{3.144}$$

Substituting the assumed solution into (3.144), we obtain

$$\frac{\Theta Z}{r} \frac{d}{dr} \left(r \frac{dR}{dr} \right) + \frac{RZ}{r^2} \frac{d^2\Theta}{d\theta^2} + R\Theta \frac{d^2 Z}{dz^2} = 0. \tag{3.145}$$

We divide both sides of equation (3.145) by $R\Theta Z$, and we get

$$\frac{1}{R} \frac{d^2 R}{dr^2} + \frac{1}{Rr} \frac{dR}{dr} + \frac{1}{r^2\Theta} \frac{d^2\Theta}{d\theta^2} = -\frac{1}{Z} \frac{d^2 Z}{dz^2} = -\beta^2, \tag{3.146}$$

where β is constant. At first, we find $Z(z)$ as a solution to the following equation

$$\frac{d^2 Z}{dz^2} - \beta^2 z = 0, \tag{3.147}$$

and hence

$$Z(z) = C_1 \cosh \beta z + C_2 \sinh \beta z. \tag{3.148}$$

Equation (3.146) implies also the next ODE of the form

$$\frac{r^2}{R} \frac{d^2 R}{dr^2} + \frac{r}{R} \frac{dR}{dr} + \beta^2 r^2 = \frac{1}{\Theta} \frac{d^2\Theta}{d\theta^2} = \alpha^2. \tag{3.149}$$

We first solve the second order ODE of the form

$$\frac{d^2\Theta}{d\theta^2} + \alpha^2\Theta = 0, \tag{3.150}$$

and its solution follows

$$\Theta(\theta) = C_3 \cos \alpha\Theta + C_4 \sin \alpha\Theta. \tag{3.151}$$

The final second order ODE is known as the *Bessel differential equation*, and has the following form

$$\frac{d^2 R}{dr^2} + \frac{1}{r}\frac{dR}{dr} + \left(\beta^2 - \frac{\alpha^2}{r^2}\right) R = 0. \tag{3.152}$$

Solutions to Bessel equations are known and are described by the Bessel functions. We have

$$R(r) = C_5 J_\alpha(\beta r) + C_6 N_\alpha(\beta r), \tag{3.153}$$

where:

$$J_\alpha(\beta r) = \sum_{m=0}^{\infty} \frac{(-1)^m (\beta r/2)^{\alpha+2m}}{m!\Gamma(\alpha + m + 1)}, \tag{3.154}$$

$$N_\alpha(\beta r) = \frac{(\cos \alpha\pi) J_\alpha(\beta r) - J_{-\alpha}(\beta r)}{\sin \alpha\pi}. \tag{3.155}$$

$J_\alpha(Br)$ stands for a Bessel function of the first kind of order α, whereas $N_\alpha(Br)$ is a Bessel function of the second kind of order α. For $\alpha = n \in N$ and large r, the Bessel functions are similar to the damped sine waves

$$J_n(r) \cong \sqrt{\frac{2}{\pi r}} \cos\left(r - \frac{\pi}{4} - \frac{n\pi}{2}\right), \tag{3.156}$$

$$N_n(r) \cong \sqrt{\frac{2}{\pi r}} \sin\left(r - \frac{\pi}{4} - \frac{n\pi}{2}\right). \tag{3.157}$$

(iv) 2D spherical product solution.
We assume that $V = V(r, \psi) = \Psi(\psi)R(r)$, and hence the truncated Laplace's equation in spherical coordinates follows

$$\frac{1}{r^2}\frac{\partial}{\partial r}\left(r^2\frac{\partial V}{\partial r}\right) + \frac{1}{r^2 \sin\psi}\frac{\partial}{\partial \psi}\left(\sin\psi\frac{\partial V}{\partial \psi}\right) = 0, \tag{3.158}$$

or equivalently

$$\frac{1}{r^2}\frac{\partial}{\partial r}\left(r^2\frac{\partial V}{\partial r}\right) + \frac{1}{r^2 \sin\psi}\left(\cos\psi\frac{\partial V}{\partial \psi} + \sin\psi\frac{\partial^2 V}{\partial \psi^2}\right) = 0. \tag{3.159}$$

Therefore, we get

$$\left(r^2\frac{d^2 R}{dr^2} + 2r\frac{dR}{dr}\right)\Psi + R\left(\frac{d^2\Psi}{d\psi^2} + \frac{1}{\tan\psi}\frac{d\Psi}{d\psi}\right) = 0 \tag{3.160}$$

or, dividing by $R\Psi$, we obtain

$$\left(\frac{r^2}{R}\frac{d^2 R}{dr^2} + \frac{2r}{R}\frac{dR}{dr}\right) + \left(\frac{1}{\Psi}\frac{d^2\Psi}{d\psi^2} + \frac{1}{\Psi\tan\psi}\frac{d\Psi}{d\psi}\right) = 0. \tag{3.161}$$

Let us introduce the following constant $n(n+1)$:

$$\frac{r^2}{R}\frac{d^2 R}{dr^2} + \frac{2r}{R}\frac{dR}{dr} = -\left(\frac{1}{\Psi}\frac{d^2\Psi}{d\psi^2} + \frac{1}{\Psi\tan\psi}\frac{d\Psi}{d\psi}\right) = n(n+1), \qquad (3.162)$$

and n is an integer. The following two separated second order ODEs are obtained:

$$r^2\frac{d^2 R}{dr^2} + 2r\frac{dR}{dr} - n(n+1)R = 0, \qquad (3.163)$$

$$\frac{d^2\Psi}{d\psi^2} + \frac{1}{\tan\psi}\frac{d\Psi}{d\psi} + n(n+1)\Psi = 0. \qquad (3.164)$$

Equation (3.163) has the following solution

$$R(r) = C_1 r^n + C_2 r^{-(n+1)}, \qquad (3.165)$$

which can be verified by direct substitution to (3.163). In order to solve (3.164) we introduce the following variable $\gamma = \cos\psi$.

3.3 Induced Electric Fields

3.3.1 *Electromotive Force*

Let us consider a straight vertical, isolated conductor, for instance a metal, where free electrons can easily move within the metal structural lattice. In the majority of (natural) cases, free electrons move in a random way, so that if one introduces an imaginary plane cutting the inductor perpendicularly, then the number of electrons passing through this plane in both directions is equal (see Fig. 3.18a). In this case, no electric field is generated by the conductor.

Let us assume now that we have introduced a uniform magnetic field, as it is shown in Fig. 3.18b. The previous random movement of free electrons is not disturbed at all. However, when the conductor starts to move with a constant velocity $\vec{v}\perp\vec{B}$, the situation changes, i.e., the electrons start moving in an ordered way, in the direction indicated by the Lorentz law (see Fig. 3.52b for the right-hand rule for vectors $\vec{B}, \vec{v}, \vec{F}$).

If one takes one electron $-q$ from the electrons flow, then this electron is subjected to action of the Lorentz force $\vec{F} = -q(\vec{v}\times\vec{B})$ pushing the electron to move vertically downward. After such a rapid transient state, one may consider the conductor AA' as a capacitor, the cross sections Π and Π' of which may be understood as the capacitor plates. As a result, we get two induced surface charges, with the top positive (Π) charge and bottom negative charge (Π'). Now, if one puts a test charge $+q_0$ inside the conductor, then the charge will move vertically downward, and hence the electric field direction $\vec{E}$ in the conductor AA' is directed downwards too, i.e., $\vec{E} = E\,\vec{j}$.

We follow here a convention used for current direction labeling. Namely, the conventional current direction is defined by the direction of flow of positive charges.

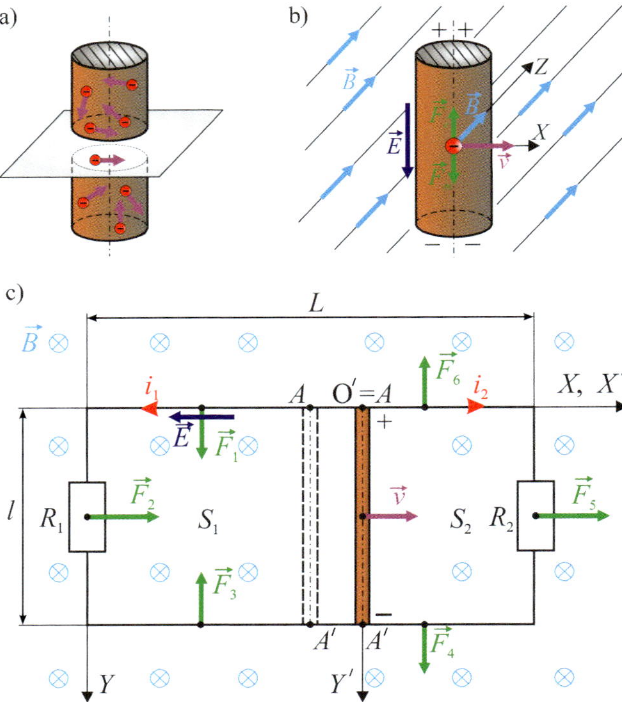

Fig. 3.18   The conductor without (a) and with (b) the action of the magnetic field, and rectangular circuit (c) with movable conductor AA'.

It means that in our case, electrons move opposite to the direction of the current. Although the current is associated with arrows indicating the direction of the flow of positive charges, it is a scalar quantity.

The potential difference between two points A, A' belonging to two cross sections can be defined by the following relationship

$$EMF \equiv \mathcal{E} = \frac{V_A - V_{A'}}{q} = \frac{1}{q} \int_{A'}^{A} \vec{F} \cdot \vec{j} \, dy = \vec{E} \cdot \vec{j} \int_{-\frac{l}{2}}^{\frac{l}{2}} dy = \vec{E} \cdot \vec{l}, \tag{3.166}$$

and is equivalent to the potential energy change. Recall that the potential is a scalar quantity (it can be positive or negative), its SI unit is volt $[\text{V}]=[\text{J/C}]$, whereas the SI unit of the electric field $[\text{E}]=[\text{N/C}]$.

The indicated electric field intensity vector $\vec{E}$ can be also derived by considering actions of two fields (magnetic and electric) on the electron. Owing to Coulomb's law, we have

$$\vec{F}_m + \vec{F}_e = \vec{0}, \tag{3.167}$$

or equivalently

$$q(\vec{E} + \vec{v} \times \vec{B}) = \vec{0}, \tag{3.168}$$

and hence

$$\vec{E} = \vec{B} \times \vec{v}. \tag{3.169}$$

Formula (3.169) shows that the electric field $\vec{E}$ can be generated by the magnetic field and the conductor velocity $\vec{v}$. This observation has been motivated by Faraday's experiments, who observed that a charge put in the magnetic field induces an electric field.

Since we have the potential energy harvested between two cylindrical conductor places, it can be treated as a battery. If we take a resistor R which closes an electrical circuit, then the potential difference between cross sections Π and Π' will generate an electromotive force, and the current i will be established. One may introduce the equivalent diagram for the *EMF* shown Fig. 3.19.

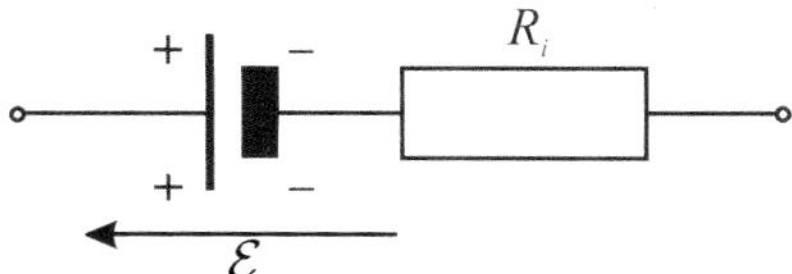

Fig. 3.19   Equivalent scheme for *EMF* $\mathcal{E}$.

It consists of a battery and an intrinsic resistance R_i. The battery, being the *EMF* source, produces the electrical energy from other form of energy like chemical, mechanical or radial ones. *EMF* is defined as $\mathcal{E} = dW/dq$, [V]=[J/C]. It should be able to perform work on charge carriers, and it is defined as the amount of work done on the unit positive charge to move it from a negative to positive terminal. The arrow of $\mathcal{E}$ shows the direction of positive charges movement through the *EMF* source *from the negative (low potential) to positive terminal (high potential)*. We have

$$\mathcal{E} = iR. \tag{3.170}$$

The electromotive force $EMF(E)$ can be also induced by the change of a magnetic flux.

In general, if we consider an open surface S being bounded by a closed contour C and assume that the magnetic flux ϕ crossing S changes in time, then the induced voltage $V = V(t, x_1, x_2, x_3)$ depends on time and space, and is defined through Faraday's law $V = -d\phi/dt$.

Observe that in the earlier considered nonconservative field, the electromotive force $EMF = V(t, x_1, x_2, x_3)$, whereas in a conservative electric field, voltage V does not depend on the position in space and is rather associated with the electrostatic potential (the electromotive force EMF is not generated in this case). When the magnetic flux ϕ changes in a circuit element, then $V = V(t)$ and $i = i(t)$, and Faraday's law states that $V(t) = -\frac{d\phi}{dt}\frac{di}{dt} = -L\frac{di}{dt}$. Now, L is the *self-inductance* of the circuit element and V is *voltage of self-inductance*.

In what follows, we show how we may derive the corresponding electric field $\vec{E}$ knowing the electric potential V, and vice versa. The general mathematical definition of a derivative of a scalar field V (physically, in spite of potential in an electric field, it may represent the height of terrain points in geology) in the direction of a unit vector $\vec{l}$ is as follows

$$\frac{\partial V}{\partial l}(M_0) = \lim_{s \to 0} \frac{f(M) - f(M_0)}{s}, \tag{3.171}$$

where: $M_0 M = \vec{l}s$, $s > 0$, or equivalently

$$\frac{\partial V}{\partial l} = \left(\vec{l} \cdot \nabla\right) V(M_0), \tag{3.172}$$

and M_0 is an arbitrary point of the field. The electric field vector $\vec{E}$ points in the decreasing potential direction, as it is shown in Fig. 3.20. We can take various directions of the unit vector $\vec{\hat{l}}$, and thus evaluate the derivatives with the use of formula (3.171). However, there exists only one vector $\vec{l}_A$, where $\vec{l}_A \cdot \vec{E} = -E$. The latter one points in the direction opposite to the vector $\vec{E}$. Since the electric field generates a potential, and vice versa, it can be understood as an action and reaction rule which satisfies classical Newton's third law.

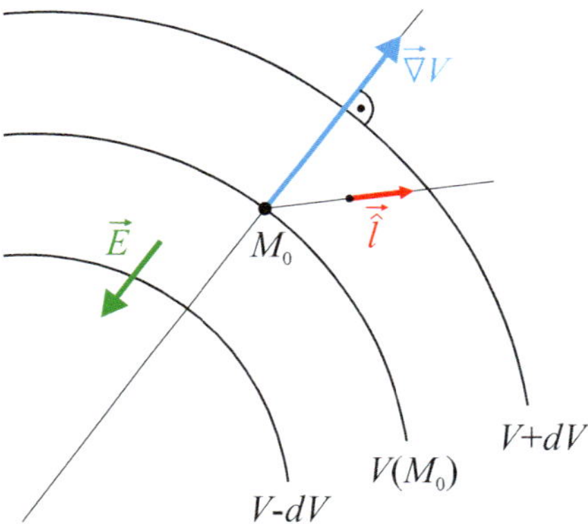

Fig. 3.20   Electric field intensity $\vec{E}$ (vector) and potential V (scalar).

In a study of either scalar or vector fields, we need to know how a field is changed due to changes of its segments. This is why the gradient (scalar field) as well as divergence and curl (vector field) operators are widely applied. We show that the change dV in the scalar function V, when traversed over a path $d\vec{r}$, is defined by the following dot product

$$dV = \nabla V \cdot d\vec{r}, \tag{3.173}$$

where $\nabla V \equiv grad V$ stands for the gradient of the scalar function V. Consider the Cartesian coordinates and take two neighboring points separated by

$$d\vec{r} = \vec{i}dx + \vec{j}dy + \vec{k}dz, \tag{3.174}$$

It follows from (3.173) and (3.174) that

$$dV = \left(\frac{\partial V}{\partial x}\vec{i} + \frac{\partial V}{\partial y}\vec{j} + \frac{\partial V}{\partial z}\vec{k}\right) \cdot \left(\vec{i}dx + \vec{j}dy + \vec{k}dz\right) V =$$
$$= \frac{\partial V}{\partial x}dx + \frac{\partial V}{\partial y}dy + \frac{\partial V}{\partial z}dz, \tag{3.175}$$

which agrees with the direct definition of the differential dV. Observe that by fixing $|d\vec{r}|$, we preserve the distance between two chosen points, and ∇V indicates the direction of the maximum increase in V. Assume that two chosen points lie on the same *equipotential surface* $V(x, y, z) = C_1$.

We have $dV = dC_1 = 0$, which means that $d\vec{r} \perp \nabla V$. Since $d\vec{r}$ is tangent to the equipotential surface, ∇V. The gradient ∇V of a potential function (here electric potential) $V(x, y, z)$ is a vector field being everywhere normal to the equipotential surfaces.

Assume that we deal with an electric field $\vec{E}$ acting on a test charge $+q$. This action generates simultaneous reaction coming from the associated potential, and hence the resultant force acting on the test charge is

$$q\left(\vec{E} + \vec{i}_A\frac{\partial V}{\partial l}\right) = 0. \tag{3.176}$$

For $\vec{l} = \vec{l}_A$, taking into account (3.172), we get

$$\vec{E} = -\nabla V(M_0), \tag{3.177}$$

where in the given Cartesian coordinates ∇ operates on the scalar function:

$$\nabla = \vec{i}\frac{\partial}{\partial x} + \vec{j}\frac{\partial}{\partial y} + \vec{k}\frac{\partial}{\partial z},$$
$$\nabla V(M_0) = \frac{\partial f(M_0)}{\partial x}\vec{i} + \frac{\partial f(M_0)}{\partial y}\vec{j} + \frac{\partial f(M_0)}{\partial z}\vec{k} = \nabla f(M_0). \tag{3.178}$$

Considering cylindrical coordinates (r, θ, z) of the point $M(r, \theta, z)$ supplemented with corresponding unit vectors $\vec{e}_r, \vec{e}_\theta, \vec{e}_z$, being tangent to coordinate curves and passing through point M, we have

$$\nabla V = \vec{e}_r\frac{\partial V}{\partial r} + \frac{1}{r}\vec{e}_\theta\frac{\partial V}{\partial \theta} + \vec{e}_z\frac{\partial V}{\partial z}, \tag{3.179}$$

whereas for the spherical coordinates of the point $M(r, \theta, \psi)$ and the corresponding unit tangent vectors $\vec{e}_r, \vec{e}_\theta, \vec{e}_\psi$ being tangent to the coordinate curves and passing through point M, we have

$$\nabla V = \vec{e}_r\frac{\partial V}{\partial r} + \frac{\vec{e}_\theta}{r\sin\psi}\frac{\partial V}{\partial \theta} + \frac{\vec{e}_\psi}{r}\frac{\partial V}{\partial \psi}. \tag{3.180}$$

Knowing $\vec{E}$, we can establish the associated potential function V. The electric field $\vec{E}$ component taken in an arbitrary direction $\vec{l}$ is equal to the negative rate of the potential change regarding the position displacement in this direction. $\vec{E}$

points opposite to the potential gradient at the point M_0, i.e., in the direction of decreasing potential value.

On the contrary, knowing a spatial distribution of a potential in the Cartesian, cylindrical and spherical coordinates, one may find the associated components of the electric field $\vec{E}$ with the use of the following formulas (see Eqs. (3.178), (3.179) and (3.180)), respectively:

$$
\begin{aligned}
E_1 &= -\frac{\partial V}{\partial x}, \quad E_2 = -\frac{\partial V}{\partial y}, \quad E_3 = -\frac{\partial V}{\partial z}, \\
E_r &= -\frac{\partial V}{\partial r}, \quad E_\theta = -\frac{1}{r}\frac{\partial V}{\partial \theta}, \quad E_z = -\frac{\partial V}{\partial z}, \\
E_r &= -\frac{\partial V}{\partial r}, \quad E_\theta = -\frac{1}{r\sin\psi}\frac{\partial V}{\partial \theta}, \quad E_\psi = -\frac{1}{r}\frac{\partial V}{\partial \psi}.
\end{aligned}
\tag{3.181}
$$

Let us now consider the case shown in Fig. 3.18c. Movement of the conductor with velocity $\vec{v}$ induces the electromotive force EMF $(\mathcal{E})$. This is due to changes of the areas located on the left- and the right-hand side of the vertical conductor AA' during its movement, and hence the magnetic field flux is defined as

$$
\phi_n(t) = \vec{B} \cdot \vec{S}_n(t), \quad n = 1, 2,
\tag{3.182}
$$

assuming that the magnetic induction field is constant and uniform, and $\vec{S}_n(t)$ denotes the surface orientation coinciding with the corresponding current.

3.3.2   *Rectangular Loop with a Conductor Moving in a Magnetic Field*

Let us consider a wire segment shown in Fig. 3.18b, and let us take the horizontal part of the wire loop of the length l (see Fig. 3.19) located on the right with respect to the moving conductor's cross section (Fig. 3.21).

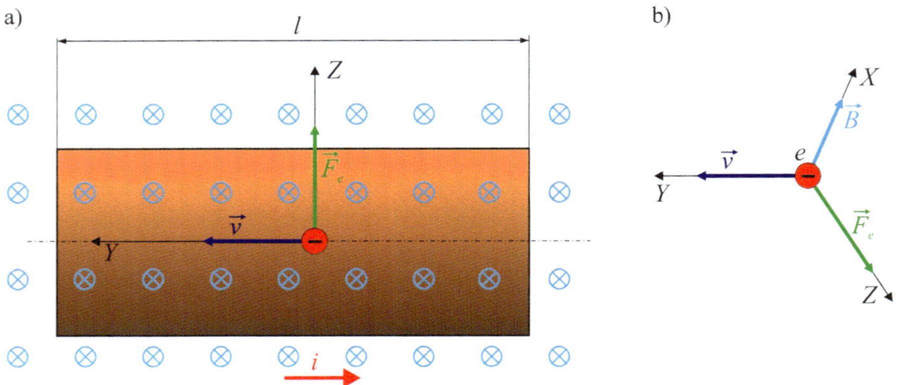

Fig. 3.21   Wire segment with the current i embedded in the magnetic field $\vec{B}$ (a) and the Lorentz force action on its electron (b).

The following Lorentz force acts on the electron moving with velocity $\vec{v}$ opposite to the direction of the current i flow

$$\vec{F}_e = q(\vec{v} \times \vec{B}) = -e(\vec{v} \times \vec{B}) = e(\vec{B} \times \vec{v}). \tag{3.183}$$

Assuming that we have N electrons moving in the wire segment of the length l, the total force

$$\vec{F} = N\vec{F}_e = eN(\vec{B} \times \vec{v}). \tag{3.184}$$

The volume of the cylindrical shape conductor of the length l and area of cross section S is connected by the relationship

$$N = nSl, \tag{3.185}$$

where n is the electron volume density. In what follows, we introduce the microscopic quantity associated with the current flow, i.e., the so-called current density $\vec{J}$, where $J=[\text{A/m2}]$. This allows to transit from a scalar (current i) to the vector $\vec{J}$ defined as follows

$$d\vec{S} \cdot \vec{J} = di, \tag{3.186}$$

where the cross section of different current is represented by the vector $d\vec{S}$. It means that the magnitude J of the density represents a current element di passing through the elementary conductor cross section area dS. The direction of the external electric field $\vec{E}$ is the same as the direction of the vector $\vec{J}$ (the electrons move opposite to the direction of $\vec{J}$). Therefore, the vectors $\vec{E}$ and $\vec{v}$ have opposite directions, and the following relationship holds

$$\vec{J} = -ne\vec{v}, \tag{3.187}$$

where n is a number of electrons per unit volume. From (3.186) one obtains

$$i = \int_S \vec{J} \cdot d\vec{S}, \tag{3.188}$$

which means that the current i can be viewed as the flux of the vector $\vec{J}$ over the surface S, and the integral is taken over the whole conductor cross section.

It follows from (3.184) that in the given Cartesian coordinates, we have

$$\vec{F} = -enSl\vec{v} \times \vec{B}) = -enSl\left(\vec{j}v \times \vec{B}\right) = enSv\left(\vec{j}l \times \vec{B}\right) = enSv\left(\vec{l} \times \vec{B}\right)$$
$$= i\left(\vec{l} \times \vec{B}\right), \tag{3.189}$$

because, due to the formula (3.187), vectors $\vec{v}$ and $\vec{J}$ have opposite directions (the direction of the current flow coincides with the direction of the vector $\vec{J}$), and $\vec{j}$ denotes the unit vector of the Cartesian axis OY.

Introducing the fixed closed conducting loop into the magnetic field $\vec{B}$, and moving the conductor AA' to the right and left with constant velocity $\vec{v}$, we produce the current $i = i_1 + i_2$, where $i_1(i_2)$ is associated with the left (right) part of the

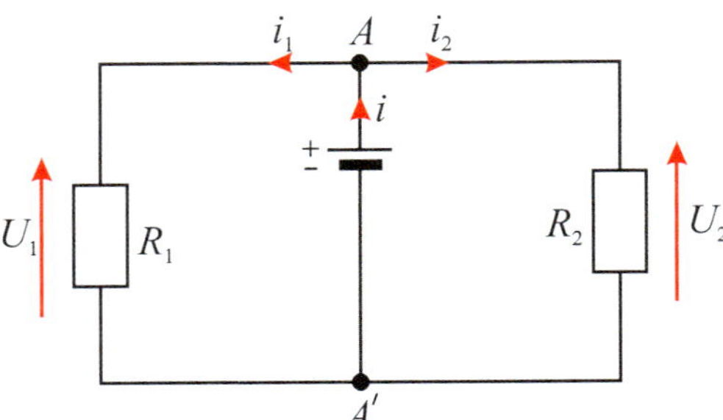

Fig. 3.22 The electric circuit with the battery AA' corresponding to the circuit shown in Fig. 3.18c at a given time instant.

loop. The potential differences $U_1 = i_1 R_1$, $U_2 = i_2 R_2$ will be registered by the voltmeters, and the situation shown in Fig. 3.18c is equivalent to that presented in Fig. 3.22.

Each of the conducting loops is affected by the magnetic flux changes ϕ_1, ϕ_2 in time by means of the following relationships:

$$\phi_1 + \phi_2 = BlL,$$
$$\phi_1 = \vec{B} \cdot \vec{S}_1(t) = Blvt \cos(\vec{B}, \vec{S}) = Blvt,$$
$$\phi_2 = \vec{B} \cdot \vec{S}_2(t) = Bl(L - vt) \cos(\vec{B}, \vec{S}) = Bl(L - vt).$$
(3.190)

In the circuit, the introduced electromotive forces EMF $(\mathcal{E}_n)$ follow Faraday's law of induction, and hence:

$$U_1 \equiv \mathcal{E}_1 = -\frac{d\phi_1}{dt} = -Blv,$$
$$U_2 \equiv \mathcal{E}_2 = -\frac{d\phi_2}{dt} = Blv,$$
$$\mathcal{E}_1 + \mathcal{E}_2 = 0.$$
(3.191)

Owing to the formula (3.189), we can derive forces $\vec{F}_n$ acting on the six corresponding parts of the length l_n of the closed loop shown in Fig. 3.18c. Namely, we have

$$\vec{F}_n = i\vec{l}_n \times \vec{B}, \quad n = 1, ..., 6,$$
(3.192)

and the vectors $\vec{F}_n$ are shown in Fig. 3.18c. Observe that $\vec{F}_1 + \vec{F}_3 = \vec{0}$, $\vec{F}_4 + \vec{F}_6 = \vec{0}$, and we have the following resultant forces:

$$F_2 = i_1 lB, \qquad F_5 = i_2 lB$$
(3.193)

which means

$$F_2 + F_5 = (i_1 + i_2)lB = ilB.$$
(3.194)

On the other hand:

$$i_1 = \frac{Blv}{R_1}, \qquad i_2 = \frac{Blv}{R_2},$$
(3.195)

and hence

$$F_2 + F_5 = B^2 l^2 v \left(\frac{1}{R_1} + \frac{1}{R_2} \right). \tag{3.196}$$

Action of the force $\vec{F}_v$ causing the movement of the conductor with velocity $\vec{v}$ causes an occurrence of the current and the forces $\vec{F}_2 + \vec{F}_5$, which are equal to the relative force $\vec{F}_R$, as it is schematically shown in Fig. 3.23.

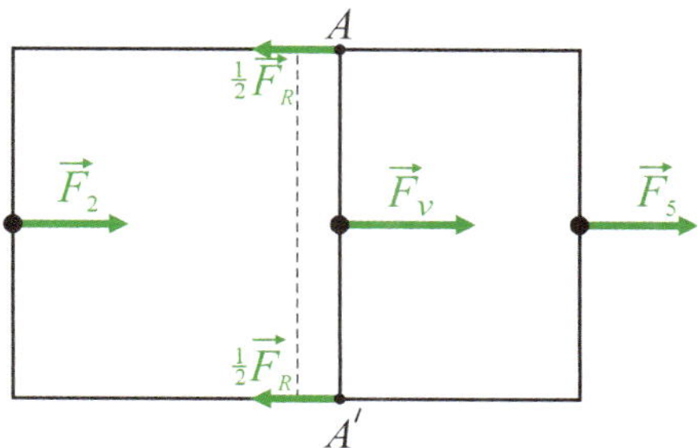

Fig. 3.23   Actions and reactions of the closed loop circuit.

Mechanical power introduced by the conductor moving with velocity $\vec{v}$ is equal to

$$P = (F_2 + F_5)\, v = B^2 l^2 v^2 \left(\frac{1}{R_1} + \frac{1}{R_2} \right), \tag{3.197}$$

and is converted into heat generated in the resistors R_1 and R_2. The studied mechatronic system includes a transition of the mechanical energy (conductor moving with the constant speed $\vec{v}$) into both electrical and heat energies.

So far, we have introduced two frames, the absolute $OXYZ$ and movable $OX'Y'Z'$ Cartesian coordinates in Fig. 3.18c.

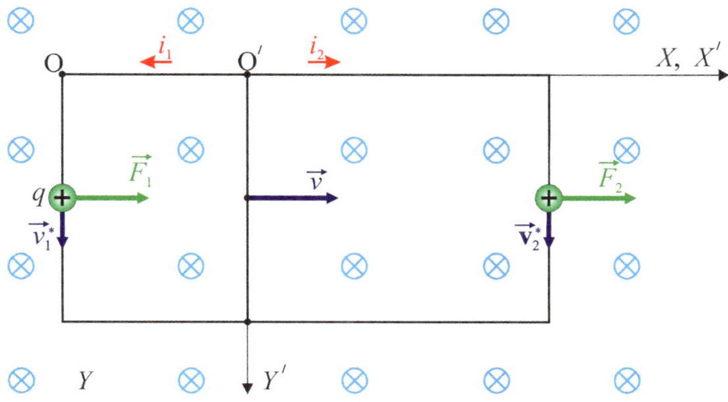

Fig. 3.24   Rectangular conducting loop in the magnetic field with movable vertical conductor (observer is fixed at the point O).

In the case of the absolute frame $(OXYZ)$, the observer is fixed in the point O, i.e., he/she is fixed with the unmovable conductor loop. In this frame of reference, the magnetic field does not move, but the positive charge moves in the direction coinciding with the corresponding current direction. The actions of the Lorentz forces are depicted in Fig. 3.24. The electromotive forces $\mathcal{E}_1$ and $\mathcal{E}_2$ are induced by the changes of the magnetic fluxes $\phi_1(t), \phi_2(t)$ in time, and they are associated with the electric field $\vec{E}_n$ in the following way:

$$\mathcal{E}_1 = \oint \vec{E}_1 \cdot \vec{l}, \qquad \mathcal{E}_2 = \oint \vec{E}_2 \cdot \vec{l}. \tag{3.198}$$

The horizontal segments of the rectangular loop produce the values $\vec{E}_n \cdot \vec{dl}$ which cancel each other, and hence we have:

$$\mathcal{E}_1 = \vec{E}_1 \cdot \vec{l}, \qquad \mathcal{E}_2 = \vec{E}_2 \cdot \vec{l}. \tag{3.199}$$

This situation is similar to that including vertical forces induced by currents i_1 and i_2 which cancel each other. Therefore, only horizontal forces $\vec{F}_1$ and $\vec{F}_2$ exist in the $OXYZ$ coordinates. It means that the induced EMF is produced in the vertical segment of the rectangular conducting loop only.

Note that for the observer fixed in the magnetic field, forces causing the movement of charges are of *purely electric origin*, and hence the induced electric fields $\vec{E}_1$, $\vec{E}_2$ are observed, and the electrically induced forces follow

$$\vec{F}_n = q\vec{E}_n. \tag{3.200}$$

Now, we consider the case when the observer is fixed in the point O', and he/she moves together with the conductor AA'. The charge q_n moves around the loop parts in directions indicated by the currents directions with velocities $\vec{v}_n$, $n = 1, 2$. It experiences the action of the velocity $\vec{v}$, because the magnetic field moves with the velocity $-\vec{v}$. Therefore, the resultant (net) velocity vectors of the charges are defined as follows

$$\vec{v}_n^* = \vec{v}_n + \vec{v}. \tag{3.201}$$

In the $O'X'Y'Z'$ coordinates, the magnetic forces

$$\vec{F}_n^* = q(\vec{v}_n + \vec{v}) \times \vec{B} = \vec{F}_{nL} + \vec{F}_D, \tag{3.202}$$

where:

$$\vec{F}_{nL} = q(\vec{v}_n \times \vec{B}), \qquad \vec{F}_D = q(\vec{v} \times \vec{B}). \tag{3.203}$$

In the $O'X'Y'Z'$ coordinates system, the observer moves with the horizontal velocity and he/she observes magnetic field forces only (see Eqs. (3.202), (3.203)). Relative motion of the loop and the magnetic field, expressed by velocity $\vec{v}$, is identical in both discussed frames, and thus the EMF is also the same in both cases

$$El = Blv, \tag{3.204}$$

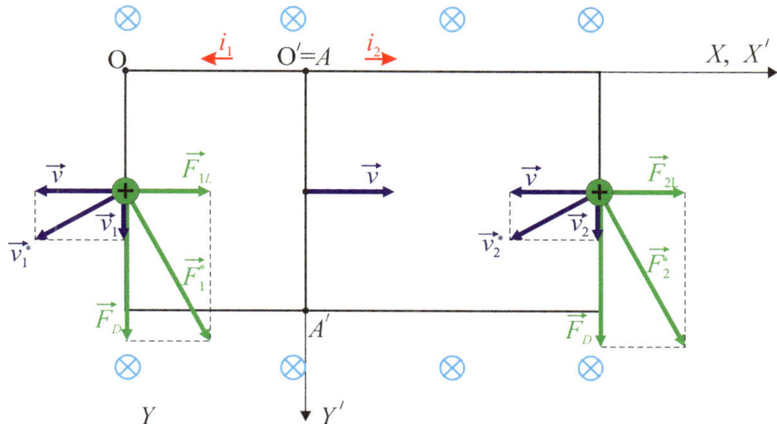

Fig. 3.25 Rectangular conducting loop in the magnetic field with movable vertical conductor (observer is fixed at the point O').

which means that in the general case,

$$\vec{E} = \vec{v} \times \vec{B}. \tag{3.205}$$

The following relationship is established between the coordinates

$$x(t) = vt + x'(t). \tag{3.206}$$

Summarizing the two analyzed cases, it has been shown that, depending on the choice of the reference frame, different $\vec{E}$ and $\vec{v}$ vectors appear. At the same time, the same magnitudes of forces $\vec{F}$ and electromotive forces are obtained.

Generally, the magnetic and electric fields depend on each other and cannot exist separately. The relationship between them follows

$$\vec{F} = q(\vec{E} + \vec{v} \times \vec{B}). \tag{3.207}$$

Assuming that the magnetic field studied thus far will move horizontally with constant speed, for an arbitrary observer fixed with the absolute (unmovable) coordinates being fixed to the Earth, the force responsible for the occurred current cannot be either purely magnetic or purely electric.

3.3.3 *Alternating Current*

We consider a rectangular current loop of the area S, embedded in a uniform magnetic field $\vec{B}$ which rotates with a constant angular velocity ω about its horizontal axis OX ($\vec{B} \perp \vec{i}$). The magnetic flux

$$\phi_B = \vec{B} \cdot \vec{S} = BS \cos \varphi = BS \cos (\omega t + \varphi_0), \tag{3.208}$$

where φ_0 is the angle between vectors $\vec{B}, \vec{S}$ at the time instant $t = 0$ (see Fig. 3.26).

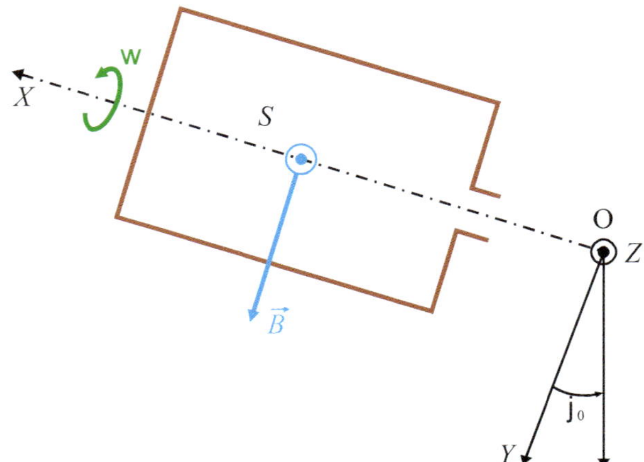

Fig. 3.26   Rectangular conductor loop rotation with the angular velocity ω.

Faraday's induction law yields the induced harmonic electromotive force

$$\mathcal{E} = -\frac{d\phi_B}{dt} = \mathcal{E}_0 \sin(\omega t + \varphi_0), \tag{3.209}$$

where $E_0 = BS\omega$. If we connect this conductor loop to the unmovable resistor R, then, owing to Ohm's law, the following current is generated

$$i = i_0 \sin(\omega t + \varphi_0), \tag{3.210}$$

where its amplitude $i_0 = \frac{\mathcal{E}_0}{R} = \frac{BS\omega}{R}$. The alternating current power is

$$P = E\,i = \mathcal{E}_0 i_0 \sin^2 \omega t, \tag{3.211}$$

and its averaged value follows

$$P_{av} = \frac{1}{T} \int_0^T P\,dt = \frac{1}{2}\varepsilon_0 i_0, \tag{3.212}$$

where: $T = \frac{2\pi}{\omega}$, $\varphi_0 = 0$.

One may also introduce the root-mean-square value of both the electromotive force and current

$$\mathcal{E}' = \frac{\mathcal{E}_0}{\sqrt{2}} \cong 0.707\varepsilon_0,$$
$$i' = \frac{i_0}{\sqrt{2}} \cong 0.707 i_0, \tag{3.213}$$

and formula (3.212) takes the form

$$P_{av} = \mathcal{E}'i'. \tag{3.214}$$

3.4 Magnetic and Electromagnetic Phenomena

3.4.1 *Magnets and Magnetic Fields*

The word *magnet* comes from Greek. Namely, the *Magnetes* were an ancient Greek tribe that lived in Thessalian Magnesia. They developed two cities in Anatolia, both named Magnesia (on the Moeander and ad Spylum). The regions colonized by the Magnetes were famous for mysterious stones which could either attract or repel each other, yielding then the terms for *magnets* and *magnetism* [Chaber and Chaber (1891)]. Lodestones (loadstones) are natural magnets which can attract pieces of iron. They were used as magnetic compasses (leading stones) employed in early navigation (for instance, in medieval China they were used until the 12th century). Lodestone's magnetic properties were firstly described in the 6th century BC in Greece [Brand *et al.* (2015)] and in the 4th century BC in China [Shu-hua (1954)] and India (see also [Carlston (189)]). A magnet is an object or a material producing a *magnetic field*. The magnetic field is a *vector field*, since in its any given point, magnetic properties are defined by a direction, magnitude and sense. Any point of the magnetic field attracts or repels other magnets, i.e., produces a magnetic force (vector). A *permanent magnet* is a material object producing its own magnetic field.

The direction of the magnetic vector is defined by a compass needle, and its magnitude (strength) is measured by the strength of the compass needle deflection and is given in teslas (T) in SI units. It follows that tesla can be defined as

$$[\mathrm{T}] = \frac{\mathrm{V} \cdot \mathrm{s}}{\mathrm{m}^2} = \frac{\mathrm{N}}{\mathrm{A} \cdot \mathrm{m}} = \frac{\mathrm{Wb}}{\mathrm{m}^2} = \frac{\mathrm{kg}}{\mathrm{C} \cdot \mathrm{s}} = \frac{\mathrm{kg}}{\mathrm{A} \cdot \mathrm{s}^2} = \frac{\mathrm{N} \cdot \mathrm{s}}{\mathrm{C} \cdot \mathrm{m}},$$

where: A – ampere, C – coulomb, kg – kilogram, m – meter, N – newton, s – second, V – volt, Wb – weber.

There exist two different models for magnetic fields: (*i*) magnetic poles and (*ii*) atomic currents.

(*i*) First of all, it should be emphasized that a magnetic monopole is an isolated magnet having only one magnetic pole, i.e., a south pole without a north pole or a north pole without a south pole. However, magnetism generated by *bar magnets* or *electromagnets* cannot be produced by magnetic monopoles. Therefore, it seems that *magnetic monopole* is a hypothetical particle coming from *particle* physics, and it is difficult to prove that it exists in our universe. Magnetic monopole is rather a mathematical term used for convenience, because in fact, there are no monopoles inside the magnet. Breaking a bar magnet into two pieces yields two bar magnets, each of them having its own north and south pole. In other words, this *magnetic monopole* concept allows to define the *magnetic vector field intensity (strength)* $\vec{H}$. The magnetic flux density $\vec{B}$ (induction) is proportional to $\vec{H}$ outside the magnet, whereas inside the magnet, the magnetization $\vec{M}$

must be added to $\vec{H}$ (this will be discussed later). The magnetization $\vec{M}$, which will be described later, is the local value of magnetic moment per unit volume M [A/m] of a magnetized material. It can be treated as a vector field rather than a vector itself, because different parts of the magnet can be magnetized with different directions and strength.

(ii) Ampére's model stands for magnetization generated by microscopic atomic circular bound currents. In the case of uniformly magnetized cylinder bar magnet, the microscopic bound currents force the magnet to behave as a macroscopic sheet of electric current flowing around the cylinder surface (see Fig. 3.27).

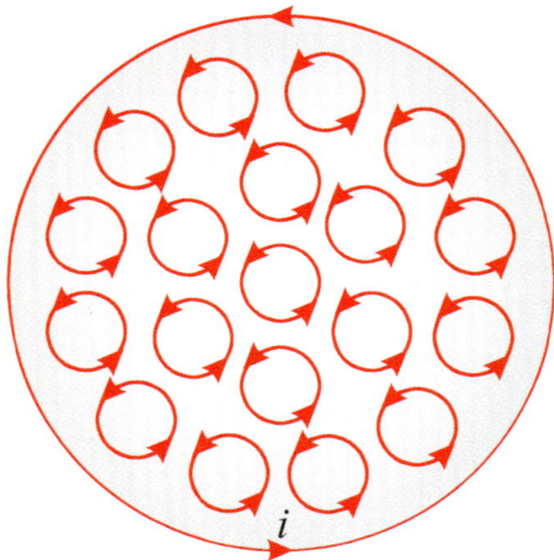

Fig. 3.27   Electric bound currents and electric current flowing around the cylinder's surface (the Ampére's hypothesis).

The magnetic moment of a material object embedded in a magnetic field is generated by the interaction between the magnetic field and the object consisting of atoms or particles. It means that each of the atoms (particles) possesses its own, either natural or inducted, dipole-type magnetic moment, denoted further by $\vec{M}$. This phenomenon can be explained using half-classical Bohr's theory assuming that electrons of the material objects move on the closed orbits around the atomic nuclei (this question will be discussed further).

In order to identify the pole of a magnet, one may look for the position of the magnet in Earth's magnetic field. Namely, a freely suspended magnet points towards the North Magnetic Pole of the Earth (located in northern Canada). However, one may also use the right-hand rule in the case of the magnetic field produced by the electric current flow (electromagnet), as it is shown in Fig. 3.28.

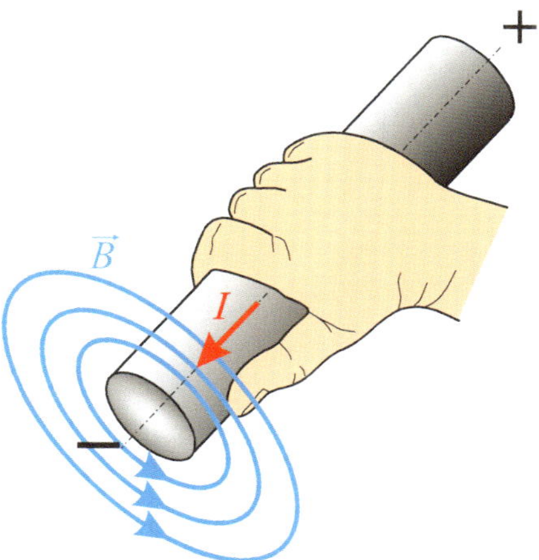

Fig. 3.28   Magnetic field induction $\vec{B}$ produced by a current I, where $\vec{B}$ direction follows the right-hand rule.

3.4.2   *Magnetic Charge and a Dipole*

Since there is an analogy between electric and magnetic fields, and one field is associated with another one following positive and negative electric charges, and it is tempting to introduce the positive and negative magnetic poles $\pm Q$ (by analogy to electric charges $\pm q$). This equivalence is illustrated in Fig. 3.29, where the current flowing around a circle loop produces the magnetic field $\vec{B}$ (a), what is modeled by a system of two magnetic monopoles (b).

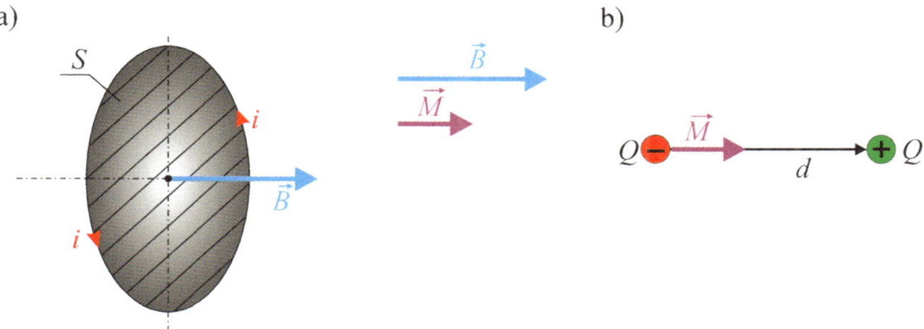

Fig. 3.29   Circular current loop versus magnetic poles $\pm Q$.

The magnetic dipole moment vector $\vec{M}$ is associated with the magnetic charge through the following relationship

$$\vec{M} = \frac{Q}{\mu_0}\vec{d},\qquad(3.215)$$

where $\vec{d}$ is the vector of magnitude equal to the distance between two charges. It points from the negative charge to the positive charge; μ_0 is the permeability constant $\mu_0 = 4\pi \cdot 10^{-7}$ [Tm/A] (the SI unit for Q is weber [Wb] = [V·s]). This constant (magnetic field) is associated with the permittivity constant ε_0 (electric field)

$$\mu_0 \varepsilon_0 = c^{-2}, \tag{3.216}$$

where c denotes the light speed. This means that, knowing one of the constants, for instance μ_0, one may find the associated constant ε_0, and vice versa. This example puts again emphasis on the correspondence between magnetic and electric fields.

The right-hand rule (or the screw rule) allows to find the direction of the magnetic moment $\vec{\mu}$, knowing the current direction.

The so-called *inverse-square law*, which is analogous to the gravitation forces for the planets interaction and to the electric charges interaction (Coulomb's law), holds also for the magnetic monopoles ("charges"). Namely, the magnetic field strength $\vec{H}$ can be associated with the magnetic charge Q in the following form

$$\vec{H} = K\frac{Q}{d^3}\vec{d}, \tag{3.217}$$

where: $K = 1/(4\pi\mu_0)$, and H [N/Wb] is understood as the force per unit magnetic charge.

Consider the magnetic force interaction between two magnetic dipoles (four magnetic charges), as shown in Fig. 3.30.

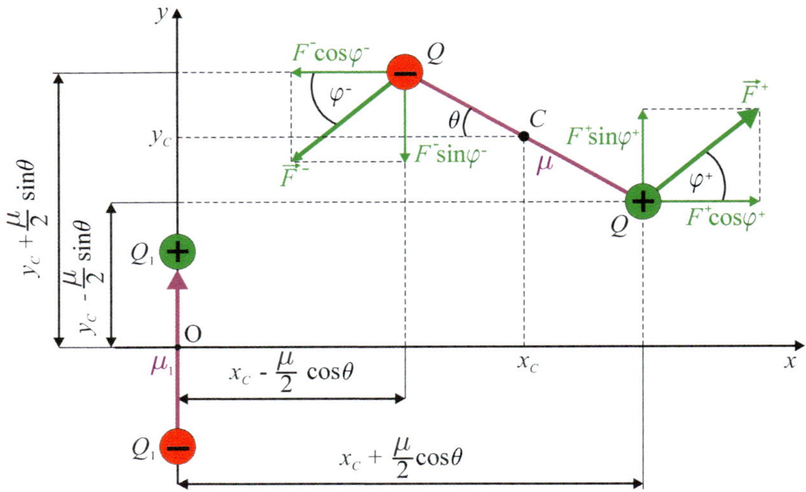

Fig. 3.30   Static configuration of two magnetic dipoles.

Actions of the magnetic forces on each of the four magnetic monopoles can be found using, for example, a geometrical approach shown in Fig. 3.2. We introduce the Cartesian coordinates by putting the dipole $Q_1\mu_1$ on the OY axis, and then define the magnetic position of the dipole $Q\mu$ with respect to the position of the first dipole. Therefore, in Fig. 3.30, only resultant forces acting on the magnetic charges $+Q$ and $-Q$ are presented (forces acting on $+Q_1, -Q_1$ are omitted).

We project two forces $\vec{F}^-$ and $\vec{F}^+$ onto the coordinates axes, and we define a torque generated by these forces with respect to the pole O. This approach yields the following equations:

$$- F^- \cos \varphi^- + F^+ \cos \varphi^+ = 0, \quad -F^- \sin \varphi^- + F^+ \sin \varphi^+ = 0,$$

$$F^- \cos \varphi^- \left(y_C + \frac{\mu}{2} \sin \theta \right) - F^- \sin \varphi^- \left(x_C - \frac{\mu}{2} \cos \theta \right) + \tag{3.218}$$

$$+F^+ \sin \varphi^+ \left(x_C + \frac{\mu}{2} \cos \theta \right) - F^+ \cos \varphi^+ \left(y_C - \frac{\mu}{2} \sin \theta \right) = 0.$$

Three algebraic equations allow to find x_C, y_C and θ, since F^-, F^+ and μ are known.

3.4.3 *Magnetic Effect of Current*

So far, we have considered the key problem of electrostatics. We have shown how, having a space set of charges (sources) $Q_1, \ldots, Q_N$ and using the superposition principle and the electrostatic laws, we can find a force acting on a test charge $+q_0$. If two wires located close and in parallel to each other are considered, then they may either attract (the currents flow in the same directions) or repel (the currents flow in opposite directions) each other (see Fig. 3.31).

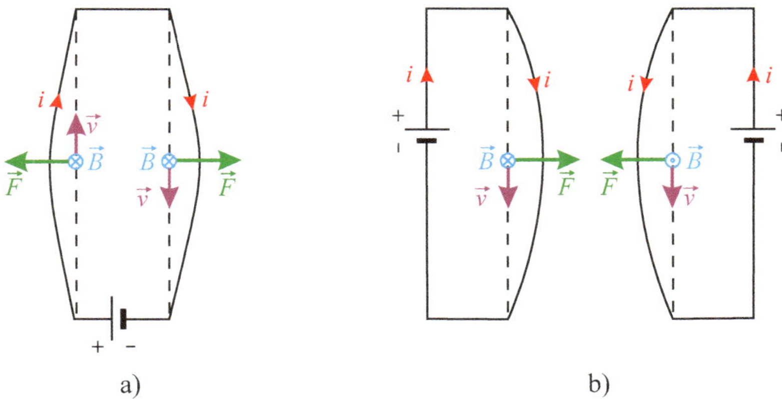

Fig. 3.31   Two parallel wires attracting each other.

The occurred forces are not electrostatic ones, because a test charge located in their vicinity is not influenced by them. As it has been already discussed, static charges create an electrostatic field of intensity the $\vec{E}$, whereas moving charges create, in addition, a magnetic field characterized by vector of magnetic induction $\vec{B}$.

The occurred either attracting or repelling forces can be easily detected with the use of a magnetic needle. It is interesting that the magnetic lines located around a wire, accompanied by the right-hand rule, allow to identify the direction of the current in a wire and the vector $\vec{B}$ (a thumb shows the current direction, whereas the remaining fingers show the direction of the magnetic field $\vec{B}$, as shown in Fig. 3.28).

In section 3.3.1, we have already introduced the Lorentz force which allows us to derive a force $\vec{F}$ acting on a charge q moving with velocity $\vec{v}$, where both electric $\vec{E}$ and magnetic $\vec{B}$ fields exist as follows

$$\vec{F} = q(\vec{E} + (\vec{v} \times \vec{B})). \tag{3.219}$$

The Lorentz force has been found experimentally, and it cannot be derived in any theoretical way.

Since magnetic forces never do any work, we take $\vec{E} = 0$ in (3.219). Let the charge q move on $d\vec{l} = \vec{v}dt$, and the work done is

$$dW_m = \vec{F}_m \cdot d\vec{l} = q(\vec{v} \times \vec{B}) \cdot \vec{v}dt = 0, \tag{3.220}$$

since $\vec{v} \times \vec{B} \perp \vec{v}$. The magnetic forces may change the direction of the moving charge, but never its velocity.

As it has been already mentioned, an *electric current* is the rate of transport of electrical charge past a specified point or across a specified surface. Movement of charges is relative, i.e., negative charges (electrons) moving to the right contribute in the same manner to the current as the positive charges (protons) moving to the left. The charges movement is characterized by a charge magnitude q, and charge velocity $\vec{v}$ (nothing will be changed when we change signs of q and $\vec{v}$ simultaneously). In practice, electrons move in the direction opposite to the current flow, which means that protons move in the current direction.

In the simple case of a linear charge of density ρ moving with velocity $\vec{v}$, a conductor part of the length $v\Delta t$ changes the charge $dq = \rho v \Delta t$, and this charge moves through a point in time interval Δt. Hence, we may define the current vector

$$\vec{i} = \rho\vec{v}. \tag{3.221}$$

If the conductor is embedded in the magnetic field, then the considered conductor part is subjected to action of the following magnetic force

$$\vec{F}_m = \int (\vec{v} \times \vec{B})dq = \int (\vec{v} \times \vec{B})\rho d\vec{l} = \int (\vec{i} \times \vec{B})d\vec{l}. \tag{3.222}$$

Since $\vec{i}$ and $d\vec{l}$ have the same direction and sense, formula (3.222) takes the following form

$$\vec{F}_m = \vec{i} \int (d\vec{l} \times \vec{B}), \tag{3.223}$$

assuming that $i = const.$

In 1820, Oersted observed that a current-carrying conductor generates the magnetic field. Let us illustrate this situation by a wire shown in Fig. 3.32, which carries

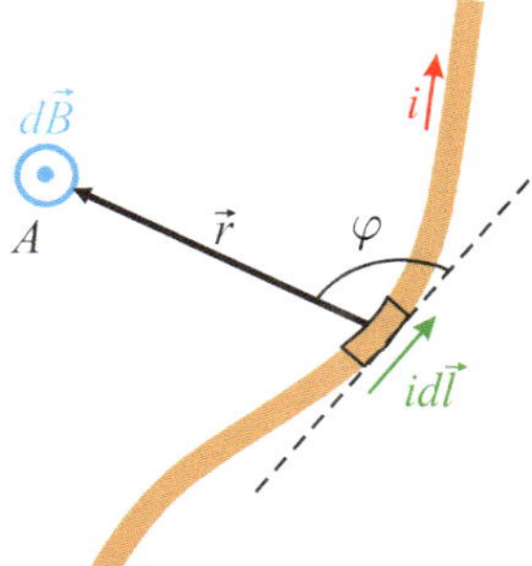

Fig. 3.32   Contribution of $d\vec{B}$ at point A by the current element $id\vec{l}$.

the current i. We are looking for the magnetic induction $d\vec{B}$ produced by the current element $id\vec{l}$ at any given point A (its position is defined by the radius vector $\vec{r}$). The Biot-Savart and Gauss' laws play a key role in *magnetostatics*, whereas Coulomb's and Ampére's laws play a key role in *electrostatics*.

The so-called Biot-Savart law defines the relationship between $d\vec{B}$ and $id\vec{l}$ in the form

$$d\vec{B} = \frac{\mu_0}{4\pi} \frac{id\vec{l} \times \vec{r}}{r^3}, \qquad (3.224)$$

the magnitude of which is

$$dB = \frac{\mu_0}{4\pi} \frac{idl}{r^2} \sin\varphi, \qquad (3.225)$$

and where $\mu_0 = 4\pi \cdot 10^{-7} \text{N/A}^2$ is the magnetic permeability of vacuum.

It should be emphasized that, although the wire with the current produces its magnetic field, only an external magnetic field may act on the wire, i.e., the wire cannot produce magnetic field which acts on the wire itself. Consider now two current-carrying wires and mutual actions of the wires through the produced magnetic forces, as it is shown in Fig. 3.33.

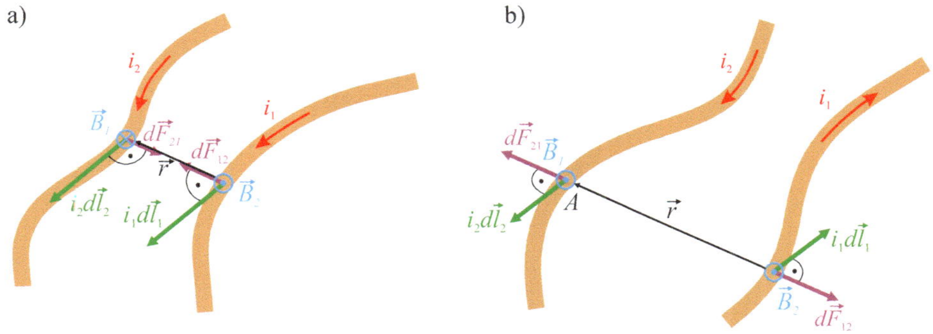

Fig. 3.33   Two current-carrying wires of arbitrary shape with the same (a) and opposite (b) current directions.

At the point A, the current element $i_1 d\vec{l}_1$ produces the magnetic field

$$d\vec{B}_1 = \frac{\mu_0}{4\pi} \frac{i_1 d\vec{l}_1 \times \vec{r}}{r^3}. \tag{3.226}$$

The total $\vec{B}_1$ can be obtained via integral evaluation over the entire length of the wire

$$\vec{B}_1 = \frac{\mu_0 i_1}{4\pi r^3} \oint_{L_1} d\vec{l}_1 \times \vec{r}. \tag{3.227}$$

Knowing $\vec{B}_1$ at the point A, one may find the magnetic force action

$$d\vec{F}_{21} = i_2 d\vec{l}_2 \times \vec{B}_1 \tag{3.228}$$

of the second wire on the point A, coming from the entire wire length L_1.

Analogously, one may find an action of $d\vec{F}_{12}$ exerted by the magnetic field $\vec{B}_2$ on the wire element $i_1 d\vec{l}_1$, which takes the form

$$\vec{B}_2 = \frac{\mu_0}{4\pi} \oint_{L_2} \frac{i_2 d\vec{l}_2 \times (-\vec{r})}{r^3}, \tag{3.229}$$

$$d\vec{F}_{12} = i_1 d\vec{l}_1 \times \vec{B}_2. \tag{3.230}$$

In the case of Fig. 3.33b, the forces $d\vec{F}_{21}$ and $d\vec{F}_{12}$ are attractive (repulsive). In what follows, we consider the case shown in Fig. 3.32, but we take a line wire and we need to find a total magnetic field $\vec{B}$ generated by this wire (see Fig. 3.34).

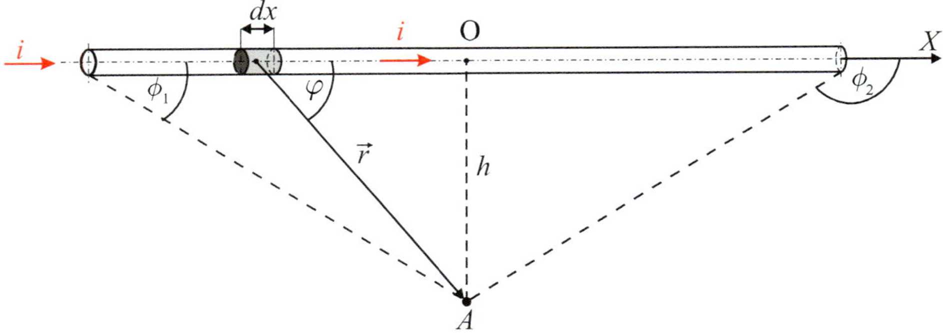

Fig. 3.34   Straight line current loop.

According to the Biot-Savart formula (3.224), we have

$$d\vec{B} = \frac{\mu_0}{4\pi} \frac{i \cdot dx \cdot \vec{i} \times \vec{r}}{r^3}, \tag{3.231}$$

and therefore

$$\mathrm{B} = \frac{\mu_0 i}{4\pi} \int_L \frac{\sin\varphi}{r^2} dx. \tag{3.232}$$

On the other hand, we have:

$$x^2 + h^2 = r^2, \quad -x = h\frac{\cos\varphi}{\sin\varphi}, \quad dx = \frac{hd\varphi}{\sin^2\varphi}, \qquad (3.233)$$

and the relation

$$r = \frac{h}{\sin\varphi} \qquad (3.234)$$

follows either from Eq. (3.234) or directly from Fig. 3.34.

From Eq. (3.233) one gets

$$B = \frac{\mu_0 i}{4\pi h} \int_{\phi_1}^{\phi_2} \sin\varphi d\varphi = \frac{\mu_0 i}{4\pi h} \left[-\cos\varphi\right]_{\phi_1}^{\phi_2} = \frac{\mu_0 i}{4\pi h} \left(\cos\phi_1 - \cos\phi_2\right). \qquad (3.235)$$

In the so far considered case, the wire length L has been defined by the angles ϕ_1 and ϕ_2. However, if $L \to \infty$, then $\phi_1 \to 0$, $\phi_2 \to -\pi$, and from (3.235) one obtains

$$B = \frac{\mu_0 i}{2\pi h}, \qquad (3.236)$$

where h denotes the distance between the linear wire and an arbitrary point A.

In what follows, we consider a magnetic field generated by a circular current loop shown in Fig. 3.35.

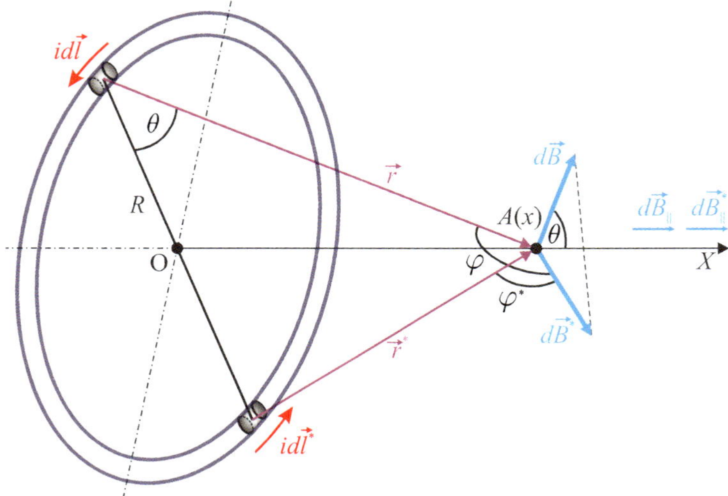

Fig. 3.35   Circular current loop and magnetic field on the axis OX.

Analogously to the previous cases, we treat the loop as a set of elements $id\vec{l}$, and hence an elementary magnetic field $d\vec{B}$ produced by the current i is given by the Biot-Savart law (3.224).

The circle surface is perpendicular to the sheet plane, and the horizontal axis OX passes through the circle center O. If two loop elementary components $id\vec{l}$ and

$id\vec{l}^*$ are located in the opposite parts of the circle, then their produced resultant contribution of $d\vec{B}$ and $d\vec{B}^*$ is

$$d\vec{B} + d\vec{B}^* = d\vec{B}_{||} + d\vec{B}_\perp + d\vec{B}^*_{||} + d\vec{B}^*_\perp = d\vec{B}_{||} + d\vec{B}^*_{||}, \tag{3.237}$$

since

$$d\vec{B}_\perp + d\vec{B}^*_\perp = \vec{0}, \tag{3.238}$$

where $||$ and $\perp$ denote vector $d\vec{B}$ of components parallel and perpendicular to the axis OX, respectively.

Therefore, the vector integral of Eq. (3.224) takes the following form

$$B = \int_L dB \cos\theta = \frac{\mu_0}{4\pi} \int_L \frac{i}{r^2} \cos\theta dl. \tag{3.239}$$

Geometric relations yield:

$$R^2 + x^2 = r^2, \qquad \cos\theta = \frac{R}{r} = \frac{R}{\sqrt{R^2 + x^2}}, \tag{3.240}$$

and hence from Eq. (3.239), one gets

$$B = \frac{\mu_0}{4\pi} \frac{iR}{(R^2 + x^2)^{3/2}} \int_0^{2\pi R} dl = \frac{\mu_0}{2} \frac{iR^2}{(R^2 + x^2)^{3/2}}. \tag{3.241}$$

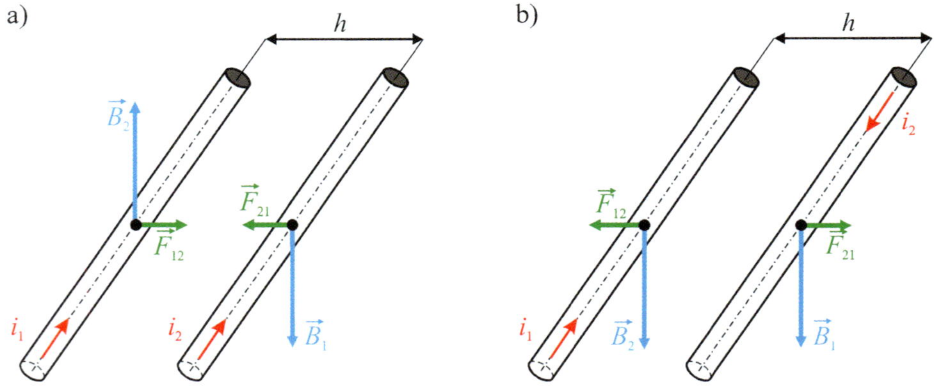

Fig. 3.36   Magnetic forces interacting between two parallel wires for the same (a) and opposite (b) current directions.

Observe that $\vec{B}$ achieves its maximum in the loop plane ($x = 0$), and its magnitude is

$$B = \frac{\mu_0 i}{2R}, \tag{3.242}$$

whereas for $x \gg R$, formula (3.241) gives

$$B = \frac{\mu_0 i R^2}{2x^3} = \frac{\mu_0}{2\pi} \frac{i\pi R^2}{x^3} = \frac{\mu_0}{2\pi} \frac{iS}{x^3}, \tag{3.243}$$

and S denotes the circle area. When we define the magnitude dipole moment M through the loop surface S, then (3.243) yields

$$B = \frac{\mu_0}{2\pi}\frac{M}{x^3}.$$
(3.244)

The magnetic dipole moment $M = iS$ is analogous to the electrical dipole moment $p = qd$.

Next, we consider the current-current interactions between two parallel wires and between two circular current loops. We apply the derived formula (3.236) to determine the mutual forces acting on the parallel wires separated by a distance h in the case of the same and opposite current directions, as it is shown in Fig. 3.36a and Fig. 3.36b, respectively.

Current $i_1(i_2)$ generates the magnetic field $B_1(B_2)$ governed by formula (3.236) of the form

$$B_n = \frac{i_n \mu_0}{2\pi h}, \qquad n = 1, 2.$$
(3.245)

Current $i_1(i_2)$ is embedded in the magnetic field $B_1(B_2)$, and hence the following mutual forces are generated:

$$\begin{aligned}
\vec{F}_{21} &= i_2 \vec{l} \times \vec{B}_1, \\
\vec{F}_{12} &= i_1 \vec{l} \times \vec{B}_2,
\end{aligned}$$
(3.246)

where l denotes the wire length, and $\vec{F}_{21}$ $(\vec{F}_{12})$ is the force exerted by the wire 2 (1) and acting on the wire 1 (2).

When the currents have the same direction, they attract each other (Fig. 3.36a). On the contrary, when the currents have opposite directions, they repel each other (Fig. 3.36b). This means that, unless the constraints are introduced, the wires will either approach (Fig. 3.36a) or separate from each other (Fig. 3.36b).

It should be emphasized that the current unit in SI system is defined through the current-current interaction of two parallel straight line wires. One ampere [A] is the current in each of long parallel wires embedded in vacuum, with the distance of 1 meter between the wires, that generate a mutual force of $2 \cdot 10^{-7}$[N] per one meter.

In Figure 3.37, actions of magnetic fields established by current loops having the same (a) and opposite attractive (c) and repelling (b) directions are presented. The magnitude of the vector $\vec{B}$ at a given point A follows:

$$B_1' + B_2' = \frac{\mu_0}{2}\left(\frac{i_1' R_1^2}{r_1} + \frac{i_2' R_2^2}{r_2}\right),$$

$$B_1'' - B_2'' = \frac{\mu_0}{2}\left(\frac{i_1'' R_1^2}{r_1} - \frac{i_2'' R_2^2}{r_2}\right),$$
(3.247)

$$B_2''' - B_1''' = \frac{\mu_0}{2}\left(\frac{i_2''' R_1^2}{r_1} - \frac{i_1''' R_2^2}{r_2}\right),$$

where $r_1 = \left(R_1^2 + x^2\right)^{3/2}$ and $r_2 = \left(R_2^2 + (x - H)^2\right)^{3/2}$.

In the cases illustrated in Fig. 3.37b and c, mutual actions of the magnetic fields may cancel each other if the following relation holds

$$\frac{i_1''}{i_2''} = \frac{R_2^2\left(R_1^2 + x^2\right)^{3/2}}{R_1^2\left(R_2^2 + (x - H)^2\right)^{3/2}}. \tag{3.248}$$

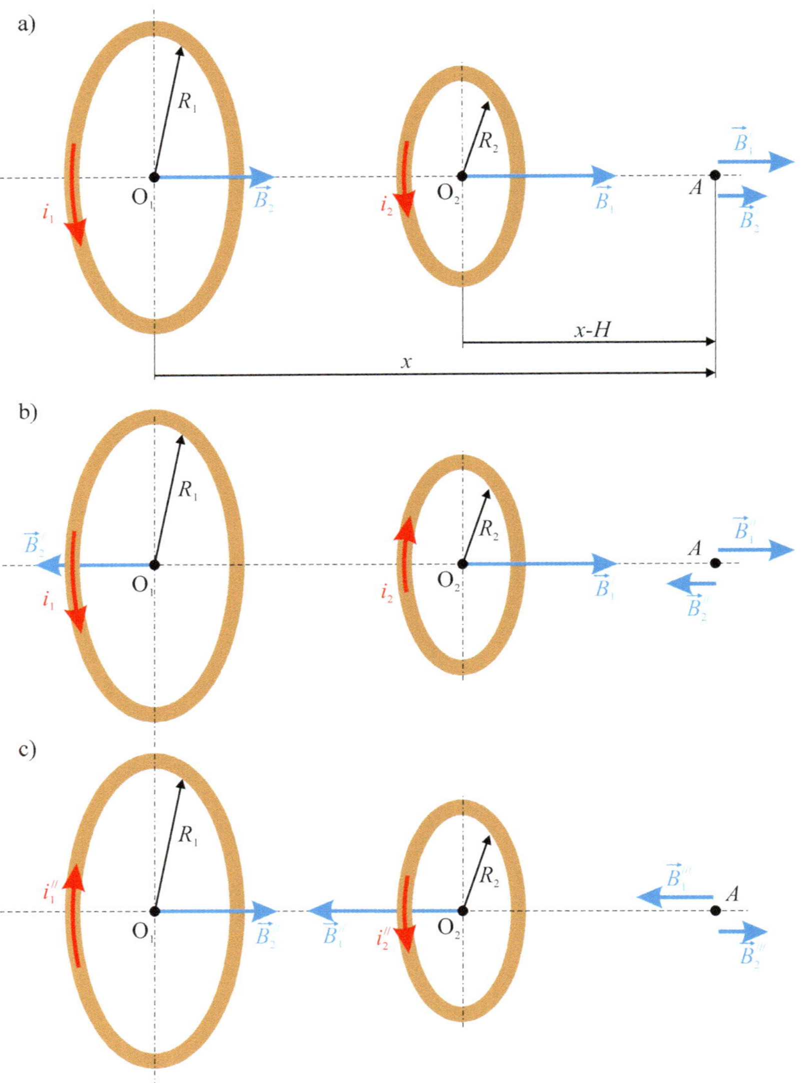

Fig. 3.37 Circular current loops having the same (a) and opposite repelling (b) and attracting (c) directions, and the contribution of $\vec{B}_1$ and $\vec{B}_2$ at point A.

3.4.4 *The Law of Ampére and its Generalization*

Ampére's law generalizes the Biot-Savart law, and it addresses the relation between the currents within the surface bounded by a closed loop and the resultant magnetic field induction along this loop (*Amperian loop*). The magnetic field $\vec{B}$ produced by its segment $d\vec{l}$ is measured with the dot product $\vec{B} \cdot d\vec{l}$. The line integral of $\vec{B} \cdot d\vec{l}$ computed along the Amperian loop is equal to the algebraic sum of the currents (net currents) that pierces the bounded surface

$$\oint \vec{B} \cdot d\vec{l} = \oint B dl \cos \varphi = \mu_0 \sum i_k, \tag{3.249}$$

where μ_0 is the permeability coefficient. Ampére's law exhibits the magnetic field vorticity (Fig. 3.38), and using the right-hand rule (see Fig. 3.28), the current i_1 is positive, whereas i_2 and i_3 are negative.

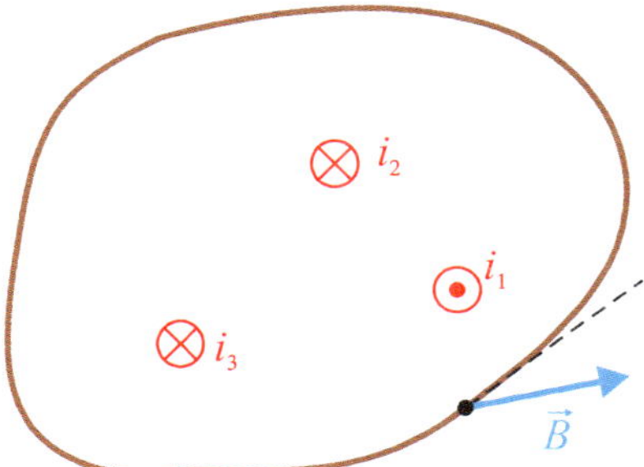

Fig. 3.38   Scheme of the Amperian loop.

In general, $\oint \vec{B} \cdot d\vec{l} \neq 0$ in any vortex field, whereas in an invortex field (for instance in the electrostatic field), we have $\oint \vec{B} \cdot d\vec{l} = 0$. However, a direct application of the Amperian loop concept is not easy.

Formula (3.245) can be presented in the equivalent form

$$\oint \vec{H} \cdot d\vec{l} = \sum i_k, \tag{3.250}$$

where $\vec{H} = \vec{B}/\mu_0$ is the magnetic field intensity ([H]=[A/m]), $\sum i_k$ stands for the magnetomotive force MMF, and in vacuum (μ_0), directions of $\vec{B}$ and $\vec{H}$ are the same. However, in general, directions of the vectors $\vec{B}$ and $\vec{H}$ can be different, and hence the full permeability tensor $[\mu]$ is introduced instead of the diagonal permeability tensor.

If the introduced Amperian loop does not comprise any piercing current, then

$$\oint \vec{H} \cdot d\vec{l} = 0, \tag{3.251}$$

which means that the electrostatic field (bar magnet, magnetic poles) is conserved, because the work done on the closed loop equals zero. A magnetic potential V_m at a given point can be defined by the radius (position) vector $\vec{r}$ in the following form

$$V_m(r) = \int\limits_{r}^{\infty} \vec{H} \cdot d\vec{r},\tag{3.252}$$

and $[\text{V}]=[\text{A}]$ (ampere). In the case of the electrostatic field, the potential difference is

$$\Delta V_{AB} = V_B - V_A = \int\limits_{A}^{B} \vec{H} \cdot d\vec{r}.\tag{3.253}$$

In what follows, we consider divergence and rotation of the magnetic induction $\vec{B}$ for a general case of a volume current. We take two points in the Cartesian coordinates, i.e., one point $\vec{r}_1(x_1, y_1, z_1)$, belonging to the volume current space V_1, and another point $\vec{r}(x, y, z)$, and we are going to find the generated induction $\vec{B}$.

The Biot-Savart law yields

$$\vec{B}(\vec{r}) = \frac{\mu_0}{4\pi} \int \frac{\vec{J}(\vec{r}_1) \times (\vec{r} - \vec{r}_1)}{|\vec{r} - \vec{r}_1|} dV_1 = \frac{\mu_0}{4\pi} \int \frac{\vec{J}(\vec{r}_1) \times (\vec{r}_0)}{|\vec{r}_0|^3} dV_1.\tag{3.254}$$

We multiply both sides of the formula (3.254) in the following way

$$\nabla \cdot \vec{B} = \frac{\mu_0}{4\pi} \int \nabla \cdot \frac{\vec{J} \times \vec{r}_0}{r_0^3} dV_1.\tag{3.255}$$

Since we have

$$\nabla \cdot \left(\vec{J} \times \frac{\vec{r}_0}{r_0^3} \right) = \frac{\vec{r}_0}{r_0^3} \cdot \left(\nabla \times \vec{J} \right) - \vec{J} \cdot \left(\nabla \times \frac{\vec{r}_0}{r_0^3} \right) = 0,\tag{3.256}$$

then

$$\nabla \cdot \vec{B} = 0.\tag{3.257}$$

It means that divergence of the magnetic induction is equal to zero.

Now we apply a cross product in the following way

$$\nabla \times \vec{B} = \frac{\mu_0}{4\pi} \int \nabla \times \left(\vec{J} \times \frac{\vec{r}_0}{r_0^3} \right) dV_1.\tag{3.258}$$

However, we have

$$\nabla \times \left(\vec{J} \times \frac{\vec{r}_0}{r_0^3} \right) = \vec{J} \left(\nabla \cdot \frac{\vec{r}_0}{r_0^3} \right) - \left(\vec{J} \cdot \nabla \right) \frac{\vec{r}_0}{r_0^3} = 4\pi\delta^3 \left(\frac{\vec{r}_0}{r_0^3} \right),\tag{3.259}$$

and from Eq. (3.258) and (3.259) we obtain

$$\nabla \times \vec{B} = \frac{\mu_0}{4\pi} \int \vec{J}(\vec{r}_1) \, 4\pi\delta^3 \left(\vec{r} - \vec{r}_1 \right) dV' = \mu_0 \vec{J}(\vec{r}).\tag{3.260}$$

The second term of Eq. (3.259) is equal zero, since $\vec{J}$ does not depend on x, y, z, whereas the first term is

$$\vec{J} \left(\nabla \cdot \frac{\vec{r}_0}{r_0^3} \right) = 4\pi\delta^3 \left(\vec{r} - \vec{r}_1 \right),\tag{3.261}$$

where we have applied a generalization of the Dirac delta to the 3D space

$$\delta^3\left(\vec{r}\right) = \delta\left(x\right)\delta\left(y\right)\delta\left(z\right),$$

$$\delta^3\left(\vec{r}\right)dV = \int\limits_{-\infty}^{\infty}\int\limits_{-\infty}^{\infty}\int\limits_{-\infty}^{\infty}\delta\left(x\right)\delta\left(y\right)\delta\left(z\right)dxdydz = 1. \tag{3.262}$$

The so far obtained formula (3.260) defines the Ampére's law in a differential form.

Its counterpart integral form is obtained using the Stokes theorem

$$\int\left(\nabla\times\vec{B}\right)\cdot d\vec{S} = \oint\vec{B}\cdot d\vec{l} = \mu_0\int\vec{J}\cdot d\vec{S} = \mu_0\sum i_k, \tag{3.263}$$

where $\sum i_k$ denotes the algebraic sum of all currents included in the Ampére's contour, and it coincides with (3.249).

The so far obtained formulas (3.257) and (3.263) have simple interpretations in the case of straight conductors (see Fig. 3.39).

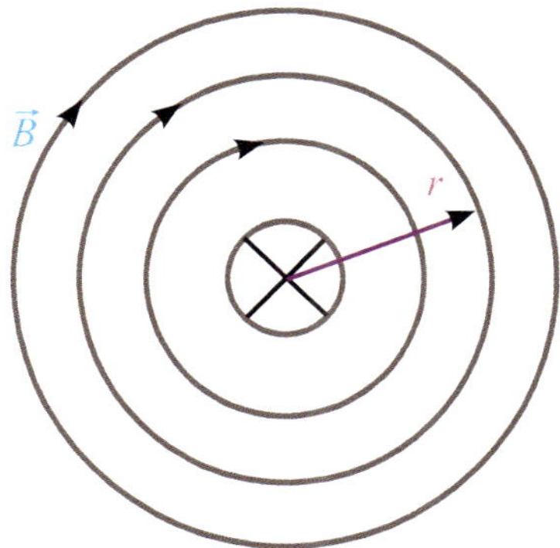

Fig. 3.39   Circles of magnetic field $\vec{B}$.

We have

$$\oint\vec{B}\cdot d\vec{l} = \oint\frac{\mu_0 i}{2\pi r}dl = \mu_0 i, \tag{3.264}$$

and the obtained result does not depend on r. Ampére's law allows to find a magnetic field for highly symmetric cases. For example, assuming that we know the constant current i in the case shown in Fig. 3.39, Ampére's law yields directly

$$\oint\vec{B}\cdot d\vec{l} = B\oint dl = B2\pi r = \mu_0 i, \tag{3.265}$$

and hence, the magnetic induction B at a distance r, measured from the conductor is

$$B = \frac{\mu_0 i}{2\pi r}. \tag{3.266}$$

Electrostatics deals with electric fields generated by *stationary charges*. *Magnetics* deals with magnetic fields generated by *constant currents*. Both introduced notions, i.e., stationary charges and constant currents are approximations to the real phenomena.

If a constant current i moves along a conductor, then it should have a constant density, and the continuity equation

$$\nabla \cdot \vec{J} = \frac{\partial \rho}{\partial t} = 0. \tag{3.267}$$

Divergence and rotation of the electrostatic field intensity are governed by the following truncated form of Maxwell's equations for *electrostatics*:

$$\nabla \cdot \vec{E} = \frac{\rho}{\varepsilon_0}, \qquad \nabla \times \vec{E} = 0. \tag{3.268}$$

Divergence and rotation of the induction of a magnetostatic field are governed by the following truncated form of Maxwell's equations for *magnetostatics*:

$$\nabla \cdot \vec{B} = 0, \qquad \nabla \times \vec{B} = \mu_0 \vec{J}. \tag{3.269}$$

In general, electric forces are larger than magnetic forces. Only if charges move with velocities comparable to the light velocity, the magnetic and electric forces are of the same order.

We consider a vertical straight line conductor with the current of intensity i. Let a torus made of a paramagnetic material lie on a horizontal plane, i.e., be situated perpendicularly to the conductor the axis of which passes through the center of the torus. Assuming that the torus axis is the circle C, we consider the circulation of the magnetic vector $\vec{M}$ and magnetic induction $\vec{B}$ along this circle. The magnetization has the same value in each point of the circle C, and hence

$$\oint_C \vec{M} \cdot d\vec{l} = 2\pi r M. \tag{3.270}$$

Circulation of the magnetization vector $\vec{M}$ is computed over the circle C with the integral, what is shown by the left-hand side of Eq. (3.270).

The entire particle bound current intensity

$$i_m = 2\pi r I = 2\pi r M, \tag{3.271}$$

and one finds that the following formula describes a relation between $\vec{M}$ circulation and the current intensity i_m

$$\oint_C \vec{M} \cdot d\vec{l} = i_m. \tag{3.272}$$

Generally, Ampére's law takes into account the microscopic molecular bound current i_m and free current i. It can be written in the following form

$$\oint_C \vec{B} \cdot d\vec{l} = \mu_0 (i + i_m), \tag{3.273}$$

and taking into account Eq. (3.272), one gets

$$\oint_C \vec{H} \cdot d\vec{l} = i, \tag{3.274}$$

where:

$$\vec{B} = \mu_0(\vec{H} + \vec{M}) = [\mu]\vec{H}, \tag{3.275}$$

and $[H] = [A/m]$. Formula (3.273) shows how, in the case of magnetic field generated by current circuits in material objects, their magnetic induction depends on both molecular current i_m and circuits free currents. The obtained Ampére's law (3.274) is valid for an arbitrary material medium including vacuum. In the case of paramagnetics and diamagnetics, vectors $\vec{M}$ and $\vec{B}$ are linearly dependent.

3.4.5 *Magnetic Dipole Moment of a Closed Planar Current-Carrying Loop*

Consider a planar current-carrying loop in the the magnetic field $\vec{B}$, as it is shown in Fig. 3.40.

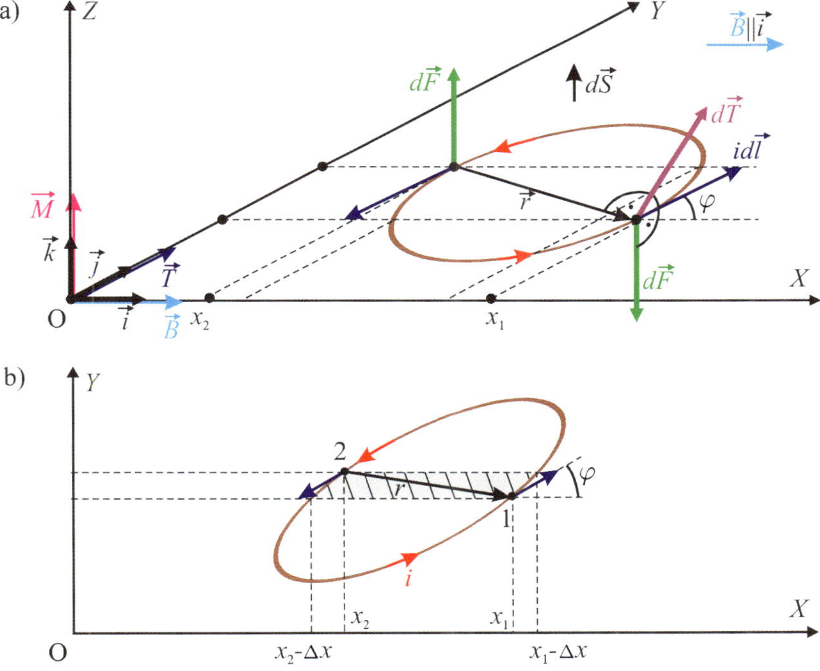

Fig. 3.40 Plane current-carrying loop in 3D space (a) and its OXY plane location with dashed parallelogram $\left|-d\vec{S}\right| = |-\vec{r} \times dl| \cong (x_1 - x_2)\,dl\sin\varphi$ (b).

The forces acting on the symmetric incremental current loop elements are

$$d\vec{F}_n = id\vec{l}_n \times \vec{B}, \quad n = 1, 2. \tag{3.276}$$

These incremental forces create a couple of forces $d\vec{F_1} + d\vec{F_2} = 0$ $(dl_1 = dl_2)$, and the torque acting on the element $dl_1 = dl$ is

$$d\vec{T} = \vec{r} \times d\vec{F} = \vec{r} \times (id\vec{l} \times \vec{B}). \tag{3.277}$$

If $\Delta x \to dx$, then vector $\vec{r}$ tends to establish the horizontal position, which forces the vector of $d\vec{T}$ to be parallel to the axis OY. It means that in the limit, we obtain

$$d\vec{T} = i(\vec{r} \times d\vec{l}) \times \vec{B} = id\vec{S} \times \vec{B}, \tag{3.278}$$

where now $d\vec{S}$ is a vector normal to the plane created by the vectors $\vec{r} \times d\vec{l}$, and

$$dS = \left| \vec{r} \times d\vec{l} \right| = rdl\sin{(\vec{r}, d\vec{l})} = rdl\sin{(\pi - \varphi)} =$$
$$= rdl\sin{\varphi} = rdy \cong (x_1 - x_2)dy. \tag{3.279}$$

In the limit, the dashed parallelogram tends to the rectangle of the sides $x_1 - x_2$ and dy, and its area represents the length of vector $d\vec{S}$.

In general, the relation

$$\vec{r} \times i(d\vec{l} \times \vec{B}) = \left(\vec{r} \times id\vec{l}\right) \times \vec{B} \tag{3.280}$$

used in (3.278) is not true for arbitrary $d\vec{l} = \vec{i}dl_x + \vec{j}dl_y$, $\vec{r} = r_x\vec{i} + \vec{j}r_y$, $\vec{B} = B\vec{i}$. However, for $\vec{r} = -r\vec{i}$, $d\vec{l} = dl\vec{j}$, $\vec{B} = B\vec{i}$, we have

$$id\vec{l} \times \vec{B} = i\begin{vmatrix} \vec{i} & \vec{j} & \vec{k} \\ 0 & dl & 0 \\ B & 0 & 0 \end{vmatrix} = -Bidl\vec{k}, \tag{3.281}$$

and the left-hand side of Eq. (3.280) is

$$\vec{r} \times \left(id\vec{l} \times \vec{B}\right) = \begin{vmatrix} \vec{i} & \vec{j} & \vec{k} \\ -r & 0 & 0 \\ 0 & 0 & -Bidl \end{vmatrix} = -irBdl\vec{j}. \tag{3.282}$$

On the other hand, we have

$$\vec{r} \times id\vec{l} = \begin{vmatrix} \vec{i} & \vec{j} & \vec{k} \\ -r & 0 & 0 \\ 0 & idl & 0 \end{vmatrix} = -ridl\vec{k}, \tag{3.283}$$

and the right-hand side of Eq. (3.280) follows

$$\left(\vec{r} \times id\vec{l}\right) \times \vec{B} = \begin{vmatrix} \vec{i} & \vec{j} & \vec{k} \\ 0 & 0 & -ridl \\ B & 0 & 0 \end{vmatrix} = -riBdl\vec{j}, \tag{3.284}$$

which proves formula (3.280) for our case.

In order to get the total torque, integration over all rectangular strips within the bounded area of the current loop should be carried out, which yields

$$\vec{T} = \int_S d\vec{T} = \int_S id\vec{S} \times \vec{B} = \left(\int_S d\vec{M}\right) \times \vec{B} = \vec{M} \times \vec{B}, \tag{3.285}$$

where $d\vec{M} = id\vec{S}$ is the incremental magnetic dipole moment, and $\vec{M} = i\vec{S}$ is the magnetic dipole moment of the coil.

Therefore, it has been shown that the action of the magnetic field vector $\vec{B} = B\vec{i}$ parallel to the closed current-carrying loop surface produces the torque $\vec{T} = T\vec{j}$, which is the vector cross product in Eq. (3.285). Under an action of the torque $\vec{T}$, the loop rotates, and the incremental potential energy required to do the work is

$$dU = Td\theta = MB\sin\theta d\theta, \tag{3.286}$$

which yields the total potential energy required to rotate the coil by the angle θ

$$U = -MB\cos\theta = -\vec{M}\cdot\vec{B} \tag{3.287}$$

staying for the magnetic potential energy $[U] = [J]$. The potential energy of the coil coming from the current i is found from Eq. (3.287):

$$U = -\int d\vec{M}\cdot\vec{B} = -i\int d\vec{S}\cdot\vec{B}. \tag{3.288}$$

One may also introduce the vector magnetic potential A in the following way. Since we deal with the constant magnetic field $\vec{B}$, then

$$\vec{\nabla}\cdot\vec{B} = \vec{\nabla}\cdot(\vec{\nabla}\times\vec{A}) = 0. \tag{3.289}$$

It follows from Eqs. (3.288) and (3.289) that

$$U = -i\int_S d\vec{S}\cdot\left(\vec{\nabla}\times\vec{A}\right). \tag{3.290}$$

We are going to transit from the surface integral to a line integral using the following Stokes theorem

$$\int_S d\vec{S}\cdot\left(\vec{\nabla}\times\vec{A}\right) = \int_L \vec{A}\cdot d\vec{l}. \tag{3.291}$$

Therefore, with the help of Eq. (3.291), the formula (3.290) yields

$$U = -i\int_L \vec{A}\cdot d\vec{l}. \tag{3.292}$$

Furthermore, let us introduce the current volume density $\vec{J}$:

$$id\vec{l} = \vec{J}dV. \tag{3.293}$$

Formula (3.292) allows to find potential with respect to $\vec{J}$ with the use of the relationship

$$U = -\int_V \left(\vec{A}\cdot\vec{J}\right)dV. \tag{3.294}$$

Let us derive the magnetic potential vector $\vec{A}$ (see Fig. 3.40a):

$$\vec{A} = \frac{1}{2}\left(\vec{B}\times\vec{r}\right), \tag{3.295}$$

then, using Eq. (3.294), the potential is

$$U = -\frac{1}{2} \int\limits_{V} \left[\left(\vec{B} \times \vec{r} \right) \cdot \vec{J} \right] dV = -\frac{1}{2} \int\limits_{V} \left[\vec{B} \cdot \left(\vec{r} \times \vec{J} \right) \right] dV =$$

$$= -\vec{B} \cdot \int\limits_{V} \frac{1}{2} \left(\vec{r} \times \vec{J} \right) dV = -\vec{B} \cdot \int\limits_{V} d\vec{M}, \tag{3.296}$$

and hence

$$d\vec{M} = \frac{1}{2} \left(\vec{r} \times \vec{J} \right) dV. \tag{3.297}$$

3.4.6 *Electromagnetic Induction*

The so far introduced two different models of magnetization (see Sec. 3.4.1) have also a strong impact on the interpretation and modeling of the electromagnetic induction phenomenon. We cannot (within the classical electromagnetic theory) use the same model for the case of a *unmovable* (fixed) closed circuit embedded in a *varying* magnetic field and the case of a movable closed circuit within a constant magnetic field.

However, in the *particular relativity theory*, a phenomenon of the so-called *electromagnetic field* appears which, depending on the choice of a reference system, can be interpreted either as the *magnetic field* or *electric field*.

Let us consider the second model, i.e., the one in which an electrical planar circuit with a linear part of length l moving with velocity $\vec{v}$ is located in a homogeneous magnetic field with induction $\vec{B}$. Vector $\vec{B}$ is perpendicular to the circuit plane and velocity $\vec{v}$, as it is shown in the Fig. 3.41.

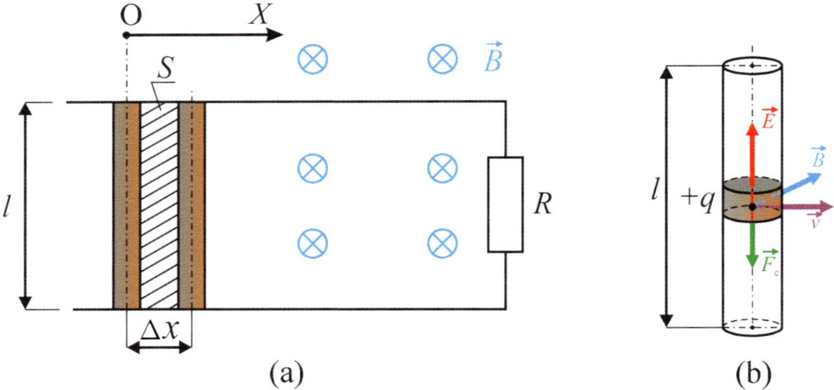

Fig. 3.41 Electrical circuit embedded in a homogeneous magnetic field with induction $\vec{B}$ (a) and the movable vertical conductor (b).

If the elementary conductor charge $+q$ is taken, then it is subjected to an action of the Lorentz force of the form

$$\vec{F} = q(\vec{v} \times \vec{B}). \tag{3.298}$$

The Lorentz force causes a movement of charge carriers, and an electrostatic field of intensity $\vec{E}$ is generated inside the circuit. The field is governed by the formula

$$\vec{F}_e = q\vec{E}. \tag{3.299}$$

The charge carriers stop if $\vec{F}_e = \vec{F}$, and thus the following relationship is obtained

$$\vec{E} = \vec{v} \times \vec{B}. \tag{3.300}$$

After the movement of the charge carried at the conductor ends, we will have different electrical potential, the difference $\mathcal{E}$ of which is defined by a scalar product of the electric field intensity $\vec{E}$ and the conductor length

$$\mathcal{E} = \vec{E} \cdot \vec{l}. \tag{3.301}$$

Since the inductor is connected to the remaining electric circuit, the generated electrical potential include an *electromotive force*, and a *current passage* occurs. Formulas (3.300) and (3.301) yield

$$\mathcal{E} = (\vec{v} \times \vec{B}) \cdot \vec{l} = \vec{l} \cdot \left(\frac{d\vec{x}}{dt} \times \vec{B} \right) = \vec{B} \cdot \left(\vec{l} \times \frac{d\vec{x}}{dt} \right)$$
$$= \frac{d}{dt} \left[\vec{B} \cdot (\vec{x} \times \vec{l}) \right] = \frac{d}{dt} \left(\vec{B} \cdot \vec{S} \right) = -\frac{d\phi_B}{dt}. \tag{3.302}$$

Formula (3.302) represents *Faraday's induction law*. Faraday experimentally proved that the so-called electromotive force (in fact, this is not a force, since it is not a vector) generated in an electric circuit is proportional to the rate of changes of the induction flux of a magnetic field enclosed by this electric circuit.

We have shown that the induction flux of a magnetic field

$$\phi_B = \vec{B} \cdot \vec{S}, \tag{3.303}$$

where the vector $\vec{S}$ of the length $S = xl$ is perpendicular to the dashed rectangle in Fig. 3.41, i.e., in our case, to the electric circuit plane. Since we deal with scalar vector product (3.303), we may get either positive or negative $\mathcal{E}$ value, depending on the angle between vectors $\vec{B}$ and $\vec{S}$.

Recall that Oersted was the first who experimentally showed that the current flowing through a conductor produces a magnetic field. On the other hand, in 1831, Faraday experimentally observed that a varying magnetic field generates a current in a conductor (see Fig. 3.42). The observed phenomenon is referred to as the *electromagnetic induction*.

In what follows, we will determine a direction of the induced current in the circuit shown in Fig. 3.42. If the magnet SN is placed inside the circuit, then the current direction generates the magnetic field which repels the approaching magnet, and vice versa.

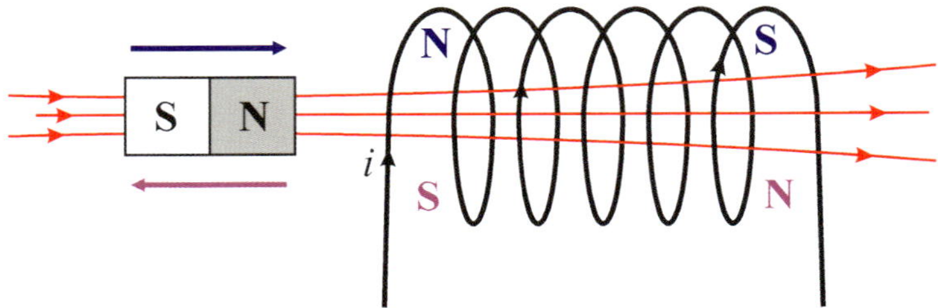

Fig. 3.42   Current induced by a variable magnetic field (magnet).

The work done during attraction or repelling of two magnetic fields is transformed into a heat energy distribution over the circuit. This phenomenon has been observed for the first time by Lenz (1834), and the direction of the current generated in a general circuit can be defined through *Lenz's law*. It says that a current induced in a circuit moves in the direction produced by the magnetic field passing through a surface enclosed by the circuit.

Let us consider the described situation taking into account a simpler example. Consider a closed conducting loop (one coil) and a magnet.

If the loop is opened, then a potential difference $V_2 - V_1 = \mathcal{E}$ appears between the loop ends, but a current is not induced. An electric field intensity $\vec{E} = -\mathrm{grad}\,\mathcal{E}$, which means that the potential difference produces an electromotive force (EMF). If the loop is closed (as it is shown in Fig. 3.43), then the induced current appears. Its direction follows the earlier described Lenz's rule. One may also consider two loops located in two parallel planes.

Assume that one of the loops is fixed (II), and its current produces magnetic effects. One may define the direction of the current induced in the second loop using the earlier described Lenz's rule. The so far mentioned examples (Figs. 3.41–3.44) indicate a dependence between purely magnetic and electric effects associated with a movement.

Knowing how to define an induced current direction, we discuss a connection between a sign of $\mathcal{E}$ direction of vectors $\vec{S}$, $\vec{E}$ and $\vec{B}_i$. Namely, we will consider the situation reported in Fig. 3.43 from a point of view of formula (3.302).

We deal with one closed loop conductor, and hence we introduce a frame of reference being at rest with respect to the loop (fixed with the loop). On the other hand, we take a general reference frame $OXYZ$ such that $\vec{k} \cdot \vec{B} > 0$ and $\vec{k} \parallel \vec{B}$. Similarly, we take also a positively directed surface vector $\vec{S}$, i.e., $\vec{k} \cdot \vec{S} > 0$ and $\vec{S} \parallel \vec{k}$.

The current movement direction presented in Fig. 3.42 illustrates validity of Faraday's induction law (3.302), where the minus sign appears.

The induced currents may occur either in conductors with small cross sections or massive conductors like rigid plates (Fig. 3.46).

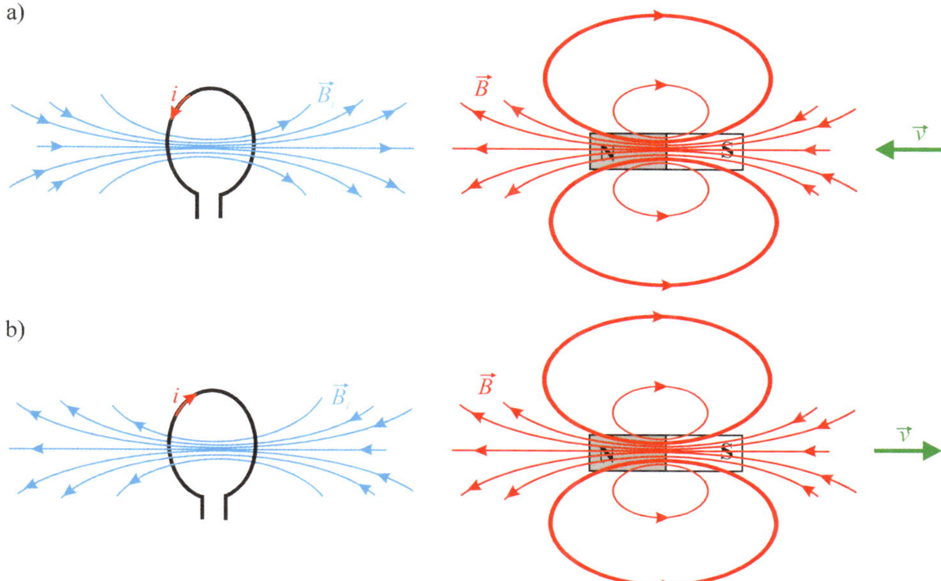

Fig. 3.43 Positive (a) and negative (b) induced ($\vec{B} > \vec{B}_i$) current direction, when the magnet NS moves towards (outwards) the conducting loop surface.

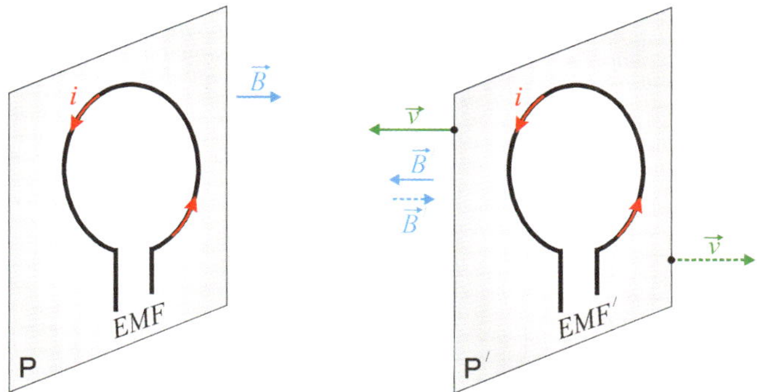

Fig. 3.44 Induced current direction in the fixed loop and in the moving loop.

Since the induced currents are within a circular vortex, they are called *eddy currents* or *Foucault currents*. They are dangerous, because they generate heat, and hence, large energy loss takes place in the electric machine elements. If a variable current flows along a conductor, the so-called skin effect may occur due to induced eddy currents. They counteract changes of the current intensity responsible for their occurrence in a neighborhood of a conductor axis, and they simultaneously cooperate with the current changes di/dt in a thin conductor layer located on its surface.

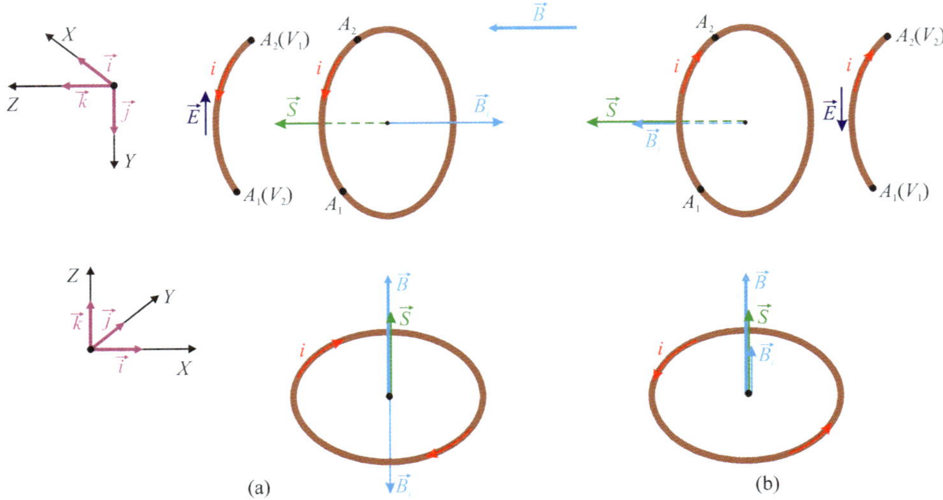

Fig. 3.45   Circled circuit surface represented by the vector $\vec{S}$, induction of the external homogeneous movable magnetic field $\vec{B}$, induction of the induced magnetic fields $\vec{B}_i$, current i directions, and electric field intensity $\vec{E}$ corresponding to the cases (a) and (b) shown in Fig. 3.43.

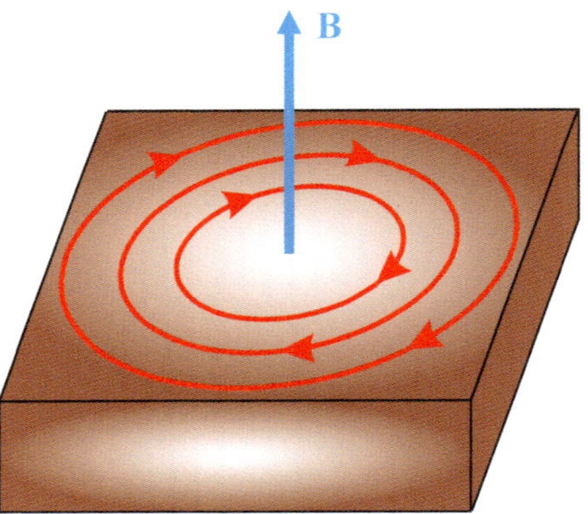

Fig. 3.46   Eddy currents induced in a rigid plate.

In other words, resistance in the inner part of the conductor is much higher in comparison to resistance close to the conductor surface. Therefore, the current density is large (small) on the conductor surface (inner part).

Owing to Ohm's law, for a closed circuit, a sum of all potentials drops along the whole circuit must be equal to the electromotive force $\mathcal{E}$, which can be quantified as a circulation of the electric field intensity $\vec{E}$ within a conductor over the entire circuit (see (3.301)). On the other hand, the entire magnetic flux ϕ_B passing through the surface S is governed by formula (3.303). Therefore, formula (3.302) can be written in the following equivalent form

$$\oint_C \vec{E} \cdot d\vec{l} = -\frac{d}{dt} \int_S \vec{B} \cdot d\vec{S}. \tag{3.304}$$

The induction law governed by (3.304) is valid for the case of an electric vortex field generated by the time-dependent magnetic field either in vacuum or nonconducting material medium. If $\vec{B}$ and $\vec{E}$ do not depend on time, then formula (3.304) yields

$$\oint_C \vec{E} \cdot d\vec{l} = 0, \tag{3.305}$$

which means that the constant (time-independent) electric field is a potential (non-vertex) field.

Let us now recall a few terms from mathematics needed for our further considerations. Assume that in a certain coordinate system, a differentiable vector field $\vec{A}(M)$ is introduced, where $M = M(x, y, z)$ is its arbitrarily taken point. Then

$$\vec{A}(M) = \vec{i} A_1(x, y, z) + \vec{j} A_2(x, y, z) + \vec{k} A_3(x, y, z).$$

Divergence of a vector field $\vec{a}$ is the scalar function

$$\operatorname{div} \vec{a} = (\vec{\nabla}, \vec{a}) = \frac{\partial a_1}{\partial x} + \frac{\partial a_2}{\partial y} + \frac{\partial a_3}{\partial z}.$$

It can be shown that $\operatorname{div} \vec{a}$ does not depend on the choice of a coordinate system.

In the majority of cases regarding a vector field analysis, it is useful to apply the so-called Hamilton's operator $\vec{\nabla}$ (known as del operator, and in many cases, the overhead arrow is omitted) which is understood as a symbolic vector. It should be emphasized that it acts for either functions or vectors standing on its right-hand side as a differential operator, whereas for the functions or vectors standing on its left-hand side, it acts as a vector which follows the rules of vector products.

We say that a scalar field f is differentiable in the point M_0 if there exists a vector $\vec{c}$ such that

$$f(M) - f(M_0) = \left(\overrightarrow{M_0 M}, \vec{c}\right) + o\left(\left|\overrightarrow{M_0 M}\right|\right) \quad \text{for} \quad M \to M_0.$$

Vector $\vec{c}$ is called a *derivative of the scalar field* f in the point M_0, and it will be denoted by $\nabla f(M_0)$.

In the Cartesian coordinate system, we have

$$\overrightarrow{M_0 M} = (x - x_0)\vec{i} + (y - y_0)\vec{j} + (z - z_0)\vec{k},$$

$$\left|\overrightarrow{M_0 M}\right| = \sqrt{(x - x_0)^2 + (y - y_0)^2 + (z - z_0)^2},$$

and hence

$$f(M) - f(M_0) = c_1(x - x_0) + c_2(y - y_0) + c_3(z - z_0) +$$

$$+ o\left(\sqrt{(x - x_0)^2 + (y - y_0)^2 + (z - z_0)^2}\right) \quad \text{for} \quad (x, y, z) \to (x_0, y_0, z_0).$$

Therefore, one may write

$$f(M) - f(M_0) = \left(\overrightarrow{M_0 M}, \nabla f(M_0)\right) + o\left(\left|\overrightarrow{M_0 M}\right|\right) \quad \text{for} \quad M \to M_0,$$

where:

$$c_1 = \frac{\partial f}{\partial x}(x_0, y_0, z_0), \quad c_2 = \frac{\partial f}{\partial y}(x_0, y_0, z_0), \quad c_3 = \frac{\partial f}{\partial z}(x_0, y_0, z_0),$$

$$\nabla = \vec{i}\frac{\partial}{\partial x} + \vec{j}\frac{\partial}{\partial y} + \vec{k}\frac{\partial}{\partial z},$$

$$\nabla f(M_0) = \vec{c} = \frac{\partial f(M_0)}{\partial x}\vec{i} + \frac{\partial f(M_0)}{\partial y}\vec{j} + \frac{\partial f(M_0)}{\partial z}\vec{k}.$$

For an arbitrary $\vec{b} = \vec{b}[b_1, b_2, b_3]$, we have

$$\vec{b} \cdot \vec{\nabla} = (\vec{b}, \vec{\nabla}) = \left(\vec{i}b_1 + \vec{j}b_2 + \vec{k}b_3\right) \cdot \left(\vec{i}\frac{\partial}{\partial x} + \vec{j}\frac{\partial}{\partial y} + \vec{k}\frac{\partial}{\partial z}\right) = b_1\frac{\partial}{\partial x} + b_2\frac{\partial}{\partial y} + b_3\frac{\partial}{\partial z}.$$

Using the introduced scalar product of $\vec{b}$ and $\vec{\nabla}$, we have

$$f(M) - f(M_0) = \left(\overrightarrow{M_0 M}, \vec{\nabla}\right) f(M_0) + o\left(\left|\overrightarrow{M_0 M}\right|\right) \quad \text{for} \quad M \to M_0.$$

Finally, let us take a unit vector $\vec{l}$ and consider a radius consisting of all points M satisfying the relationship $\overrightarrow{M_0 M} = \vec{l}S$ for $S > 0$.

A derivative of a scalar field f in the direction $\vec{l}$ in the point M_0 is defined by the following formula

$$\frac{\partial f}{\partial l}(M_0) = \lim_{S \to 0} \frac{f(M) - f(M_0)}{S},$$

or equivalently

$$\frac{\partial f}{\partial l} = (\vec{l} \cdot \vec{\nabla}) f(M_0).$$

There are two mathematical tools for quantifying how the vector field $\vec{a}$ will change from point to point: *divergence* (scalar) and *curl* (vector). Divergence of the vector field $\vec{a}$ at the point M_0 is defined as

$$\operatorname{div}\vec{a} = \lim_{\Delta\vartheta \to 0} \frac{\oint \vec{a} \cdot d\vec{S}}{\Delta\vartheta}, \tag{3.306}$$

where $\Delta\vartheta$ is an infinitesimal volume surrounding M_0, and integration is carried out over the area S of this volume. Formula (3.306) can be explicitly derived in Cartesian, cylindrical and spherical coordinates.

We derive an expression for divergence in the Cartesian coordinates (Fig. 3.47) now.

The vector field defined at M_0 is

$$\vec{a}(x, y, z) = a_1\vec{i} + a_2\vec{j} + a_3\vec{k},$$

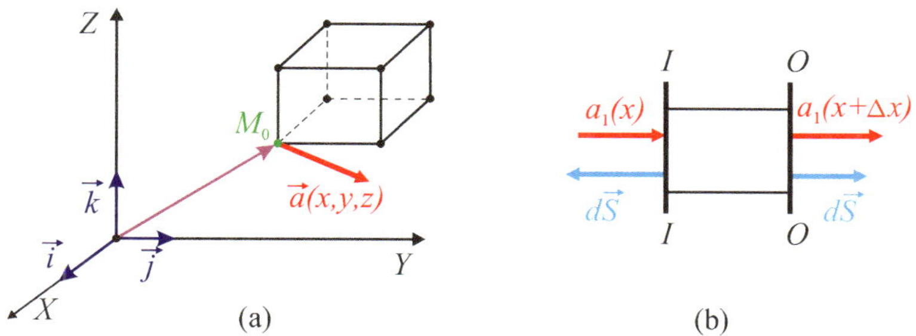

Fig. 3.47   Cube with the edges $\Delta x, \Delta y, \Delta z$ and the vector field $\vec{A}(x,y,z)$ at $M_0(x,y,z)$ (a) and two faces I (input) and O (output) of the cube (b).

and M_0 is the cube corner with the lowest values of x, y, z. We cover all three pairs of the cube faces while expressing $\oint \vec{a} \cdot d\vec{S}$, where $d\vec{S}$ direction is outward. We have

$$\int_I \vec{a} \cdot d\vec{S} \cong -a_1 \Delta y dz,$$

$$\int_O \vec{a} \cdot d\vec{S} \cong a_1(x + \Delta x)\Delta y dz \cong \left(a_1 + \frac{\partial a_1}{\partial x}\Delta x\right)\Delta y dz, \tag{3.307}$$

and hence

$$\int_I \vec{a} \cdot d\vec{S} + \int_O \vec{a} \cdot d\vec{S} = \frac{\partial a_1}{\partial x}\Delta x \Delta y dz. \tag{3.308}$$

Taking into account pairs of faces, we get

$$\oint \vec{a} \cdot d\vec{S} \cong \left(\frac{\partial a_1}{\partial x} + \frac{\partial a_2}{\partial y} + \frac{\partial a_3}{\partial z}\right)\Delta\vartheta, \tag{3.309}$$

where $\Delta\vartheta = \Delta x \Delta y \Delta z$, and from Eqs. (3.306) and (3.309) we obtain

$$\text{div}\vec{a} = \nabla \cdot \vec{a} = \frac{\partial a_1}{\partial x} + \frac{\partial a_2}{\partial y} + \frac{\partial a_3}{\partial z}, \tag{3.310}$$

where the del operator ∇ is defined explicitly only in Cartesian coordinates, $\nabla = \frac{\partial}{\partial x}\vec{i} + \frac{\partial}{\partial y}\vec{j} + \frac{\partial}{\partial z}\vec{k}$.

In cylindrical and spherical coordinates, we have:

$$\text{div}\vec{a} = \frac{1}{r}\frac{\partial}{\partial r}(ra_r) + \frac{1}{r}\frac{\partial a_\theta}{\partial\theta} + \frac{\partial a_z}{\partial z}, \tag{3.311}$$

$$\text{div}\vec{a} = \frac{1}{r^2}\frac{\partial}{\partial r}(r^2 a_r) + \frac{1}{r\sin\psi}\frac{\partial}{\partial\psi}(a_\psi \sin\psi) + \frac{1}{r\sin\psi}\frac{\partial a_\theta}{\partial\theta}, \tag{3.312}$$

respectively.

Maxwell's equations for static $\vec{E}$ and $\vec{D}$ fields are as below:

$$div\vec{D} = \rho,$$

$$div\,\vec{E} = \frac{\rho}{\varepsilon}, \tag{3.313}$$

where the electric field $\vec{E}$ is a function of the permittivity ε, whereas the electric flux density $\vec{D}$ does not depend on ε.

It follows directly from Gauss' law that

$$\oint \vec{D} \cdot d\vec{S} = Q, \tag{3.314}$$

where Q denotes the enclosed charge. Taking the neighborhood $\Delta\vartheta$ and going to the limit, from Eq. (3.314) we obtain:

$$\lim_{\Delta\vartheta \to 0} \frac{\oint \vec{D} \cdot d\vec{S}}{\Delta\vartheta} = div\ \vec{D} = \rho, \tag{3.315}$$

where

$$\rho = \lim_{\Delta\vartheta \to 0} \frac{Q}{\Delta\vartheta}. \tag{3.316}$$

Since for any electric field, for a medium characterized by a variable $\varepsilon = \varepsilon(x, y, z)$, we have $\vec{D} = \varepsilon\vec{E}$, hence

$$div\ \vec{E}\varepsilon = \rho, \tag{3.317}$$

and for a medium, for which $\varepsilon = const$ (isotropic medium), we get Eq. (3.313).

A key role in physics, mechanics, hydrodynamics, magneto-electrodynamics and mechatronics is played by Gauss' formula which allows to transform various conservation rules to the equivalent modeling through the differential equations.

Let $\vartheta \subset R^3$ be a bounded space the boundary S of which is a piecewise smooth surface. Assume that we have a continuous differential vector field $\vec{a} = (a_1, a_2, a_3)$ in $\overline{\vartheta} = \vartheta \cup S$. Then, a flow of the vector field $\vec{a}$ passing through the space boundary S, defined as $\iint_S (\vec{a}, \vec{n})dS$, is equal to the triple integral of $div\vec{a}$ with respect to the space S, i.e., the following relationship holds

$$\iint_S (\vec{a}, \vec{n})dS = \iiint_\vartheta div\ \vec{a}d\vartheta, \tag{3.318}$$

or equivalently

$$\iint_S a_1 dydz a_2 dxdz a_3 dxdy \iiint_\vartheta \left(\frac{\partial a_1}{\partial x} + \frac{\partial a_2}{\partial y} + \frac{\partial a_3}{\partial z} \right) dxdydz. \tag{3.319}$$

Let us assume that in an oriented Euclidean space, a simple surface Σ is defined in the following way:

$$\vec{r} = \vec{r}(u, v), \qquad (u, v) \in \Omega \subset R^2. \tag{3.320}$$

A closed surface has a positively oriented smooth (or piecewise smooth) contour, i.e., if one walked around the boundary $\partial\Omega$, the surface Ω would be located on the left side. Let the boundary $\partial\Omega$ be described by equations:

$$u = u(S), \quad v = v(S), \quad \alpha < S \leq \beta. \tag{3.321}$$

Therefore, $\partial\Sigma$ is understood as a positively oriented boundary of the surface Σ. The orientation of the surface Σ is defined by the field of normal vector $\vec{N} = [\vec{r}_u, \vec{r}_v]$, and it coincides with the positive orientation of the surface boundary, what is in agreement with the right-handed screw rule.

It follows from Eq. (3.319) that for any sufficiently regular vector field A, we have

$$\oint_S \Delta \cdot d\vec{S} = \int_\vartheta (\nabla \cdot \vec{a}) d\vartheta, \tag{3.322}$$

and the obtained formula is called the *divergence (Gauss') theorem*.

Consider a field with known charge, and assume the charge density ρ is known within the volume ϑ enclosed by the surface S.

It follows directly from Gauss' law that

$$\oint \vec{D} \cdot d\vec{S} = \int \rho d\vartheta = \int \nabla \cdot \vec{D}. \tag{3.323}$$

The divergence theorem is applied while converting the volume integral into a closed surface integral and vice versa. It is applied either to time-varying (dynamics) or static fields.

Curl (rotation) of a vector field $\vec{a}$ is defined by the formula

$$curl\,\vec{a} = [\vec{\nabla}, \vec{a}] = \begin{vmatrix} \vec{i} & \vec{j} & \vec{k} \\ \frac{\partial}{\partial x} & \frac{\partial}{\partial y} & \frac{\partial}{\partial z} \\ a_1 & a_2 & a_3 \end{vmatrix} = \vec{\nabla} \times \vec{a} =$$

$$= \vec{i}\left(\frac{\partial a_3}{\partial y} - \frac{\partial a_2}{\partial z}\right) + \vec{j}\left(\frac{\partial a_1}{\partial x} - \frac{\partial a_3}{\partial z}\right) + \vec{k}\left(\frac{\partial a_2}{\partial x} - \frac{\partial a_1}{\partial y}\right). \tag{3.324}$$

One can show that, in an arbitrary counterclockwise coordinate system, the curl of a vector field remains unchanged, whereas it changes its sign during a transition from the counterclockwise coordinate (right) to a clockwise (left) coordinate system. This is why a curl is called a *pseudo-vector*.

It follows from (3.324) that the curl of a vector field $\vec{a}$ produces another vector field. Geometric interpretation of the curl is presented in Fig. 3.48.

Projection of the curl of $\vec{a}$ onto vector $\Delta\vec{S}$ normal to the surface ΔS is defined by the formula

$$(curl\ \vec{a})\left(\frac{\Delta\vec{S}}{\Delta S}\right) = \lim_{\Delta S \to 0} \frac{\oint \vec{a} \cdot d\vec{l}}{\Delta S}, \tag{3.325}$$

where $d\vec{l}$ stands for infinitesimal part of the contour C traversed so that the area ΔS is on the left, and hence $\Delta\vec{S}$ is determined by the right-hand rule. We use the geometric approach to validate formula (3.324) in Cartesian coordinates by using Eq. (3.325).

Since by definition $\vec{a} = a_1\vec{i} + a_2\vec{j} + a_3\vec{k}$, we consider the rectangular area $\Delta y \Delta z$ for $x = const$, as it is shown in Fig. 3.49.

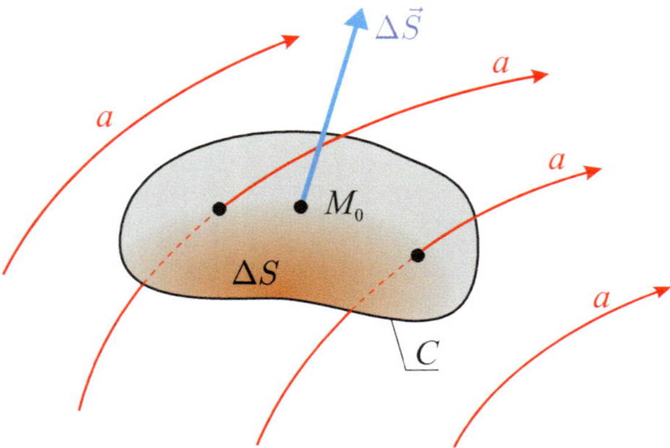

Fig. 3.48   Area aS enclosed by the closed contour C surrounding the point M_0.

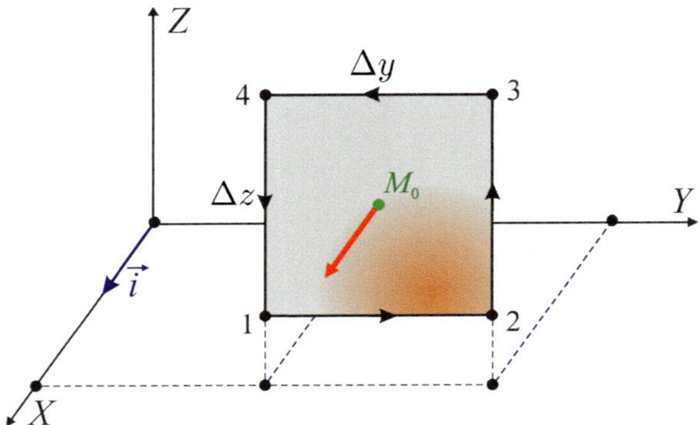

Fig. 3.49   Rectangular area $\Delta y \Delta z$ with the middle point M_0 fixed at $x = const$.

Formula (3.325) takes the form

$$(\text{curl } \vec{a}) \cdot \vec{i} = \lim_{\Delta x \Delta z \to 0} \frac{\oint \vec{a} \cdot d\vec{l}}{\Delta x \Delta z}, \tag{3.326}$$

and

$$\oint = \int_1^2 + \int_2^3 + \int_3^4 + \int_4^1 = a_2 \Delta y + \left(a_3 + \frac{\partial a_3}{\partial y} \Delta y \right) \Delta z +$$

$$+ \left(a_2 + \frac{\partial a_2}{\partial z} \Delta z \right) (-\Delta y) + a_3 (-\Delta z) = \left(\frac{\partial a_3}{\partial y} - \frac{\partial a_2}{\partial z} \right) \Delta y \Delta z. \tag{3.327}$$

Combining the y and z components of the curl $\vec{a}$, we obtain (3.324). In cylindrical and spherical coordinates, we obtain

$$\text{curl } \vec{a} = \left(\frac{1}{r} \frac{\partial a_3}{\partial \theta} - \frac{\partial a_\theta}{\partial z} \right) \vec{e}_r + \left(\frac{\partial a_r}{\partial z} - \frac{\partial a_z}{\partial r} \right) \vec{e}_\Theta + \frac{1}{r} \left(\frac{\partial (r a_\theta)}{\partial r} - \frac{\partial a_r}{\partial \theta} \right) \vec{e}_z, \tag{3.328}$$

$$\text{curl } \vec{a} = \frac{1}{r \sin \psi} \left(\frac{\partial (a_\theta \sin \psi)}{\partial \psi} - \frac{\partial a_\psi}{\partial \theta} \right) \vec{e}_r + \frac{1}{r} \left(\frac{1}{\sin \psi} \frac{\partial a_r}{\partial \theta} - \frac{\partial (r a_\theta)}{\partial r} \right) \vec{e}_\psi +$$
$$+ \frac{1}{r} \left(\frac{\partial (r a_\psi)}{\partial r} - \frac{\partial a_r}{\partial \psi} \right) \vec{e}_\theta. \tag{3.329}$$

In many cases, the following field properties are used:

(i) The divergence of a curl is a zero scalar

$$\nabla \cdot (\text{curl } \vec{a}) = \nabla \cdot \left(\vec{\nabla} \times \vec{A} \right) = 0. \tag{3.330}$$

In the Cartesian coordinates, we have

$$\left(\frac{\partial}{\partial x} \vec{i} + \frac{\partial}{\partial y} \vec{j} + \frac{\partial}{\partial z} \vec{k} \right) \cdot \left[\left(\frac{\partial a_3}{\partial y} - \frac{\partial A_2}{\partial z} \right) \vec{i} \right.$$
$$\left. + \left(\frac{\partial a_1}{\partial x} - \frac{\partial a_3}{\partial z} \right) \vec{j} + \left(\frac{\partial a_2}{\partial x} - \frac{\partial a_1}{\partial y} \right) \vec{k} \right] = \frac{\partial}{\partial x} \left(\frac{\partial a_3}{\partial y} - \frac{\partial a_2}{\partial z} \right) \tag{3.331}$$
$$+ \frac{\partial}{\partial y} \left(\frac{\partial a_1}{\partial x} - \frac{\partial a_3}{\partial z} \right) + \frac{\partial}{\partial z} \left(\frac{\partial a_2}{\partial x} - \frac{\partial a_1}{\partial y} \right) = 0.$$

(ii) The curl of a gradient is a zero vector

$$\nabla \times (\nabla f(M_0)) = 0. \tag{3.332}$$

In the Cartesian coordinates, we obtain

$$\begin{vmatrix} \vec{i} & \vec{j} & \vec{k} \\ \frac{\partial}{\partial x} & \frac{\partial}{\partial y} & \frac{\partial}{\partial z} \\ \frac{\partial f}{\partial x} & \frac{\partial f}{\partial y} & \frac{\partial f}{\partial z} \end{vmatrix} = \vec{i} \left(\frac{\partial^2 f}{\partial y \partial z} - \frac{\partial^2 f}{\partial y \partial z} \right)$$
$$+ \vec{j} \left(\frac{\partial^2 f}{\partial x \partial z} - \frac{\partial^2 f}{\partial x \partial z} \right) + \vec{k} \left(\frac{\partial^2 f}{\partial x \partial y} - \frac{\partial^2 f}{\partial x \partial y} \right) = 0. \tag{3.333}$$

Finally, let us define the Laplacian of a scalar V (it should not be confused with the potential) and a vector $\vec{a}$. In the Cartesian coordinates, we have

$$\nabla^2 V = \nabla \cdot (\nabla V) = \left(\frac{\partial}{\partial x} \vec{i} + \frac{\partial}{\partial y} \vec{j} + \frac{\partial}{\partial z} \vec{k} \right) \cdot \left(\frac{\partial V}{\partial x} \vec{i} + \frac{\partial V}{\partial y} \vec{j} + \frac{\partial V}{\partial z} \vec{k} \right)$$
$$= \left(\frac{\partial^2 V}{\partial x^2} + \frac{\partial^2 V}{\partial y^2} + \frac{\partial^2 V}{\partial z^2} \right). \tag{3.334}$$

Laplacian in the cylindrical coordinates

$$\nabla^2 V = \frac{1}{r} \frac{\partial}{\partial r} \left(r \frac{\partial V}{\partial r} \right) + \frac{1}{r^2} \frac{\partial^2 V}{\partial \theta^2} + \frac{\partial^2 V}{\partial z^2}, \tag{3.335}$$

whereas in the spherical coordinates

$$\nabla^2 V = \frac{1}{r^2} \frac{\partial}{\partial r} \left(r^2 \frac{\partial V}{\partial r} \right) + \frac{1}{r^2 \sin \psi} \frac{\partial}{\partial \psi} \left(\sin \psi \frac{\partial V}{\partial \psi} \right) + \frac{1}{r^2 \sin \psi} \frac{\partial^2 V}{\partial \theta^2}. \tag{3.336}$$

In the Cartesian coordinate system, we have

$$\nabla^2 \vec{a} = \nabla^2 \left(a_1 \vec{i} + a_2 \vec{j} + a_3 \vec{k} \right) = \nabla^2 a_1 \vec{i} + \nabla^2 a_2 \vec{j} + \nabla^2 a_3 \vec{k}. \tag{3.337}$$

It is easy to verify that the following identity holds

$$\nabla^2 \vec{a} = \vec{\nabla} \left(\vec{\nabla} \cdot \vec{a} \right) - \vec{\nabla} \times \left(\vec{\nabla} \times \vec{a} \right). \tag{3.338}$$

3.4.7　*Electric and Magnetic Susceptibility and Permeability*

Taking magnetization into consideration, there are three types of materials in nature. *Dielectrics* usually exhibit polarization in the direction of a field $\vec{E}$. *Paramagnetics (diamagnetics)* are polarized in the direction of the magnetic induction $\vec{B}$, with the same (opposite) sense regarding $\vec{B}$. There exist also *ferromagnetics* (iron) which keep magnetization when the external field is removed.

In the case of *paramagnetic materials*, the dipoles associated with spins of electrons are subjected to the action of a moment of forces which tries to move the dipoles parallel to the field lines. In *diamagnetic materials*, orbital velocity of electrons changes in a way that the orbital dipole moment moves in the direction opposite to the field direction. In general, paramagnetic and diamagnetic phenomena are very weak. A very strong field repels diamagnetic material and attracts paramagnetic one.

In what follows, we illustrate and discuss natural reasons for occurrence of polarization $\vec{P}$. Many materials (mainly dielectrics), if subjected to the action of an electric field, change positions of atoms and dipoles within the material and exhibit the so-called polarization governed by the equation

$$\vec{P} = \varepsilon_0 \chi_e \vec{E}, \tag{3.339}$$

where χ_e is the electric susceptibility, ε_0 is a term introduced to keep a linear dependence between $\vec{P}$ and $\vec{E}$. The electric susceptibility may depend on temperature and internal material lattice structure. If we apply an electric induction field $\vec{D}$, then for a linear medium

$$\vec{D} = \varepsilon_0 \vec{E} + \vec{P} = \varepsilon \vec{E}. \tag{3.340}$$

where

$$\varepsilon = \varepsilon_0 (1 + \chi_e). \tag{3.341}$$

The proportional coefficient ε is called an *electrical permeability* of a medium. In vacuum, the electric susceptibility $\chi_e = 0$ (there is no matter), and from Eq. (3.341) we obtain that the electrical permeability in vacuum is $\varepsilon = \varepsilon_0$.

Using Eq. (3.341), we may also define a relative electric permeability

$$\varepsilon_r = \frac{\varepsilon}{\varepsilon_0} = 1 + \chi_e. \tag{3.342}$$

Here, we give exemplary values of the relative electric permeability for a few materials, i.e., $\varepsilon_r \cong 1$ for helium, argon, hydrogen, air, nitrogen; $\varepsilon_r = 80.1$ for water, $\varepsilon_r = 99$ for ice, and $\varepsilon_r = 5.7$ for diamond.

In *crystals*, electric susceptibility χ_{ce} depends on the polarization direction, and we use a tensor of electric susceptibility to define the following linear dependence

$$\vec{P} = \varepsilon_0 \chi_{ce} \vec{E}, \tag{3.343}$$

which means that nine coefficients $\chi_{e^{i,j}}$ $(i,j = 1,2,3)$ must be used (in general) to define the susceptibility tensor.

A nondimensional quantity χ_m is negative for diamagnetics (gold: $-3.4 \cdot 10^{-5}$, silver $-2.4 \cdot 10^{-5}$, copper $-9.7 \cdot 10^{-6}$, water: $-9.0 \cdot 10^{-6}$) and positive for paramagnetics (oxygen $1.9 \cdot 10^{-6}$; wolfram $7.8 \cdot 10^{-5}$; platinum $2.8 \cdot 10^{-4}$).

Magnetization of a medium is equivalent to a magnetic field coming from internal volume currents and surface currents of the magnetized medium. The current density $\vec{J}$ is a sum of two vectors

$$\vec{J} = \vec{J}_\vartheta + \vec{J}_f, \tag{3.344}$$

where $\vec{J}_\vartheta$ represent the volume currents yielded by the medium magnetization as a result of a polarization of atoms and dipoles, whereas $\vec{J}_f$ (called a free current) is yielded by an external source (like a battery) to transport the charges.

It can be shown that

$$\vec{J}_\vartheta = \nabla \times \vec{M}, \tag{3.345}$$

where $\vec{M}$ is the magnetic dipole moment per unit volume. It is also known as the magnetization or the magnetic polarization and plays a similar role in magnetostatics as the electric polarization $\vec{P}$ plays in electrostatics.

In order to distinguish magnetic materials (paramagnetic and diamagnetic ones), we need to introduce a dimensionless constant χ (the term comes from Latin and denotes receptiveness) characterizing the proportion between a *magnetic moment* $\vec{M}$ per unit volume and *magnetic field strength* $\vec{H}$. The magnetic moment points from the south to the north pole of a magnet, and it is a vector of an electric current or the torque of magnetic field acting on the magnet.

The volume magnetic *susceptibility* χ_m is defined by the following formula

$$\vec{M} = \chi_m \vec{H}, \tag{3.346}$$

where M [A/m] is the material magnetization (net magnetic dipole moment per unit volume); $\vec{H}$ is the magnetic field strength.

In a homogeneous and linear medium, the volume current density $\vec{J}_\vartheta$ and the volume free current density are coupled by the following formula

$$\vec{J}_\vartheta = \nabla \times \vec{M} = \nabla \times (\chi_m \vec{H}) = \chi_m \vec{J}_f. \tag{3.347}$$

Ampére's law yields

$$\frac{1}{\mu_0}(\nabla \times \vec{B}) = \vec{J}_f + (\nabla \times \vec{M}), \tag{3.348}$$

or equivalently

$$\nabla \times \vec{H} = \vec{J}_f, \tag{3.349}$$

where the magnetic field intensity

$$\vec{H} = \frac{1}{\mu_0}\vec{B} - \vec{M}. \tag{3.350}$$

The magnetic field intensity $\vec{H}$ is important in magnetostatics (in electrostatics,, a similar role is played by the electric field induction $\vec{D}$). We have distinguished the free currents $\vec{J}_f$, since we can control them. In contrast, we cannot change the action of the volume current $\vec{J}_\vartheta$, since it depends only on the magnetization process. Owing to the applied Gauss' law, the application of $\vec{H}$ allows to use only the controllable currents $\vec{J}_f$, which is preferable by electrical engineers and designers. This also yields a motivation to use $\vec{H}$ in Eq. (3.347) instead of the induction $\vec{B}$.

Equation (3.350), taking into account (3.347), can be recast to the following form

$$\vec{B} = \mu\vec{H} = \mu_0(1 + \chi_m)\vec{H} = \mu_0\mu_r\vec{H}, \tag{3.351}$$

where μ is the magnetic permeability of a medium, μ_r is the relative magnetic permeability, and μ_0 is the vacuum magnetic permeability.

In what follows, we combine the so far described two different models for magnetic fields. Let us assume that we have a 3D space of homogeneous external magnetic field with the induction $\vec{B}_e$. If we introduce a cylinder made either of a paramagnetic or diamagnetic material (magnetic materials will be described later), this cylinder experiences its own internal magnetic induction $\vec{B}_i$ owing to ordering of its atom magnetic moments, and $\vec{B}_e \parallel \vec{B}_i$. As a result, the following resultant magnetic field induction is obtained

$$\vec{B} = \vec{B}_e + \vec{B}_i, \tag{3.352}$$

what is shown in Fig. 3.50.

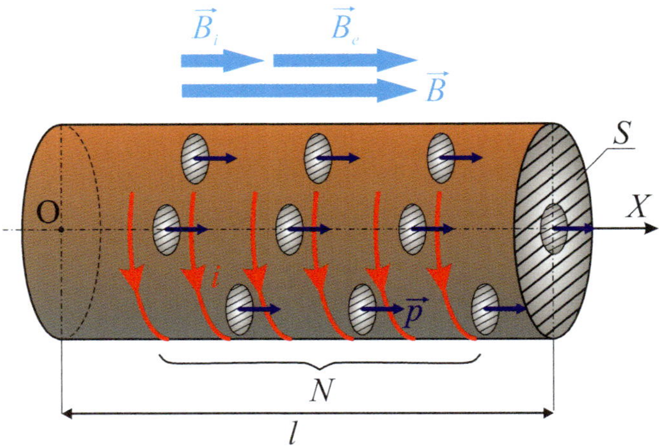

Fig. 3.50   Resultant magnetic field induction $\vec{B}$.

As it has been mentioned above (see Fig. 3.50), in any given long cylinder with a cross section perpendicular to the OX axis, all bound currents cancel each other. Only the current flows on the cylindrical surface are active and an internal magnetic field

$$B_i = \mu_0 I \tag{3.353}$$

is induced in the cylinder body, where $I = iN/l$ and $[I]=[\text{A/m}]$. Formula (3.353) comes from the standard computation of the magnetic field of a long solenoid, and iN denotes the entire bound currents intensity, whereas I stands for the density of the entire current intensity per cylinder length unit, and μ_0 is the proportionality coefficient. One may generate a magnetic field in a different surrounding medium like gas, fluid or vacuum. It is useful to introduce the so-called *relative magnetic permeability coefficient* μ.

If one takes a conductor with the current (or a magnet) and puts it into a medium different from the vacuum, one observes a different magnetic induction vector at a given point of the magnetic field (see (3.351):

$$\vec{B} = \mu_r \vec{B}_0 = (\chi_m + 1)\vec{B}_0, \tag{3.354}$$

where $\vec{B}_0$ refers to vacuum. Hence, a linear dependence between induction in a material object and a vacuum is assumed.

In general, the induction of the external magnetic field may change both magnetic permeability μ and magnetic susceptibility χ in a nonlinear way. However, in many cases, χ and μ will not depend on the magnetic field induction. Equivalently, if one experimentally finds that for a given material object we have $\vec{J}\tilde{\ }\vec{B}$, then χ and μ can be treated as constants (I stands for the current density).

Let us assume that we have N particles of a studied material object located in the unit volume part ΔV. Each particle has its own dielectric polarization vector $\vec{p}_n$, $n= 1, \ldots, N$, and then, the magnetization vector of the object unit volume follows

$$\vec{M} = \sum_{n=1}^{N} \vec{p}_n = \vec{P}. \tag{3.355}$$

In the case when $\vec{p}_1 = \cdots = \vec{p}_N = \vec{p}$, we have

$$\vec{M} = N\vec{p}. \tag{3.356}$$

Observe that $[M]=[I]=[\text{A/m}]$, and actually, $M = I$. The entire magnetic moment of the solenoid of the infinite length (see Fig. 3.50) is equal to the sum of its N magnetic moments iS, where S is the cylinder cross section area, i.e., we have

$$P = iSN. \tag{3.357}$$

Therefore, the magnetic moment of the solenoid per its unit volume should be equal to the magnetization M, which means that

$$M = \frac{P}{V} = \frac{iSN}{lS} = I. \tag{3.358}$$

In other words, the magnetization is equal to the density of the entire current intensity. Therefore, from Eqs. (3.347) and (3.351), we obtain

$$\vec{M} = \frac{\chi_m}{\mu\mu_0}\vec{B}, \tag{3.359}$$

assuming that χ_{m} and μ do not depend on B.

Magnetic permeability is either a scalar (isotropic medium) or a second rank tensor (anisotropic medium) – see the next subsection.

One may extend this definition to the so-called *incremental permeability* defined as the ratio $\Delta B/\Delta H$ used in a local linearization procedure. In SI units, $[\mu] =$ H/m $= $ N/(A)^2, $[B] = $ Wb $= $ V/(m)$^2 = $ T.

We define an *electric charge* as a *physical property* of matter which generates a force when it appears near other electrically charged matter. There exist *positive* and *negative* electric charges. One positively charged substance repels other positively charged substance but attracts other negatively charged substance, and vice versa (see Fig. 3.51).

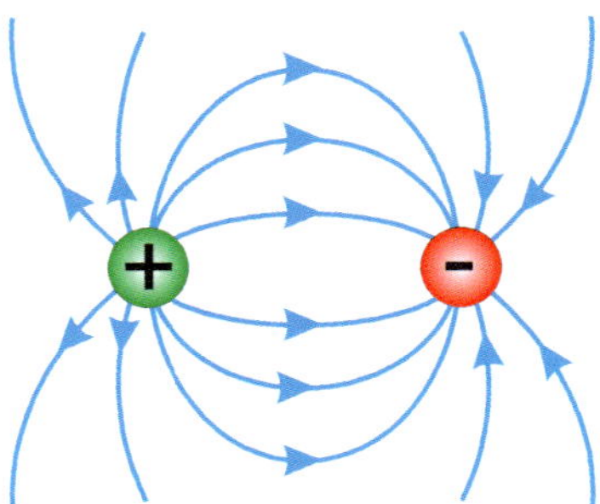

Fig. 3.51   Electric field generated by a positive $(+)$ and negative $(-)$ point charge.

The electric charge is often defined by a symbol q [C = A·s]. A particle of charge q, being in an electric field intensity $\vec{E}$ and a magnetic field of induction $\vec{B}$, moves with instantaneous velocity $\vec{v}$ due to action of the Lorentz force

$$\vec{F} = q[\vec{E} + (\vec{v} \times \vec{B})], \tag{3.360}$$

where $q(\vec{v} \times \vec{B})$ is called the electric (magnetic) force. The magnetic Lorentz force component is the force acting on a current-carrying wire in a magnetic field.

If the Cartesian coordinates and $\vec{r}$ denoting the *charged particle* vector position (see Fig. 3.52) are introduced, then the Lorentz force (3.360) can be written as follows

$$\vec{F}(\vec{r}, \dot{\vec{r}}) = q[\vec{E}(\vec{r}, t) + \dot{\vec{r}} \times \vec{B}(\vec{r}, t)], \tag{3.361}$$

where the dot denotes a time derivative.

One may extend the so far introduced concept to a *continuous charge distribution*. In the latter case, we have

$$d\vec{F} = dq(\vec{E} + \vec{v} \times \vec{B}), \tag{3.362}$$

where $d\vec{F}$ is the force acting on a small piece of the charge distribution with the charge quantity dq. Assuming that a volume of the small piece of the charge distribution is dV, one may introduce $\vec{f} = d\vec{F}/dV$ (force density), $\rho = dq/dV$ (charge

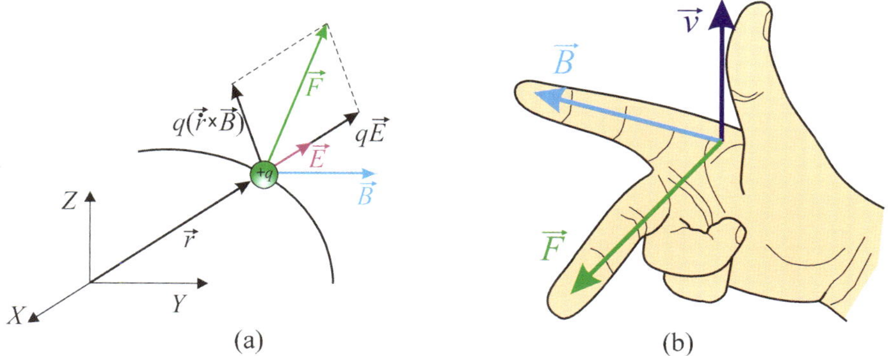

Fig. 3.52   Charged particle moving along a path under the Lorentz force $\vec{F}$ for a given time instant t (a) and the right-hand rule for vectors $\vec{B}, \vec{v}, \vec{F}$ (b).

density), and Eq. (3.361) takes the form

$$\vec{f} = \rho\vec{E} + \vec{J} \times \vec{B},\tag{3.363}$$

where $\vec{J} = \rho\vec{v}$ is the current density corresponding to the motion of the charge continuum. Now, the total force $\vec{F}$ is the volume integral of formula (3.363):

$$\vec{F} = \iiint\limits_{V} (\rho\vec{E} + \vec{J} \times \vec{B})d\vec{v}.\tag{3.364}$$

In the case of a straight wire carrying an electrical current and being placed in a magnetic field, one may think of the charges moving along the wire creating a microscopic force on the wire, and the Lorentz force (in the described situation it is sometimes referred to as Laplace's force) is governed by the relationship

$$\vec{F} = I\vec{l} \times \vec{B},\tag{3.365}$$

where $\vec{l}$ is the unit wire length vector, having the wire direction and the sense of the current flow. In the case of a curved wire, one may apply the formula (3.365) to the *infinitesimal* wire segment, which yields

$$\vec{F} = I \int d\vec{l} \times \vec{B}.\tag{3.366}$$

In order to apply the Lagrange equations and extend a classical mechanics approach to study the Lorentz forces, both the electric $\vec{E}$ and magnetic $\vec{B}$ fields are replaced by the magnetic potential $\vec{A}$ (vector) and electrostatic potential ϕ (scalar) in the following way:

$$\vec{B} = \nabla \times \vec{A},\tag{3.367}$$

$$\vec{E} = -\nabla\phi - \frac{\partial\vec{A}}{\partial t},\tag{3.368}$$

where $\nabla\phi$ stands for a *gradient*, and $\nabla\times$ is a curl. Equation (3.360), taking into account (3.363) and (3.364), yields

$$\vec{F} = q\left[-\left(\nabla\phi + \frac{\partial\vec{A}}{\partial t}\right) + \vec{v}\times(\nabla\times\vec{A})\right],\qquad(3.369)$$

or equivalently

$$\vec{F} = q\left[-\left(\nabla\phi + \frac{\partial\vec{A}}{\partial t}\right) + \nabla(\vec{v}\cdot\vec{A}) - (\vec{v}\cdot\nabla)\vec{A}\right].\qquad(3.370)$$

One may introduce now the Lagrangian L for a charged particle of the mass m, and the charge q, located in an electromagnetic field of the following form

$$L = \frac{m}{2}\dot{r}^2 + q\vec{A}\cdot\dot{\vec{r}} - q\phi.\qquad(3.371)$$

3.4.8   *Permeability and Susceptibility as Tensors and Dyadics*

In the MKS (meter, kilogram, second) units, Maxwell's equations require the following constructive equations (see Eqs. (3.340) and (3.350), i.e.,

$$\begin{aligned}
\vec{B} &= \mu\vec{H} = \mu_0\left(\vec{H} + \vec{M}\right),\\
\vec{D} &= \varepsilon\vec{E} = \varepsilon_0\vec{E} + \vec{P},
\end{aligned}\qquad(3.372)$$

where $\vec{B}$ is the magnetic flux density [Wb/m^2], $\vec{H}$ is the magnetic field intensity [A/m], $\vec{D}$ is the electric displacement [C/m^2], $\vec{M}$ is the magnetization [A/m], $\vec{P}$ is the electric polarization [C/m^2]. In addition, ε is the permittivity constant [F/m], μ is the permeability constant [H/m], whereas ε_0 is the dielectric constant of free space (vacuum) $(1/36\pi)\cdot 10^{-9}$ [F/m], and μ_0 is the permeability of vacuum which is equal $4\pi\cdot 10^{-7}$ [H/m]. If we introduce the magnetic χ_m and electric χ_e susceptibilities, then:

$$\begin{aligned}
\mu &= \mu_0(1 + \chi_m),\\
\varepsilon &= \varepsilon_0(1 + \chi_e).
\end{aligned}\qquad(3.373)$$

Equations (3.372) can be written in their equivalent tensor forms:

$$\begin{aligned}
\begin{bmatrix} B_1 \\ B_2 \\ B_3 \end{bmatrix} &= \begin{bmatrix} \mu_{11} & \mu_{12} & \mu_{13} \\ \mu_{21} & \mu_{22} & \mu_{23} \\ \mu_{31} & \mu_{32} & \mu_{33} \end{bmatrix}\begin{bmatrix} H_1 \\ H_2 \\ H_3 \end{bmatrix},\\[2mm]
\begin{bmatrix} D_1 \\ D_2 \\ D_3 \end{bmatrix} &= \begin{bmatrix} \varepsilon_{11} & \varepsilon_{12} & \varepsilon_{13} \\ \varepsilon_{21} & \varepsilon_{22} & \varepsilon_{23} \\ \varepsilon_{31} & \varepsilon_{32} & \varepsilon_{33} \end{bmatrix}\begin{bmatrix} E_1 \\ E_2 \\ E_3 \end{bmatrix},
\end{aligned}\qquad(3.374)$$

where all vectors $\vec{B}$, $\vec{H}$, $\vec{D}$ and $\vec{E}$ are expressed with the unit vectors $\vec{i}, \vec{j}, \vec{k}$. Treating Eq. (3.373) in tensor notation (3.374), we obtain:

$$\begin{bmatrix} B_1 \\ B_2 \\ B_3 \end{bmatrix} = \mu_0 \left\{ \begin{bmatrix} 1 & 0 & 0 \\ 0 & 1 & 0 \\ 0 & 0 & 1 \end{bmatrix} + \begin{bmatrix} \chi_{m11} & \chi_{m12} & \chi_{m13} \\ \chi_{m21} & \chi_{m22} & \chi_{m23} \\ \chi_{m31} & \chi_{m32} & \chi_{m33} \end{bmatrix} \right\} \begin{bmatrix} H_1 \\ H_2 \\ H_3 \end{bmatrix},$$

$$\begin{bmatrix} D_1 \\ D_2 \\ D_3 \end{bmatrix} = \varepsilon_0 \left\{ \begin{bmatrix} 1 & 0 & 0 \\ 0 & 1 & 0 \\ 0 & 0 & 1 \end{bmatrix} + \begin{bmatrix} \chi_{e11} & \chi_{e12} & \chi_{e13} \\ \chi_{e21} & \chi_{e22} & \chi_{e23} \\ \chi_{e31} & \chi_{e32} & \chi_{e33} \end{bmatrix} \right\} \begin{bmatrix} E_1 \\ E_2 \\ E_3 \end{bmatrix}. \tag{3.375}$$

Introducing a column vector as below

$$\vec{u} = \begin{bmatrix} \vec{i} \\ \vec{j} \\ \vec{k} \end{bmatrix}, \tag{3.376}$$

one may use a dyadic vector

$$\overset{\leftrightarrow}{\mu} = \vec{u}^T [\mu] \vec{u} =$$

$$= \begin{bmatrix} \vec{i} & \vec{j} & \vec{k} \end{bmatrix} \left(\begin{bmatrix} \mu_{11} & \mu_{12} & \mu_{13} \\ \mu_{21} & \mu_{22} & \mu_{23} \\ \mu_{31} & \mu_{32} & \mu_{33} \end{bmatrix} \begin{bmatrix} \vec{i} \\ \vec{j} \\ \vec{k} \end{bmatrix} \right) = \begin{bmatrix} \vec{i} & \vec{j} & \vec{k} \end{bmatrix} \begin{bmatrix} \mu_{11}\vec{i} & \mu_{12}\vec{j} & \mu_{13}\vec{k} \\ \mu_{21}\vec{i} & \mu_{22}\vec{j} & \mu_{23}\vec{k} \\ \mu_{31}\vec{i} & \mu_{32}\vec{j} & \mu_{33}\vec{k} \end{bmatrix} =$$

$$= \mu_{11}\vec{i}\vec{i} + \mu_{12}\vec{j}\vec{i} + \mu_{13}\vec{k}\vec{i} + \mu_{21}\vec{i}\vec{j} + \mu_{22}\vec{j}\vec{j} + \mu_{23}\vec{k}\vec{j} + \mu_{31}\vec{i}\vec{k} + \mu_{32}\vec{j}\vec{k} + \mu_{33}\vec{k}\vec{k}. \tag{3.377}$$

Therefore, equations (3.374) can be presented also in the corresponding dyadic forms:

$$\vec{B} = \overset{\leftrightarrow}{\mu} \cdot \vec{H},$$
$$\vec{D} = \overset{\leftrightarrow}{\varepsilon} \cdot \vec{E}. \tag{3.378}$$

The previously introduced dyadic permeability μ and dyadic permittivity ε allow also to introduce the corresponding *dyadic susceptibilities*. We introduce the following cross product action in the first equation of (3.372):

$$\nabla \times \left(\overset{\leftrightarrow}{\mu} \cdot \vec{H} \right) = \nabla \times \mu_0 \left(\vec{H} + \vec{M} \right), \tag{3.379}$$

and hence we get

$$\nabla \times \left[\overset{\leftrightarrow}{\mu} \cdot \vec{H} - \mu_0 \left(\vec{H} + \vec{M} \right) \right] = 0. \tag{3.380}$$

The following dyadic susceptibility is defined

$$\vec{M} = \overset{\leftrightarrow}{\chi}_m \vec{H}, \tag{3.381}$$

and then from Eq. (3.380), one gets

$$\nabla \times \left[\overset{\leftrightarrow}{\mu} - \mu_0 \left(\overset{\leftrightarrow}{I} + \overset{\leftrightarrow}{\chi} \right) \right] \cdot \vec{H} = 0, \tag{3.382}$$

which is satisfied if

$$\overset{\leftrightarrow}{\mu} = \mu_0 \left(\overset{\leftrightarrow}{I} + \overset{\leftrightarrow}{\chi} \right), \tag{3.383}$$

where the unit dyadic tensor

$$\overset{\leftrightarrow}{I} = \vec{i}\vec{i} + \vec{j}\vec{j} + \vec{k}\vec{k}. \tag{3.384}$$

Note that the following relations hold

$$\overset{\leftrightarrow}{I} \cdot \vec{H} = \left[\vec{i}\vec{i} + \vec{j}\vec{j} + \vec{k}\vec{k} \right] \cdot \begin{bmatrix} H_1\vec{i} \\ H_2\vec{j} \\ H_3\vec{k} \end{bmatrix} = H_1\vec{i} + H_2\vec{j} + H_3\vec{k} = \vec{H}. \tag{3.385}$$

Equation (3.383) exhibits the relationship between dyadic permeability and susceptibility.

In the case when H_i is collinear with B_i (semi-infinite isotropic magnetic medium), the tensor $[\mu]$ takes a simple diagonal form

$$[\mu] = diag\,[\mu]. \tag{3.386}$$

Scalar permeability can be represented by a diagonal matrix with three different terms

$$[\mu] = \begin{bmatrix} \mu_{11} & 0 & 0 \\ 0 & \mu_{22} & 0 \\ 0 & 0 & \mu_{33} \end{bmatrix}, \tag{3.387}$$

where

$$\mu_{ii} = \frac{B_i}{H_i}. \tag{3.388}$$

In general, when we have finite samples, H_i is not collinear with B_i. However, we may compute the diagonal terms of the permeability tensor.

From the equation

$$[\mu]\vec{H} = \mu_0 \left(\vec{H} + \vec{M} \right) \tag{3.389}$$

we get

$$\begin{bmatrix} \mu_{11} & \mu_{12} & \mu_{13} \\ \mu_{21} & \mu_{22} & \mu_{23} \\ \mu_{31} & \mu_{32} & \mu_{33} \end{bmatrix} \begin{bmatrix} H_1\vec{i} \\ H_2\vec{j} \\ H_3\vec{k} \end{bmatrix} = \mu_0 \begin{bmatrix} (H_1 + M_1)\vec{i} \\ (H_2 + M_2)\vec{j} \\ (H_3 + M_3)\vec{k} \end{bmatrix}, \tag{3.390}$$

and consequently

$$H_1\mu_{11}\vec{i} + H_2\mu_{12}\vec{j} + H_3\mu_{13}\vec{k} = \mu_0(H_1 + M_1)\vec{i},$$

$$H_1\mu_{21}\vec{i} + H_2\mu_{22}\vec{j} + H_3\mu_{23}\vec{k} = \mu_0(H_2 + M_2)\vec{j}, \tag{3.391}$$

$$H_1\mu_{31}\vec{i} + H_2\mu_{32}\vec{j} + H_3\mu_{33}\vec{k} = \mu_0(H_3 + M_3)\vec{k}.$$

Compression of the terms with subscripts „ii" at unit vectors $\vec{i}, \vec{j}, \vec{k}$ in Eq. (3.391) yields

$$\mu_{ii} = \mu_0 \left(\frac{H_i + M_i}{H_i} \right). \tag{3.392}$$

The left-hand side of the equation (3.389) is

$$[\mu]\vec{H} = \begin{bmatrix} \mu_{11} & \mu_{12} & \mu_{13} \\ \mu_{21} & \mu_{22} & \mu_{23} \\ \mu_{31} & \mu_{32} & \mu_{33} \end{bmatrix} \begin{bmatrix} H_1\vec{i} \\ H_2\vec{j} \\ H_3\vec{k} \end{bmatrix} = \begin{bmatrix} \mu_{11}H_1 + \mu_{12}H_2 + \mu_{13}H_3 \\ \mu_{21}H_1 + \mu_{22}H_2 + \mu_{23}H_3 \\ \mu_{31}H_1 + \mu_{32}H_2 + \mu_{33}H_3 \end{bmatrix} \begin{bmatrix} \vec{i} \\ \vec{j} \\ \vec{k} \end{bmatrix}. \tag{3.393}$$

Taking into account Eqs. (3.393) and (3.390), we obtain

$$\begin{aligned} \mu_{11}H_1 + \mu_{12}H_2 + \mu_{13}H_3 &= \mu_0(H_1 + M_1), \\ \mu_{21}H_1 + \mu_{22}H_2 + \mu_{23}H_3 &= \mu_0(H_2 + M_2), \\ \mu_{31}H_1 + \mu_{32}H_2 + \mu_{33}H_3 &= \mu_0(H_3 + M_3). \end{aligned} \tag{3.394}$$

Relationships (3.394) and (3.392) yield the following set of algebraic equations

$$\begin{aligned} \mu_{12}H_2 + \mu_{13}H_3 &= 0, \\ \mu_{21}H_1 + \mu_{23}H_3 &= 0, \\ \mu_{31}H_1 + \mu_{32}H_2 &= 0. \end{aligned} \tag{3.395}$$

Equations (3.395) are obtained from the equation (3.389):

$$\left([\mu] - \mu_0[I]\right)\vec{H} = \mu_0\vec{M}, \tag{3.396}$$

which gives the following matrix notation

$$\begin{bmatrix} \mu_{11} - \mu_0 & \mu_{12} & \mu_{13} \\ \mu_{21} & \mu_{22} - \mu_0 & \mu_{23} \\ \mu_{31} & \mu_{32} & \mu_{33} - \mu_0 \end{bmatrix} \begin{bmatrix} H_1 \\ H_2 \\ H_3 \end{bmatrix} = \mu_0 \begin{bmatrix} M_1 \\ M_2 \\ M_3 \end{bmatrix}, \tag{3.397}$$

where $\mu_{ii} = \mu_0(1 + M_i/H_i)$.

The sought nondiagonal terms of the permeability tensor can be found from the formula

$$\mu_{ji} = \mu_0\left(\frac{M_i}{H_j}\right), \qquad i \neq j, \tag{3.398}$$

assuming that the following relationships hold:

$$\frac{M_2}{M_3} = -\frac{H_3}{H_2}, \qquad \frac{M_1}{M_3} = -\frac{H_3}{H_1}, \qquad \frac{M_1}{M_2} = -\frac{H_2}{H_1}. \tag{3.399}$$

The following final remarks are important:

(i) when H_i is collinear with B_i (semi-infinite isotropic magnetic medium), we have $\mu = \mu_{11} = \mu_{22} = \mu_{33}$;

(ii) when H_i is not collinear with B_i (finite-size magnetic media of various shapes, then Eqs. (3.392) and (3.398) hold for $M_i = M\cos\beta_i$, and β_i is the angle between M and H_i;

(iii) B_i are different for each direction of $\vec{H}$ for $\mu_{11} \neq \mu_{22} \neq \mu_{33}$.

3.4.9 *Diamagnetic Materials*

Diamagnetic materials are those materials whose permanent magnetic dipole moments equal zero. This configuration state of the particles is implied by zero vector sum of orbital and spin magnetic moments. The following characteristic features are typical for this materials group

(i) lack of permanent atom magnetic dipole;
(ii) magnetic susceptibility χ is small and negative $(-10^{-6}; -10^{-5})$;
(iii) induced magnetic moments generate the magnetization opposite to the applied external field;
(iv) diamagnetic object is repelled by the magnetic field.

Typical examples of the diamagnetic materials are: inactive gases, the majority of organic compounds, water, glass and metals such as Ag, Au, Bi, Cu, Hg, Zn.

Introduction of a diamagnetic sample into the magnetic external field $\vec{B}_0$ results in an induction effect. Namely, the magnetic field flux passing through the loops (orbits) with the current induced by the electron movements changes the previous orbital electrons movement, i.e., the electrons are accelerated or slowed down (Faraday's theory) and, owing to Lenz's rule, the induced magnetic field opposes the increase in the magnetic flux.

The action of an external magnetic field $\vec{B}_0$ implies the so-called *Larmor precession of the electron*, i.e., the previous loop vector of the orbit being normal to the loop surface is rotated with additional angular velocity about $\vec{B}_0$. This precession of electrons induces dipole moments of atoms against the field. Since the induced magnetization opposes the external field, diamagnetic objects are repelled by the magnetic field.

Although one may reasonably suspect that the so far described diamagnetic phenomena are typical for all materials, because the induction process affects all atoms, in practice, such phenomena are rarely observed. The reason is that they are much weaker than paramagnetic and ferromagnetic phenomena. In other words, the diamagnetic phenomena occur also in paramagnetic and ferromagnetic materials, but their influence on magnetic properties of the two mentioned groups of the magnetic material is negligible.

The distribution of magnetic dipole moments in a diamagnetic material is schematically shown in Fig. 3.53.

It is known that a change in the resultant magnetic moment $\Delta\vec{p}$ of an atom having Z electrons and being subjected to the action of external magnetic field $\vec{B}_0$ is estimated by the following equation

$$\Delta\vec{p} = -\frac{e^2 r^2 Z \vec{B}}{4m}, \tag{3.400}$$

where m is the electron mass, Z is the electrons number and r^2 denotes the average value of projections of radii of orbits of Z electrons onto the direction perpendicular

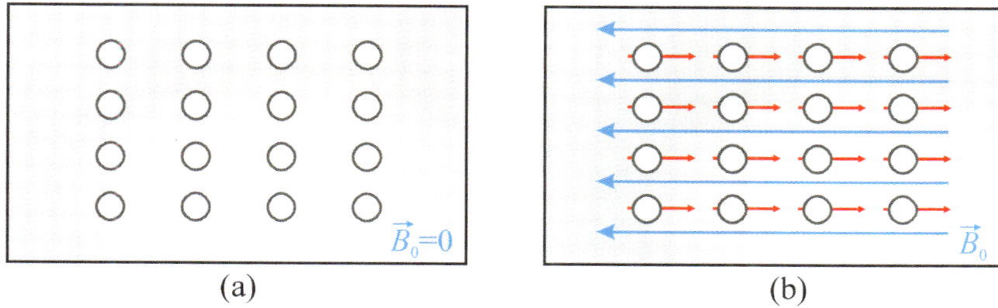

Fig. 3.53   Magnetic dipole moments in the absence (a) and existence (b) of the external field B_0 exhibited by a diamagnetic material sample.

to $\vec{B}$. The magnetization

$$\vec{M} = n_0 \Delta \vec{p} = -\frac{e^2 r^2 Z n_0}{4m} \vec{B}, \qquad (3.401)$$

where n_0 denotes the number of particles in the volume unit. Since we have

$$M = \frac{\chi B}{\mu \mu_0} \approx \frac{\chi B}{\mu_0}, \qquad (3.402)$$

then, from Eqs. (3.401) and (3.402), we get

$$\chi = -\frac{\mu_0 n_0 e^2 r^2 Z}{4m}. \qquad (3.403)$$

This formula yields small values of magnetic susceptibility of diamagnetics estimated in item (ii).

3.4.10   *Paramagnetic Materials*

Materials consisting of atoms or particles possessing permanent nonzero magnetic dipole moments μ are called *paramagnetic*. The resultant vector sum of orbital and spin magnetic moments is zero, since dipole moments are randomly distributed. However, when we apply an external uniform magnetic field $\vec{B}_0$, the dipoles rotate to follow the magnetic field alignment (see Fig. 3.54).

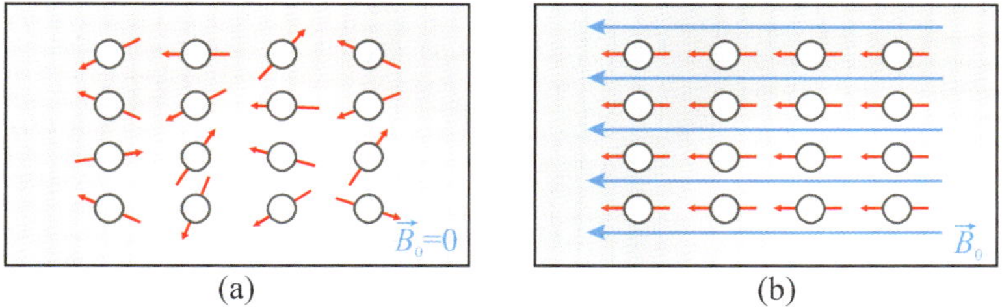

Fig. 3.54   Magnetic dipole moments in the absence (a) and presence (b) of the external field $\vec{B}_0$ exhibited by a paramagnetic material sample.

Paramagnetics have the following typical properties:

(i) permanent atom magnetic dipole moment exists;

(ii) magnetic susceptibility is small (10^{-7}-10^{-2}) and positive;

(iii) external field generates the magnetization being slightly higher than that of an external field;

(iv) paramagnetic material is attracted by a magnetic field.

The group of paramagnetic materials includes, for instance, oxygen, nitric oxide, platinum, aluminum, alkaline earth (metal) family and alkali metals.

After the action of $\vec{B}_0$ on a sample of paramagnetic material, its total magnetic field exhibits two components: external field $\vec{B}_0$ and induced field $\mu_0 \vec{M}$. Since $\vec{B} \parallel \vec{M}$, then the resultant induced field $\vec{B} = \aleph \vec{B}_0$. The permeability factor $\aleph$ slightly exceeds 1 and depends on the temperature T. The magnetization decreases with an increase in temperature, since the electron motions are disturbed by thermal actions, and hence also the previous alignment of dipoles is disturbed. This phenomenon is governed by Curie's law for paramagnetics (P. Curié in 1895) through the following formula

$$M = C\frac{B_0}{T},\tag{3.404}$$

where C stands for the Curie constant and depends on the material properties. The linear part of formula (3.404) is valid only for small values of B_0/T, as it is shown in Fig. 3.55.

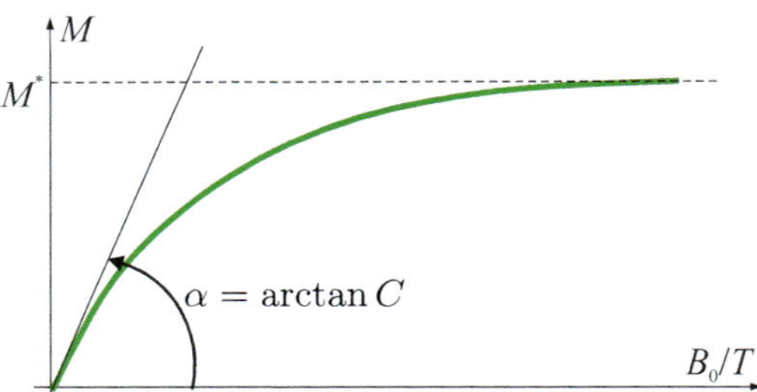

Fig. 3.55   Magnetization versus B_0/T for a paramagnetic material sample.

For large values of B_0/T, the so-called *saturation effect* is observed, where plateau M^* is achieved. This phenomenon is associated with parallel alignment of the dipoles when magnetization achieves its maximal value due to thermally excited motion of electrons. The removal of B_0 causes random distribution of directions of dipoles again, and the magnetic forces of atoms are too small to conserve the previous alignment.

The theory of paramagnetism was developed by D. Langevin in 1905. The average value of the atom magnetic moment projected onto the magnetic field B depends on the ratio γ of the atom energy devoted to its interaction with the external magnetic field B and thermal energy of the form

$$\gamma = \frac{p_m B}{kT},\tag{3.405}$$

where p_m is the magnetic moment.

Magnetization process of a paramagnetic sample depends essentially on γ, i.e.,

$$M = \frac{n_0 p_m^2 B}{3kT} = \chi_m \frac{B}{\mu}, \qquad \gamma \ll 1, \tag{3.406}$$

$$M^* = n_0 p_m, \qquad \gamma \gg 1, \tag{3.407}$$

where n_0 denotes a number of atoms per unit volume.

This means that for $\gamma \ll 1$, the dependence of $M(B)$ is linear, whereas for $\gamma \gg 1$, we have a saturation phenomenon. According to Eq. (3.406):

$$\chi_m = \frac{\mu_0 n_0 p_m^2}{3kT}, \tag{3.408}$$

which yields the earlier discussed Curie's law stating that $\chi_m \sim \frac{1}{T}$.

3.4.11 *Ferromagnetic Materials*

Materials having permanent dipole moments and exhibiting strong interactions between neighboring atomic dipole moments are called *ferromagnetic*. They possess the following properties:

(i) magnetic susceptibility χ_m is very high (10^5) and is nonlinear regarding the field parameters;

(ii) permanent atom magnetic dipole moments exist;

(iii) external field generates huge magnetization;

(iv) remanent magnetization and coercive field occur;

(v) ferromagnetism vanishes at temperatures higher than those of Curie points.

Ferromagnetic materials include iron, nickel, cobalt, and some lanthanide series. A temperature increase implies a decrease in the ferromagnetic material properties, and when the temperature exceeds the so-called threshold values defined by the Curie points, the paramagnetic properties occur instead of the ferromagnetic ones. The Curie point T_C may vary from 300 K (gadolinium) to 970 K (iron). When temperature $T > T_C$, the magnetization formula follows the Curie-Weiss law governed by the relationship

$$M = \frac{CB}{T - T_C}. \tag{3.409}$$

Large values of the susceptibility χ_m affect nonlinear behavior of the magnetization M and the resultant field B versus the applied field B_0. The permeability μ becomes also field-dependent.

Applying magnetization and demagnetization to a ferromagnetic material sample, the so-called magnetic hysteresis curve is obtained in Fig. 3.56.

An increase in the magnetic field B_0 yields an increase in the ferromagnetic sample magnetization up to the saturation level M^* (point B). A decrease in B_0 results in the transition via another path, and for $B_0 = 0$, the so-called remanent

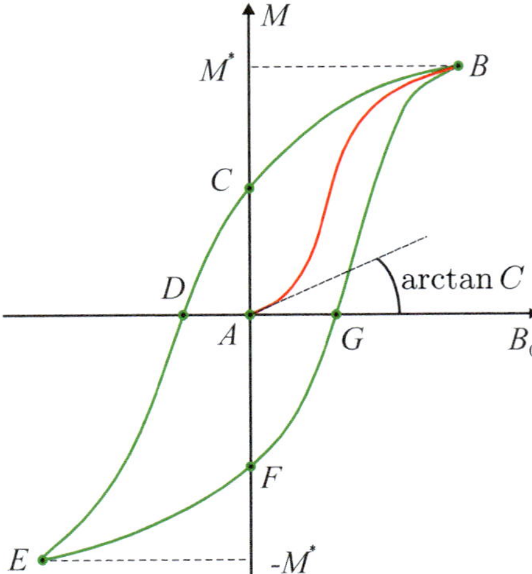

Fig. 3.56　Magnetization a ferromagnetic sample in external field B_0.

magnetization occurs (point C). In order to remove magnetization, we must change the direction of B_0 to achieve $M = 0$ in point D. Continuing the process, we observe the reversed magnetization of the sample which is again saturated at point E. Before closing a circle of external field magnetization, we pass through the permanent magnetization at point F, and the coercive field (G) finally achieves the original magnetization (B). The earlier illustrated closed orbit of magnetization/demagnetization is a characteristic feature of the ferromagnetic samples. Furthermore, we say that a ferromagnetic is hard (soft), when the coercive field is high (low).

3.5　An Introduction to Electromagnetic Fields

As it is well known, there exist four fundamental forces in our universe, i.e., gravity, weak, strong and electromagnetic ones. In mechatronics, we deal with magnetic, electric and mechanical forces. Mechanical forces are well known and described in various mechanical books [Awrejcewicz and Koruba (2012); Awrejcewicz (2012a,b)], and they obey three Newton's laws.

The electric charge is quantified as a multiple of the electron ($-$) or proton ($+$) charges ($\pm e = \pm 1.60210^{-19}$C), where C (coulomb) is the unit of the electric charge. The charge of 1 C is the charge which flows through a 120 W light bulb in 1 second. Theoretically, two charges of 1 C each, being separated by 1 meter, should repel each other with a force around 10^9 N. However, this will never happen, since the nature will never collect the charge of 1 C in one point. Coulomb's law allows to compute charges q_1 and q_2 (recall that charges of the same (different) sign repel (attract) each other) as follows

$$\vec{F} = k \frac{q_1 q_2}{r^3} \vec{r}, \tag{3.410}$$

where r denotes the distance between them, $k = \frac{1}{4\pi\varepsilon_0}$ and ε_0 is the value of free space (vacuum) permittivity ($\varepsilon_0 = 8.85 \cdot 10^{12}$ [F/m]), where Farad [F] is the unit of electrical capacitance; magnitude of charge Q and voltage V applied to parallel plates of the capacitor is the capacitance ($C = Q/V$).

In order to show that Coulomb's law satisfies Newton's third law, equation (3.410) has been presented in the vector notation. Coulomb's law is similar to that of the gravitational law governed by the equation

$$\vec{F} = G\frac{m_1 m_2}{r^3}\vec{r},\tag{3.411}$$

where $G = 6.67 \cdot 10^{-11} \mathrm{Nm}^2/\mathrm{kg}^2$ is the gravitational constant, and the gravity is the weakest of the previously mentioned gravitational forces.

The Lorentz force law unifies both electric and magnetic fields. As it is known, the magnetic field may be induced by electric currents, which may be understood as currents produced by electrons in atomic orbits or in terms of macroscopic currents in wires. The interaction of a magnetic field (sometimes called also magnetic flux densities) $\vec{B}$ with a charge q and an electric field intensity $\vec{E}$ is governed by the Lorentz equation

$$\vec{F} = q\vec{E} + q\vec{v} \times \vec{B},\tag{3.412}$$

where $\vec{v}$ is the charge of q velocity. In SI notation $[B] = [T]$ (tesla). The Gauss (G) magnetic field units is $1\ [T] = 10^4\ [G]$, and $1\ [T] = 1\ [Wb/m^2]$, where weber (Wb) is the unit of the magnetic flux.

Simple schemes of both electric $q \cdot E$ and magnetic $q\vec{v} \times \vec{B}$ forces acting on a moving positive charge are shown in Fig. 3.57.

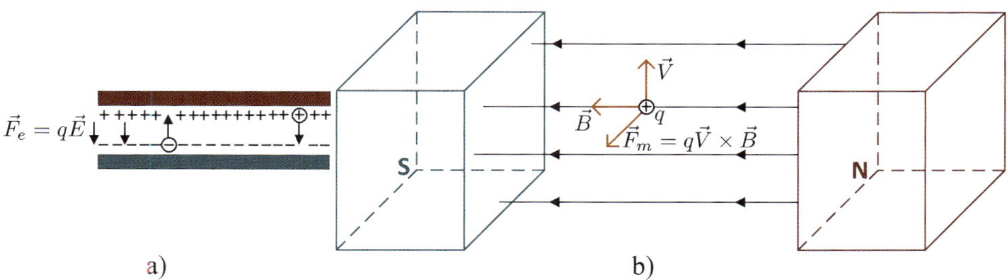

Fig. 3.57 Electric and magnetic constituents of the electromagnetic forces (1) (S/N) denote the south north pole of a magnet.

In the case of applications to current-carrying wires, a force acting on a straight wire of the length l is given by the following formula

$$\vec{F} = \vec{I}l \times \vec{B},\tag{3.413}$$

where $\vec{I}$ is the vector of the electric current and l denotes the length of the wire.

In the world of atomic scales, the electromagnetic forces are dominant over the other mentioned fundamental forces, and they are responsible for building the material atomic and molecular structure. However, the magnetic part of the electromagnetic force can be also neglected since it appears only at high resolutions. On the other hand, the so far discussed gravity and electromagnetic forces obey the so-called inverse-square law (coming from geometrical considerations and being applied to diverse physical phenomena). They are infinite in range and they have similar mathematical forms. This observation motivated Einstein to look for one unified force being responsible for all possible interactions in our universe matter. However, it appears that the common form of the electromagnetic and gravity forces is due to existence of an exchange particle of zero mass rather than because of an inherent symmetry.

The electric force (in Newtons) $\vec{F}_e$ per charge q (in Coulombs) defines the electric field $\vec{E}$ in the following way

$$\vec{E} = \frac{\vec{F}_e}{q},\tag{3.414}$$

and $[E] = [N/C]$ or $[E] = [N/m]$. It points radially outwards from a positive charge towards a negative point charge, as it is schematically depicted in Fig. 3.58.

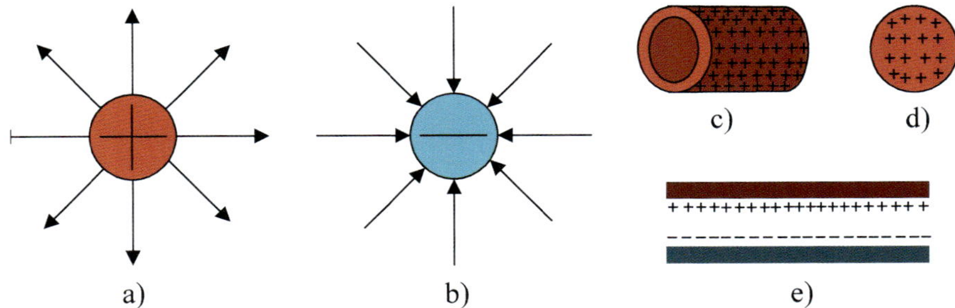

Fig. 3.58   Electric field produced by a positive (a) and negative (b) point charge, a charged cylinder (c), a charged sphere (d) and charged parallel plates (e).

In the case of the point charge, the electric field is

$$E = \frac{F}{q} = \frac{1}{q}\frac{kQq}{r^2} = \frac{kQ}{r^2},\tag{3.415}$$

and radius r determines the electric potential of a point charge as follows

$$V = \frac{kQ}{r} = \frac{Q}{4\pi\varepsilon_0 r}.\tag{3.416}$$

Therefore, the equipotential surfaces are spheres of radii r, and the equipotential lines are circles. In the case of parallel conducting plates (see Fig. 3.58e), the equipotential lines are parallel to the plates and are perpendicular to the electric field lines.

In a general physics, three constants are used to describe magnetic and electric fields propagation, i.e., the speed of light $c = 299,792,458$ [m/s], the electric permittivity of free space ε_0 and the magnetic permeability of (free) space $\mu_0 = 4\pi \cdot 10^{-7}$ N/A^2, where A denotes ampere. Note that the permeability of the majority of materials is close to μ_0, since they are either *paramagnetic* or *diamagnetic*. However, sometimes the permeability of ferromagnetic materials is very large, and therefore a relative permeability μ_r is introduced. For example, in the case of the magnetic iron $\mu_r = 200$, whereas in the case of permalloy (78.5% nickel, 21.5% iron) $\mu_r = 8000$. A value of the free space permittivity ε_0 can be computed from the following formula

$$c = \frac{1}{\sqrt{\mu_0 \varepsilon_0}}, \tag{3.417}$$

which gives $\varepsilon_0 = 8.85 \cdot 10^{-12}$ F/m. In any dielectric material, the orientation of molecules is random. However, when an electric field is applied, the material starts to exhibit orientation of the dipole moments of polar molecules. Observe that even though the total molecule charge is zero, its positive and negative charges do not overlap due to the nature of chemical bonds, i.e., they have permanent *dipole moments*. If we consider an ideal pair of opposite charges of magnitudes $+q$ and $-q$ located within a distance r, it produces the electric dipole moment being defined as follows

$$\vec{p} = q\vec{r}, \tag{3.418}$$

where q is the magnitude of the charge and r is the distance between charges towards the positive charge. An electric field $\vec{E}_d$ produced by a dipole is

$$\vec{E}_d = -\frac{1}{4\pi\varepsilon_0} \frac{\vec{p}}{d^4} \vec{d}, \tag{3.419}$$

where $d \gg r$ and $\vec{d} \perp \vec{r}$. The *potential* (voltage or electric potential energy) of the point charge can be estimated in the following way. We take zero potential in infinity and we compute the work needed to bring a test charge q from its infinite location to that at the distance r. The work done by the Coulomb force acting along a straight radial line r from r_a to r_b follows (see Fig. 3.59)

$$W = V_A - V_B = \int_{r_A}^{r_B} \frac{kq}{r^2} dr = kq\left(\frac{1}{r_A} - \frac{1}{r_B}\right), \tag{3.420}$$

where $k = \frac{1}{4\pi\varepsilon_0}$ is the Coulomb constant. Hence, we obtain Eq. (3.416), assuming that $r_B \to \infty$.

On the other hand, the potential of the electric dipole is estimated through formula (3.420) in the following way

$$V = kq\frac{r_- - r_+}{r_- r_+}, \tag{3.421}$$

where $r_A = r_+$, $r_B = r_-$. In Fig. 3.59, equipotential lines are also shown. They form circles, and scaling of voltage at equal increments shows how these circles get further apart with increasing r.

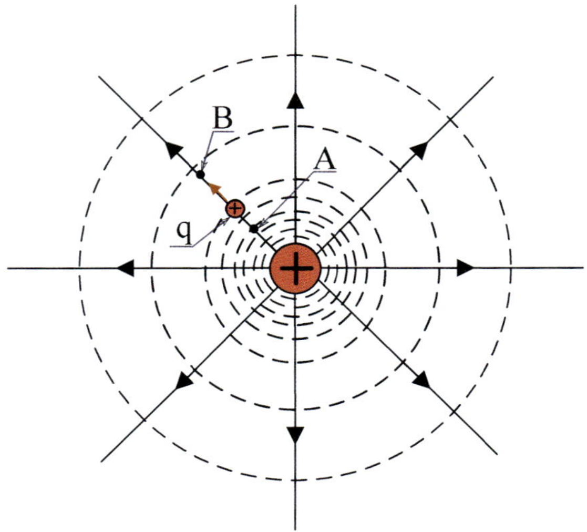

Fig. 3.59   Test charge q moving radially and the equipotential lines (dashed).

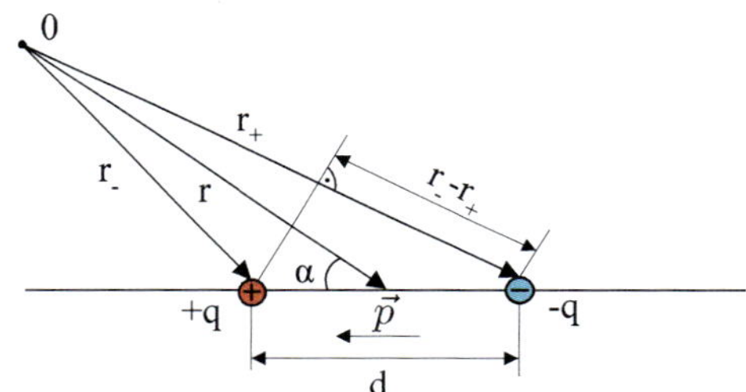

Fig. 3.60   Scheme for electric dipole potential estimation.

Owing to the scheme shown in Fig. 3.60, the following observations hold

$$r_- - r_+ \cong d\cos\alpha, \quad r_+ r_- \cong r^2, \quad \vec{p} = q\vec{d}. \tag{3.422}$$

Hence, taking into account Eq. (3.422) in (3.421):

$$V = \frac{kp\cos\alpha}{r^2}. \tag{3.423}$$

The electric potential of the dipole is reported in Fig. 3.61 and it shows the mirror symmetry about the center point of the dipole. Equipotential lines are everywhere perpendicular to the electric field lines.

Electric and magnetic fields store the energy. It is usually computed per volume unit $\mathcal{U}$, and hence it may be presented as the energy densities of electric/magnetic field of the forms:

$$\frac{H_E}{\mathcal{U}} = \frac{1}{2}\varepsilon E^2, \tag{3.424}$$

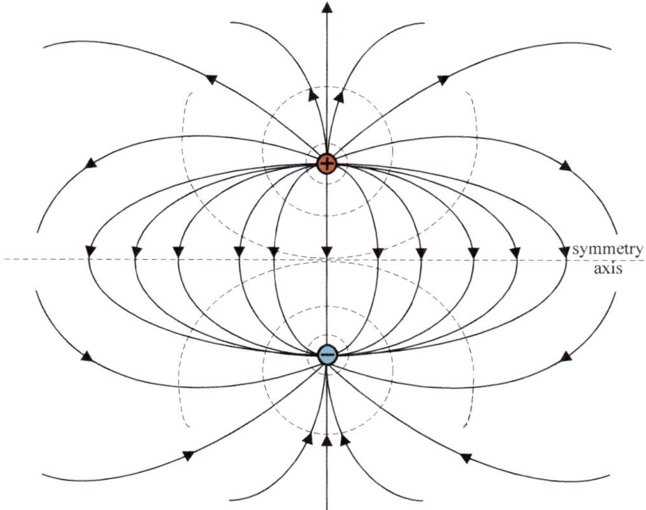

Fig. 3.61   Equipotential lines of a dipole.

$$\frac{H_B}{\mathcal{U}} = \frac{1}{2}\frac{B^2}{\mu}. \tag{3.425}$$

Enthalpy in Greek means "to put heat into". Nowadays, enthalpy represents rather a measure of the total energy of a system, not necessarily related to a heat transfer. This general word is applied to a variety of mechatronic, chemical and biological systems in which the energy can be transferred. Formal definition of the enthalpy follows

$$H = U + pV, \tag{3.426}$$

where H [J] denotes the system enthalpy, U [J] is the system internal energy, p [Pa] is the pressure on the boundary between the system and the surrounding environment and $\mathcal{U}$ is the system volume. Dividing both sides of Eq. (3.426) by $\mathcal{U}$, one obtains the enthalpy density functions being also denoted by H.

Let us consider a parallel-plate capacitor of the plate area A, the distance between the plates r, where $Q = CV$ and $V = Er$. Its electric energy can be directly computed from (3.424), i.e.,

$$H_E = \frac{1}{2}\varepsilon E^2 Ar = \frac{1}{2}\varepsilon\frac{A}{r}V^2 = \frac{1}{2}CV^2. \tag{3.427}$$

Let us recall that a long straight coil of a wire produces a nearly uniform magnetic field $\vec{B}$, assuming that the current $\vec{I}$ flows through that wire, i.e., a *solenoid*. It is well known that the magnetic field induced by the electric current $\vec{I}$ in a solenoid coil is similar to that produced by a bar magnet, which has been presented in Fig. 3.62.

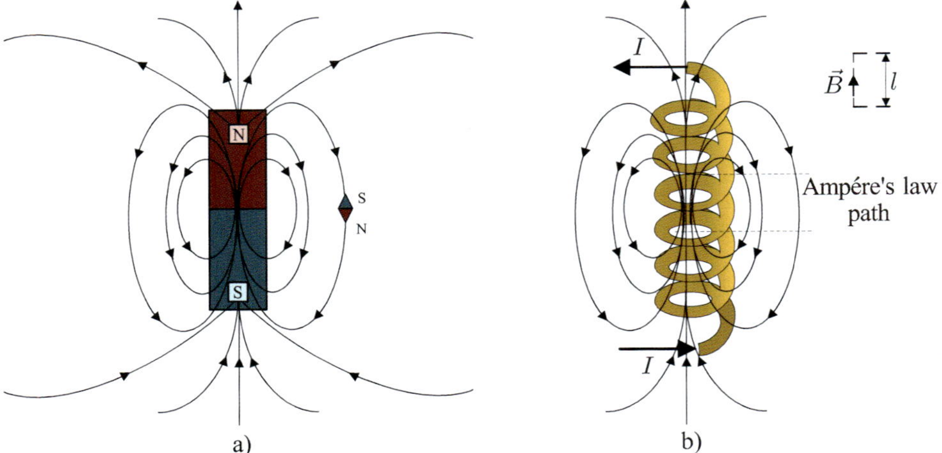

Fig. 3.62   Magnetic field produced by a bar magnet (a) and a solenoid (b) with the north (N) and south (S) poles and the compass needle marked.

The magnetic field produced in the solenoid (see Fig. 3.62b) is negligible far from the coil. Now, taking a rectangular path such that its length side l is parallel to $\vec{B}$, for the idealized case, Ampére's law yields

$$\vec{B}l = \mu n\vec{I}, \tag{3.428}$$

and hence

$$\vec{B} = \mu N\vec{I}, \tag{3.429}$$

where $n/l = N$.

Formula (3.429) describes the long solenoid field approximation in its center, where $\mu = \mu_0\mu_r$, $\mu_0 = 4\pi \cdot 10^{-7}$ T/Am, and μ_r is called the *relative permeability*.

So far, we have considered the static cases. If one moves a rectangular wire (conductor) into a stationary magnetic field with velocity v, then voltage, called also a *motional EMF*, governed by *Faraday's law* will be induced (see Fig. 3.63).

Since the EMF can be presented as the work done per unit charge q, then

$$EMF = \frac{W}{q} = \frac{Fl}{q} = \frac{qvBI}{q} = vBl = Bl\frac{dx}{dt} = \frac{d\,(BA)}{dt} = \frac{d\phi}{dt}, \tag{3.430}$$

where $\phi = BA$ is the magnetic flux, B denotes the external magnetic field, and $A = xl$ is the rectangular coil area. Observe the similarity to Newton's third law, i.e., the induced current I creates a magnetic field B_I which opposes the magnetic field B created in the rectangular coil. If one takes a coil with the number of N turns, then formula (3.430) can be cast to the following form

$$EMF = -N\frac{d\phi}{dt}, \tag{3.431}$$

and the minus sign comes from Lenz's law. Equation (3.431) represent Faraday's law. Lenz's law says that if an EMF is generated by a change in the magnetic flux

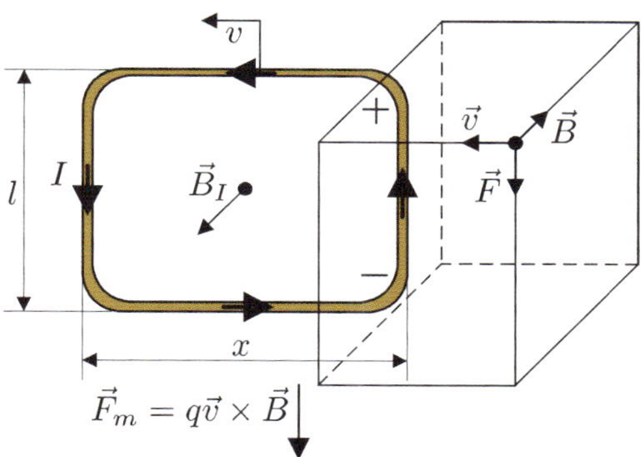

Fig. 3.63   Scheme of a motional *EMF* (electromotive force).

associated with B due to Faraday's law, then the induced EMF polarity produces a current I whose magnetic field B_I opposes the change of B. Lenz's law is illustrated in Fig. 3.64, where the action of a bar magnet with south and north poles is shown (taken from R. Nave/hyperphysics.phy-astr.gsu.edu).

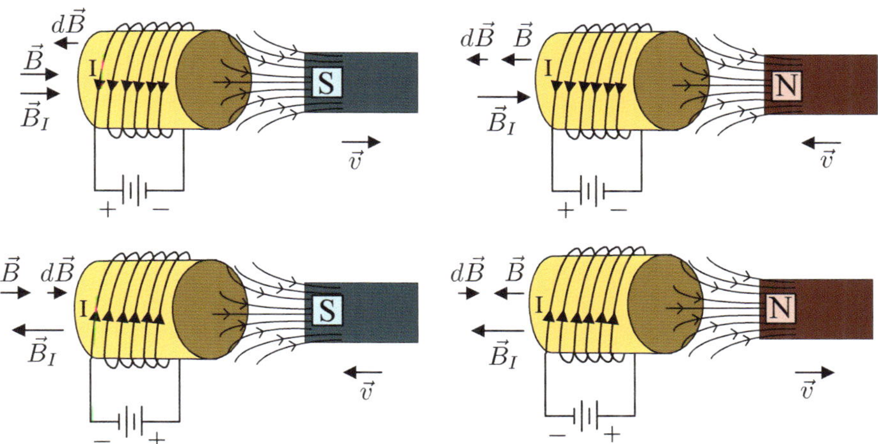

Fig. 3.64   Illustration of Lenz's law.

The so far discussed electric and magnetic laws can be presented in a more compact, generalized form. *Ampére's law* can be described by the line integral around a closed loop by the following equation

$$\oint \vec{B} \cdot d\vec{s} = \mu_0 I + \frac{1}{c^2}\frac{\partial}{\partial t}\int \vec{E} \cdot d\vec{A},\qquad(3.432)$$

or by the following differential equation

$$\nabla \times \vec{B} = \frac{4\pi k}{c^2}\vec{J} + \frac{1}{c^2}\frac{\partial \vec{E}}{\partial t},\qquad(3.433)$$

where $k = \frac{1}{4\pi\varepsilon_0}$, $c = \frac{1}{\sqrt{\mu_0\varepsilon_0}}$, and $\vec{J}$ is the current density. The line and the surface are shown in Fig. 3.65a,b, respectively.

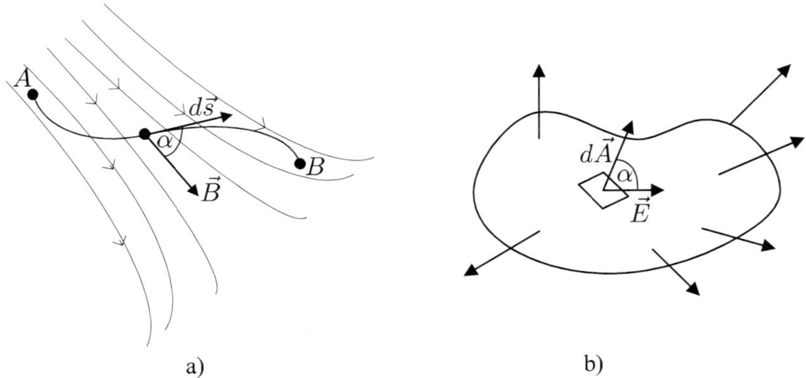

Fig. 3.65   Scheme of the line integral $\int\limits_{A}^{B} \vec{B} \cdot d\vec{s} = \int\limits_{A}^{B} B\cos\alpha ds$ (a) and of the surface integral corresponding to the electric flux $\phi = \int \vec{E} \cdot d\vec{A} = \int B\cos(\alpha)dA$ (b) computations.

The total electric flux ϕ through a closed surface A is equal to the enclosed charge divided by permittivity ε_0, i.e.,

$$\phi = \frac{q}{\varepsilon_0} = \oint \vec{E} \cdot d\vec{A}, \qquad \vec{E} = \frac{\Delta\phi}{\Delta\vec{A}}, \tag{3.434}$$

and $\vec{E} \perp \Delta\vec{A}$. For a point charge q, the electric flux is (see Fig. 3.66a)

$$\phi = EA = 4\pi r^2 E = \frac{q}{\varepsilon_0}, \tag{3.435}$$

whereas for a conducting sphere with charge Q, the electric flux is (see Fig. 3.66b)

$$\phi = 4\pi r^2 E = \frac{Q}{\varepsilon_0} \quad \text{for} \quad r > R,$$
$$\phi = 0 \quad \text{for} \quad r < R. \tag{3.436}$$

Recall that if there are two point charges q_1 and q_2 located at a distance r, then the force acting on each point is governed by formula (3.410). There are two Gauss' laws. The integral and differential forms of Gauss' law for magnetism follow:

$$\oint \vec{B} \cdot d\vec{A} = 0, \tag{3.437}$$

$$\nabla \cdot \vec{B} = 0. \tag{3.438}$$

The surface integral (3.437) is equal to zero, which means that the net magnetic flux through any closed surface is zero (it is nonzero if there is a magnetic monopole

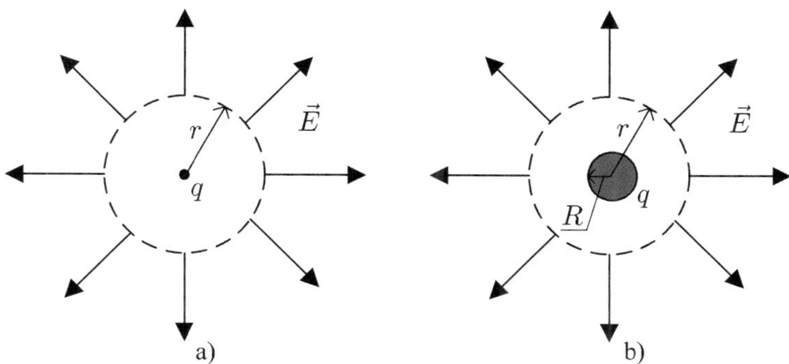

Fig. 3.66   Electric field of a point charge (a) and of a conducting sphere (b).

source). If there is no monopole source, the divergence of a vector field is proportional to the point source density. Gauss' law for electricity also has either integral form

$$\oint \vec{E} \cdot d\vec{A} = \frac{q}{\varepsilon_0} = 4\pi k q,$$

(3.439)

or differential form

$$\nabla \cdot \vec{E} = \frac{\rho}{\varepsilon_0} = 4\pi k \rho.$$

(3.440)

It represents the following observation: the electric flux through any closed surface is proportional to the total charge of the surface. The electric field divergence (see Eq. (3.440)) allows to measure the density of sources, whereas the surface integral (see Eq. (3.439)) yields a measure of the enclosed net charge.

Faraday's law for induction has the following form of the line integral

$$\oint \vec{E} \cdot d\vec{s} = -\frac{d\phi}{dt},$$

(3.441)

i.e., the line integral of the electric field E around a closed loop is equal to the negative of the rate of change of the magnetic flux ϕ through the surface bounded by the loop.

Its corresponding differential representation is

$$\nabla \times \vec{E} = -\frac{\partial \vec{B}}{\partial t}.$$

(3.442)

Maxwell's equations represent a set of mathematical formulas of the so far illustrated and discussed Gauss' electric and magnetic laws as well as of Faraday's and Ampére's laws.

They will be repeated below in their more general differential representation valid for polarizable magnetic isotropic linear materials.

(i) Gauss' laws for electricity:

$$\nabla \cdot \vec{D} = \rho,$$

(3.443)

$$\vec{D} = \varepsilon \vec{E} + \vec{P}, \tag{3.444}$$

where $\vec{D}$ denotes the electric displacement, ρ is the charge density, ε is the permittivity, $\vec{P}$ is the polarization vector (in the case if free space $\vec{D} = \varepsilon_0 \vec{E}$).

(*ii*) Gauss' law for magnetism

$$\nabla \cdot \vec{B} = 0, \tag{3.445}$$

where B is the magnetic field vector.

(*iii*) Faraday's law of induction

$$\nabla \times \vec{E} = -\frac{\partial \vec{B}}{\partial t}. \tag{3.446}$$

(*iv*) Ampére's law

$$\nabla \times \vec{H} = \vec{J} \times \frac{\partial \vec{D}}{\partial t}, \tag{3.447}$$

where

$$\vec{B} = \mu_0 \left(\vec{H} + \vec{M} \right), \tag{3.448}$$

and $\vec{H}$ denotes the magnetic field strength vector, whereas $\vec{M}$ is the magnetization vector (for a free space $\vec{B} = \mu_0 \vec{H}$). Formula (3.448) requires explanation. Owing to Ampére's law, when magnetic fields generated by currents pass through magnetic materials, it is difficult to distinguish which part of the field comes from the material and which from the external currents. Therefore, the magnetic field strength $\vec{H}$ has been introduced. Note that $\vec{H}$ and $\vec{M}$ have the same unit [A/m]. Sometimes, instead of Eq. (3.448) we take $\vec{B} = \vec{B}_0 + \mu_0 \vec{M}$, where μ_0 is the magnetic permeability of free space and $\vec{B}_0$ is the externally applied magnetic field.

Chapter 4

Modeling of Piezoelectric Phenomena

4.1 Piezoelectric Materials, Materials Laws and Constitutive Equations

Piezoelectric materials (piezoceramics) play a key role in mechatronic systems. As it has been pointed out in [Kamlah (2001); von Wagner (2003)], the piezoelectric ceramic lead (Pb)-zirconate (Zr)-titanate (Ti), denoted by PZT, is often used in manufacturing of sensors and actuators. PZT is a mixture of $PBTiO_3$ and $PbZrO_3$. Since both $PbTiO_3$ and $PbZrO_3$ are ferroelectric, their lattice structure depends on temperature, where the key role is played by the Curie temperature T_c. In Fig. 4.1, cubic (a) and tetragonal (b) unit cells of $PbTiO_3$ piezoelectric ceramic are shown.

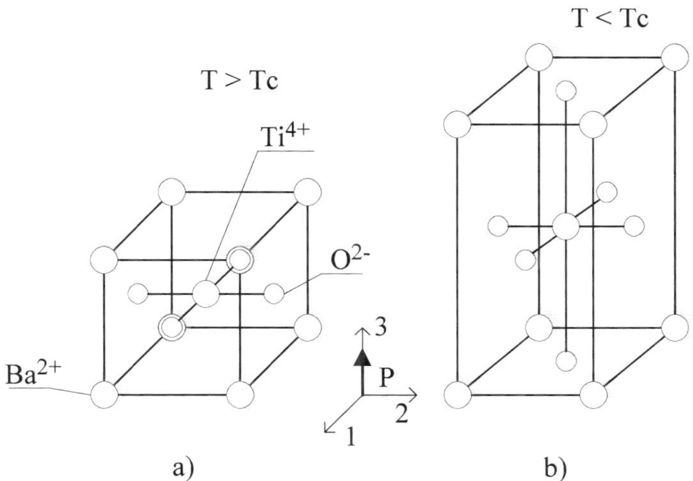

Fig. 4.1   Regular hexahedron for $T > T_c$ (a) and deformed rectangular prism $T < T_c$ (b) of cells of $BaTiO_3$ piezoelectric ceramic (Ba^{2+} O^2 Ti^{4+}) with polarization vector P marked.

Piezoceramics are manufactured of isotropic ferroelectric ceramics using a strong, homogeneous, external electric field. In general, Zr and Ti, as well as their compounds, exist in a form of randomly oriented grains belonging to different

123

crystal classes, i.e., tetragonal pyramids or rhombohedra prisms. The introduced electric field allows for a rearrangement of domain walls in the grains, yielding a macroscopic polarizations. After the strong electric field is removed, a transversely isotropic piezoelectric material is built and a net remnant polarization axis is perpendicular to the isotropy plane. In what follows, the obtained bulk material possesses properties equivalent to that of a single hexagonal crystal of 6 mm (however, be aware that crystals of the given piezoceramic material are not of the same size). The Curie temperature (150-350°C) is responsible for piezoelectric unit behavior. Namely, below this threshold, the previous symmetric structure (see Fig. 4.1a) is forced due to the displacement of Ti (or Zr). In what follows, a tetragonal (of possibly rhombohedral) dipole emerges giving a rise for coupling of previously uncoupled mechatronic, i.e., mechanical and electrical, properties.

For $T > T_c$, the cuboid preserves all symmetry properties and the dipoles are not produced. The dipole effects induced by strong electric fields in each elementary cell are polarized in a chaotic manner, i.e., their directions are randomly distributed. However, once the temperature $T < T_c$ and the electric field is applied in a proper (polarization) direction, the elementary dipoles structure is organized in a synchronized (oriented) manner, exhibiting the piezoelectric microscopic effect.

The so far described polarization phenomena are associated with the dielectric and butterfly hysteresis effects, i.e., they can be presented on the P-E and S-E planes (see Figs. 4.2, 4.3). In the initial I (origins), the dipoles are randomly organized. Introduction of E and its quasi static increase yields an ordering process of dipoles orientation and, finally, in the state II, all dipole domains of the piezoelectric test sample are ordered in a way coinciding with the applied E direction.

A decrease in E does not influence the ordered piezoelectric state which is conserved for $E = 0$. This corresponds to P_s and S_s states, and point III corresponds to the system configuration state, in which the working regime of the piezoelectric devices used in various industrial application plays a key role. The linear approximation is applied in this point, presented in Figs. 4.2b and 4.3b, respectively. The same holds for point VI, although in this case, the vector of electric field is E. In point III, the direction of the vector E is changed, and between point III and IV, dipoles are disordered. Further decrease in E (after crossing of point IV) results in the ordering occurrence, which is achieved in point V. An increase in E preserves the order dipoles state up to the point VI $(0, -P_s)$, and then the movement from point VI to VII is accompanied by disorder occurrence, which is finally kept in point VII. Then, the so far described hysteresis cycle repeats, and the mentioned behavior has been experimentally verified.

The so-called piezoelectric effect couples normal (S_1, S_2, S_3) and shear (S_4, S_5, S_6) strains (in fact, in commonly manufactured units, S_6 is uncoupled) with the electric fields E_1, E_2, E_3. Three different piezoelectric effects d_{33}, d_{31} and d_{15} are displayed in Fig. 4.4, using an example of electrically excited piezoceramic rod with a square cross section.

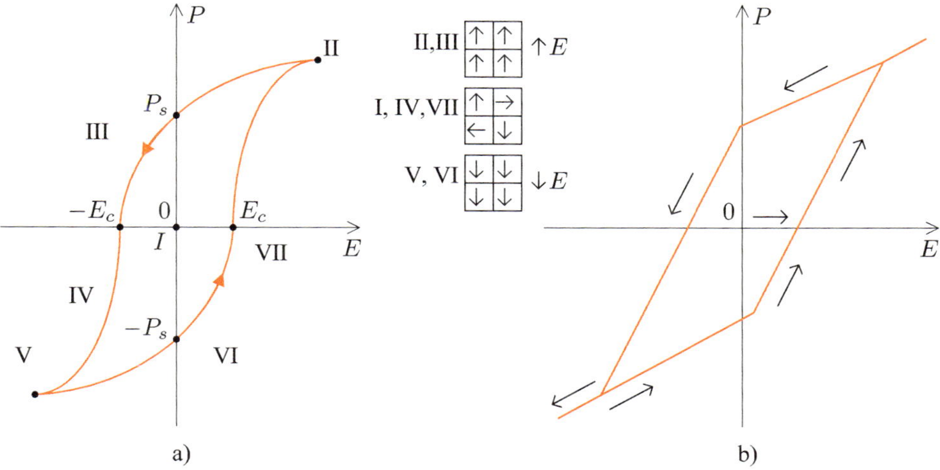

Fig. 4.2 The dielectric hysteresis (a) and its piecewise approximation in the *P-E* plane (b).

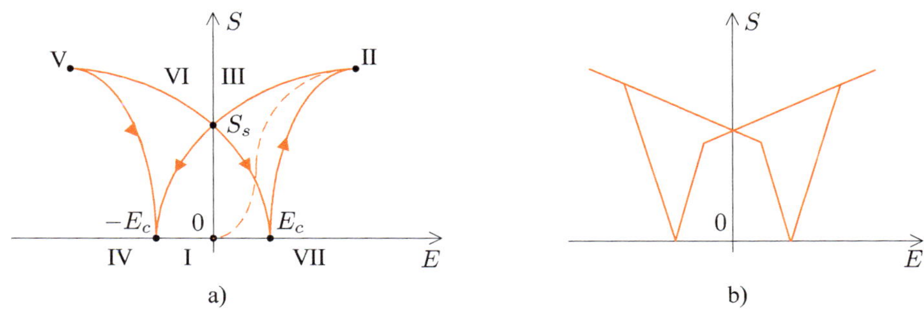

Fig. 4.3 The dielectric hysteresis (a) and its piecewise approximation in the *S-E* plane (b).

The piezoelectric rod deformations (configurations) reported in Fig. 4.4 depend on the used electrode geometry, boundary condition and excitations.

The strongest piezoelectric effect d_{33} couples S_3 and E_3 in the polarization direction P. On the other hand, the d_{31}-effect couples S_1 and E_3, and the material is elongated in the direction of the polarization vector P_3 and simultaneously contracted in the plane perpendicular to P_3. Finally, the d_{15}-effect, i.e., the shear effect, occurs in the plane perpendicular to P.

The d_{33}-effect is widely applied in quasi-statically driven positioners and in modern combustion motors. The d_{31}-effect is used in pneumatic valves in excitation of bending vibrations in bimorph structures or to guide threats in textile machines. The d_{15}-effect may even achieve the magnitude of d_{33}-effect coupling and is widely applied in multiaxial positioners and acceleration sensors.

The fundamental piezoelectric relations can be derived from the following two

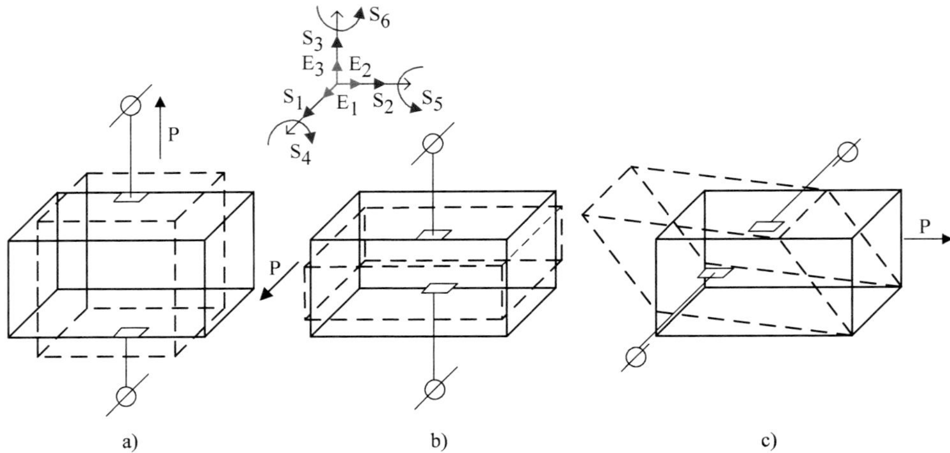

Fig. 4.4   Piezoelectric effects d_{33} (a), d_{31} (b) and d_{15} (c) (solid line denotes the unit configuration after deformation and for simplicity a direction normal to the electrode surface is parallel to E_i).

coupled vector equations:

$$f_1(S, E, T) = 0,$$
$$f_2(S, E, T) = 0, \tag{4.1}$$

where mechanical variables $S(T)$ refer to strain (stress), whereas the electrical variables represent the electric flux density D and the electric field E (sometimes, polarization P is used instead of D).

Assuming that the point (S_0, E_0, T_0) satisfies both vector equations (4.1), the expansion into the Taylor series gives the following linear constitutive equations (see [Ikeda (1990)] for more details):

$$T = k^E S + c_T^E \dot{S} - e^1 E,$$
$$D = eS + c_D \dot{S} + \varepsilon^S E. \tag{4.2}$$

In fact, mechanical elasticity (stiffness) $k = k^E$ and damping coefficients of coupling tensors c_T^E, as well as dielectric coefficients $\varepsilon = \varepsilon^s$ and damping coefficients c_D, denote the constants measured either at constant electric field E or at constant strain, respectively.

Remark 1. It should be emphasized that equations (4.2) are strongly simplified in comparison to equations (4.1), and their engineering motivation, identification and validation is required in practice.

Remark 2. In general, equations (4.2) hold for linear homogeneous elastic and piezoelectric materials.

Remark 3. Equations (4.2) are given in the so-called *mixed form*, since the mixed tensors of strain and electric field are taken as independent variables. This form is mainly used in applications, since the equations correspond to direct estimation of the boundary conditions expressed mainly through displacements and voltage.

Remark 4. Neglecting the damping coefficient, equations (4.22) have the following index form notation:

$$T_{ij} = k_{ijkl}^{E} s_{kl} - e_{kij} E_k,$$
$$D_i = e_{ikl} s_{kl} + \varepsilon_{ik}^{S} E_k. \tag{4.3}$$

Taking D and T as independent values, we get:

$$s_{ij} = s_{ijkl}^{D} T_{kl} + g_{kij} D_k,$$
$$E_i = -g_{ikl} T_{kl} + \beta_{ik}^{T} D_k. \tag{4.4}$$

However, one may use also the independent extensive variables D_k and S_{kl}, and the following index form constitutive equations are obtained:

$$T_{ij} = k_{ijkl}^{D} s_{kl} - h_{kij} D_k,$$
$$E_i = -h_{ikl} s_{kl} - \beta_{ik}^{S} D_k. \tag{4.5}$$

Finally, if one takes the intensive variables E_k and T_{kl} as independent ones, the following constitutive equations are obtained:

$$s_{ij} = s_{ijkl}^{E} T_{kl} + d_{kij} E_k,$$
$$D_i = d_{ikl} T_{kl} + \varepsilon_{ik}^{T} E_k. \tag{4.6}$$

Above, k_{ijkl} are stiffness coefficients, s_{ijkl} are compliance coefficients, ε_{ik} are dielectric permeability constants, and β_{ik} are susceptibility constants. Remaining constant coefficients $h_{ijkl}, d_{ijkl}, e_{ikl}$ and g_{ikl} govern coupling between electric and mechanical fields.

For convenience, material constant for PIC-181 ceramics and their corresponding units are taken from reference [Ceramics (2008)] and are reported in Table 4.1.

For instance, considering equations (4.5) and taking into account a linear planar piezoelectric ceramic, we have different material constants, including $k_1 \ldots k_5$ (material stiffness), h_1, h_2, h_3 (electromechanical coupling) and the dielectric material properties β_1, β_2. Fortunately, the known materials exhibit symmetry properties allowing, in general, for application of a matrix notation due to introduction of new stiffness coefficients $k_{ijkl} = k_{pq}$ and new coupling terms $e_{ijkl} = e_{ip}$. In other words, we have applied the following replacements: $ij \to p, kl \to q$, and the indices replacement follows: $11 \to 1$, $22 \to 2$, $33 \to 3$, 23 or $31 \to 4$, 13 or $31 \to 5$, 12 or $21 \to 6$ (see also [Meitzler *et al.* (1988)]).

Table 4.1   PIC-181 ceramics constants corresponding to coefficients in equations (4.3)-(4.6) (see [von Wagner and Hagedorn (2002)]).

Constant	Unit	Value
Density ρ	kg·m^{-3}	7850
Compliance s_{11}^E	m^{2}N^{-1}	$1.175 \cdot 10^{-11}$
s_{12}^E	"	$-4.07 \cdot 10^{-12}$
s_{13}^E	"	$-4.996 \cdot 10^{-12}$
s_{33}^E	"	$1.411 \cdot 10^{-11}$
s_{44}^E	"	$3.533 \cdot 10^{-11}$
s_{66}^E	"	$2(s_{11}^E - s_{12}^E)$
Compliance s_{11}^D	m^{2}N^{-1}	$1.058 \cdot 10^{-11}$
s_{12}^D	"	$-5.235 \cdot 10^{-12}$
s_{13}^D	"	$-2.268 \cdot 10^{-12}$
s_{33}^D	"	$1.137 \cdot 10^{-11}$
s_{44}^D	"	$2.134 \cdot 10^{-11}$
s_{66}^D	"	$2(s_{11}^D - s_{12}^D)$
Stiffness c_{11}^E	N·m^{-2}	$152.3 \cdot 10^9$
c_{12}^E	"	$89.1 \cdot 10^9$
c_{13}^E	"	$85.5 \cdot 10^9$
c_{33}^E	"	$134 \cdot 10^9$
c_{44}^E	"	$28.3 \cdot 10^9$
c_{66}^E	"	$1/2(c_{11}^E - c_{12}^E)$
Stiffness c_{11}^D	N·m^{-2}	$155 \cdot 10^9$
c_{12}^D	"	$91.8 \cdot 10^9$
c_{13}^D	"	$70.6 \cdot 10^9$
c_{33}^D	"	$166.4 \cdot 10^9$
c_{44}^D	"	$46.9 \cdot 10^9$
c_{66}^D	"	$1/2(c_{11}^D - c_{12}^D)$
Piezoelectric coupl. d_{31}	m·V^{-1}	$-1.08 \cdot 10^{-10}$
d_{33}	"	$2.53 \cdot 10^{-10}$
d_{15}	"	$3.89 \cdot 10^{-10}$
Piezoelectric coupl. e_{31}	N·V^{-1}m^{-1}	-4.5
e_{33}	"	14.7
e_{15}	"	11
Permittivity ε_{11}^T	$8.854 \cdot 10^{-12}$F·m^{-1}	12241
ε_{33}^T	"	1135
ε_{11}^S	"	740
ε_{33}^S	"	634

We have
$$\mathbf{T} = [T_{11} \quad T_{22} \quad T_{33} \quad T_{23} \quad T_{13} \quad T_{12}]^T = [T_1 \quad T_2 \quad T_3 \quad T_4 \quad T_5 \quad T_6]^T,$$
$$\mathbf{S} = [s_{11} \quad s_{22} \quad s_{33} \quad 2s_{23} \quad 2s_{13} \quad 2s_{12}]^T = [S_1 \quad S_2 \quad S_3 \quad S_4 \quad S_5 \quad S_6]^T,$$
$$\mathbf{D} = [D_1 \quad D_2 \quad D_3]^T, \mathbf{E} = [E_1 \quad E_2 \quad E_3]^T.$$

For instance, in the case of a transverse isotropy piezoceramic material (rotation about axis 3 does not change material properties), equations (4.3) take the following form

$$
\begin{bmatrix} T_1 \\ T_2 \\ T_3 \\ T_4 \\ T_5 \\ T_6 \end{bmatrix} =
\begin{bmatrix}
k_{11}^E & k_{12}^E & k_{13}^E & 0 & 0 & 0 \\
k_{12}^E & k_{11}^E & k_{13}^E & 0 & 0 & 0 \\
k_{13}^E & k_{13}^E & k_{33}^E & 0 & 0 & 0 \\
0 & 0 & 0 & k_{44}^E & 0 & 0 \\
0 & 0 & 0 & 0 & k_{44}^E & 0 \\
0 & 0 & 0 & 0 & 0 & k_{66}^E
\end{bmatrix}
\begin{bmatrix} S_1 \\ S_2 \\ S_3 \\ S_4 \\ S_5 \\ S_6 \end{bmatrix} +
\begin{bmatrix}
0 & 0 & -e_{31} \\
0 & 0 & -e_{31} \\
0 & 0 & -e_{33} \\
0 & -e_{15} & 0 \\
-e_{15} & 0 & 0 \\
0 & 0 & 0
\end{bmatrix}
\begin{bmatrix} E_1 \\ E_2 \\ E_3 \end{bmatrix},
$$

$$
\begin{bmatrix} D_1 \\ D_2 \\ D_3 \end{bmatrix} =
\begin{bmatrix}
0 & 0 & 0 & 0 & e_{15} & 0 \\
0 & 0 & 0 & e_{15} & 0 & 0 \\
e_{31} & e_{31} & e_{33} & 0 & 0 & 0
\end{bmatrix}
\begin{bmatrix} S_1 \\ S_2 \\ S_3 \\ S_4 \\ S_5 \\ S_6 \end{bmatrix} +
\begin{bmatrix}
\varepsilon_{11}^S & 0 & 0 \\
0 & \varepsilon_{11}^S & 0 \\
0 & 0 & \varepsilon_{33}^S
\end{bmatrix}
\begin{bmatrix} E_1 \\ E_2 \\ E_3 \end{bmatrix},
$$

$$(4.7)$$

where: $k_{66}^E = 0.5(k_{11}^E - k_{12}^E)$ and a transverse isotropy piezoceramic material has been analyzed.

Sometimes, in addition to the used notation, coefficients $a_i = \delta_{ri}$ are introduced, which indicate the direction r of polarization Pr. A similar-like notation reduction can be applied to the elastic compliance coefficients S_{ijkl}, and piezoelectric coupling coefficients d_{ikl}, g_{kij}.

The main advantage of piezoelectric materials used in various application is their ability to change electrical energy to mechanical energy, and vice versa. In other words, both energies are coupled, and then a naturally motivated idea of estimation of a coupling coefficient occurs. If one takes a transformer with primary L_1 and secondary L_2 inductance and mutual inductance M_{12}, then $L_1 L_2 = k^2 M_{12}^2$. Now, if one applies the geometric approach to the elastic

$$
u_k = \frac{1}{2} T_i s_{ij}^E T_j,
\tag{4.8}
$$

and to dielectric

$$
u_d = \frac{1}{2} E_m \varepsilon_{mn}^T E_n,
\tag{4.9}
$$

as well as to mutual

$$
u_d = \frac{1}{2} E_m d_{ni}^T E_n
\tag{4.10}
$$

energy densities, then one finds the so-called Berlincourt, Currand and Joffe coupling coefficient k_{BCJ} of the form [Mason (1950, 1964)]:

$$
k_{BJC} = \frac{u_m}{\sqrt{u_e u_d}}.
\tag{4.11}
$$

Other widely used definitions based on a concept of the so-called *quasi-static conversion cycle* owing to the IEEE standard on piezoelectricity and the Ulitko definitions [Ulitko (1977)] are briefly sketched in Fig. 4.5a, b and Fig. 4.5c, d, respectively.

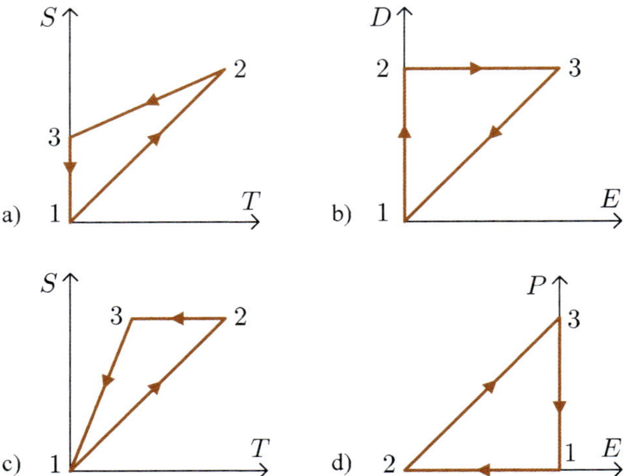

Fig. 4.5 Quasi-static piezoelectric conversion cycles according to the IEEE standards (a, b) and the Ulitko definition (c, d).

Owing to the definition introduced by the IEEE, a particular stress field T is applied to an unstressed piezotransducer having short-circuited electrodes (state 1). The transducer state is represented now by mechanical (T_2, S_2) and electrical $(0, D_2)$ fields, and the energy introduced to the transducer W corresponds to the $1 \to 2$ cycle part. In the state point 2, the electrodes are disconnected and the $2 \to 3$ cycle part corresponds to stress releasing and to the W_2 energy part. The remaining energy part W_1 is removed (for instance, is transferred into heat) by electrical loading applied to the electrodes ($3 \to 1$ part of the conversion cycle).

The coupling coefficient k_{IEEE} defines a ratio of the converted total stored energy in the following way

$$k_{IEEE}^2 = \frac{W_1}{W} = \frac{W_1}{W_1 + W_2}. \tag{4.12}$$

If one takes $W_1 = 0$, then a cycle is degenerated to a line segment, and the piezoelectric coupling disappears ($k_{IEEE} = 0$) (see Fig. 4.6a, b).

On the other hand, if $W_2 = 0$ (see Fig. 4.6c, d), then the points 2 and 3 overlap ($k_{IEEE} = 1$).

The Ulitko definition concept, reported in Fig. 4.5c, d, allows to introduce the following coupling coefficient

$$k_u^2 = \frac{u_0 - u_s}{u_0}, \tag{4.13}$$

where $u_0(u_s)$ correspond to the internal energy with open/short-circuited electrodes. It should be emphasized that the part $2 \to 3$ of the quasi-static conversion cycle corresponds to the energy converted by a quasi-static discharge of the electric fields keeping the strain fields constant.

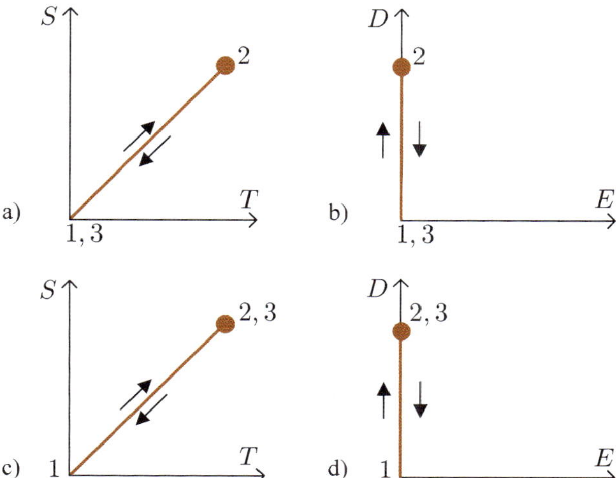

Fig. 4.6   Degenerated cases of the IEEE standard piezoelectric conversion cycles for $k_{IEEE} = 0$ (a, b) and $k_{IEEE} = 1$ (c, d).

In order to give more physical insight to k_{IEEE}, the following simple example is considered. Namely, we apply only stress T_4 to piezoelectric element and, assuming its homogeneous fields in the piezoelement, equations (4.6) have the following fully developed form:

$$S_1 = s_{11}^E T_1 + s_{12}^E T_2 + s_{13}^E T_3 - d_{31} E_3,$$
$$S_2 = s_{21}^E T_1 + s_{22}^E T_2 + s_{23}^E T_3 - d_{31} E_3,$$
$$S_3 = s_{31}^E T_1 + s_{32}^E T_2 + s_{33}^E T_3 - d_{33} E_3,$$
$$S_4 = s_{44}^E T_4 - d_{15} E_2, \tag{4.14}$$
$$S_5 = s_{44}^E T_5 - d_{15} E_1, S_6 = s_{66}^E T_6,$$
$$D_1 = d_{15} T_5 + \varepsilon_{11}^T E_1, D_2 = d_{15} T_4 + \varepsilon_{11}^T E_2,$$
$$D_3 = d_{31} T_4 + d_{31} T_2 + d_{33} T_3 + \varepsilon_{11}^T E_3,$$

and one finds from Eq. (4.14) that

$$S_4 = s_{44}^E T_4, D_2 = d_{15} T_4. \tag{4.15}$$

The total stored specific energy is

$$W_1 + W_2 = \frac{1}{2} s_{44}^E T_4^2. \tag{4.16}$$

After unloading the piezoelement (the electrodes are disconnected), full energy related to stress approaches zero ($T_4 = 0$), and electric and strain fields appear instead. Equation (4.6) yields:

$$E_2 = \frac{D_2}{\varepsilon_{11}^T} = \frac{d_{15} T_4}{\varepsilon_{11}^T},$$
$$S_4 = d_{15} E_2 = \frac{d_{15}^2}{\varepsilon_{11}^T} T_4. \tag{4.17}$$

The strain S_4 pumps the following specific energy

$$W_1 = \frac{1}{2} \frac{d_{15}^2}{\varepsilon_{11}^T} T_4^2.$$

(4.18)

Let us verify the units of $[W]$. We have

$$[W] = \left(\frac{\mathrm{m}}{\mathrm{V}}\right) \frac{\mathrm{m}}{\mathrm{F}} \left(\frac{\mathrm{N}}{\mathrm{m}^2}\right)^2 = \frac{\mathrm{N}^2}{\mathrm{V} \cdot \mathrm{m} \cdot \mathrm{C}} = \frac{\mathrm{J}}{\mathrm{m}^3},$$

(4.19)

which is in agreement with Table 4.1. Therefore, the coupling coefficient is defined as follows

$$k_{IEEE}^2 = \left(\frac{W_1}{W_1 + W_2}\right) = \frac{d_{15}^2}{\varepsilon_{11}^T s_{44}^E}.$$

(4.20)

Now, coming back to the k_{BCJ} definition (see Eq. (4.11)), we take nonzero stress T_4 and electric field E_1 quantities. The corresponding energies follow

$$u_m = \frac{1}{2} d_{15} E_2 T_4, u_e = \frac{1}{2} s_{44}^E T_4^2, u_d = \frac{1}{2} \varepsilon_{11}^T E_1^2,$$

(4.21)

and one finds that $k_{BJC} = k_{IEEE}$, although (in general), the k_{BJC} definition includes only a material property. The similar observation holds also for the Ulitko definition. Namely, using Eqs. (4.13) and (4.17) one finds

$$T_4 = \frac{\varepsilon_{11}^T}{\varepsilon_{11}^T s_{44}^E + d_{15}^2} S_4,$$

(4.22)

and hence

$$u_0 = \frac{1}{2} T_4 S_4 = \frac{1}{2} \frac{\varepsilon_{11}^T}{\varepsilon_{11}^T s_{44}^E - d_{15}^2} S_4^2.$$

(4.23)

One the other hand, for $E_2 = 0$ (a short-circuit case) one gets

$$u_s = \frac{1}{2} \frac{1}{s_{44}^E} S_4^2,$$

(4.24)

and

$$k_u = k_{BCJ}.$$

The so far considered examples have shown that various definitions of the coupling coefficients overlap, assuming that both uniform stress-strain and electric fields have been applied. However, the situation dramatically changes when a material with nonhomogenous fields is studied. It has been reported by [Chang *et al.* (1995)] that the most suitable deformation, remaining valid also for piezoelectric transducers with nonuniform fields structure, is that proposed by Ulitko, whereas the Berlincourt et al. is valid rather only for homogeneous transducer fields distributions. Furthermore, the input quantities refer rather to strain fields T and electric displacement D, whereas the stress and electric fields S and E are computed. On the other hand, the variables S and E are directly associated with the given electrode voltage. Furthermore, the inertial energy density function follows

$$z_u = S^T k^E S + E^T \varepsilon^S E.$$

(4.25)

Here, we encounter a problem similar to the one that has been discussed in reference [Meitzler *et al.* (1988)] while estimating the inertial tensor coefficients. Positive definiteness of the quadratic forms $S^T k^E S > 0$ and $E^T \varepsilon E > 0$ allows to yield the following restrictions of the coefficient k^E and e^S reported in [Schonecker (2009)]:

$$k_{ij}^E > 0, \qquad \varepsilon_{jj} > 0,$$
$$|k_{11}^E| > |k_{12}^E|, \tag{4.26}$$

$$2\left(k_{13}^E\right) < \left(k_{11}^E + k_{12}^E\right) k_{33}^E, \tag{4.27}$$

$$d_{31}^2 < s_{11}^E \varepsilon_{33}^T, d_{33}^2 < s_{33}^E \varepsilon_{33}^T, d_{15} < s_{44}^E \varepsilon_{11}^T, \tag{4.28}$$

$$s_{jj}^E > 0, \qquad \varepsilon_{jj}^T > 0, \tag{4.29}$$

$$|s_{11}^E| > |s_{12}^E|,$$
$$2\left(s_{13}^E\right) < \left(s_{11}^E + s_{12}^E\right) s_{33}^E, \tag{4.30}$$

$$d_{31}^2 < s_{11}^E \varepsilon_{33}^T, d_{33}^2 < s_{33}^E \varepsilon_{33}^T, d_{15} < s_{44}^E \varepsilon_{11}^T. \tag{4.31}$$

Inequalities (4.29), (4.30) refer to an elastic factor, whereas inequalities (4.31) give bounds for the coupling parameters.

4.2 One-dimensional Rod Polarized Along its Axis – an Example of an Actuator

We consider a one-dimensional (1D) rod of the length L and axial displacement $w(x_3)$ (see Fig. 4.7).

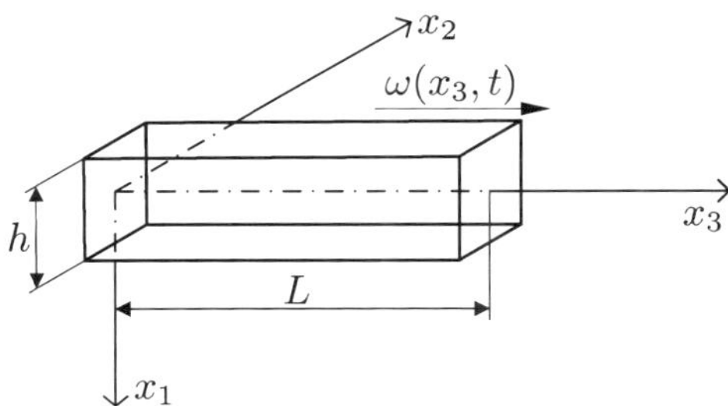

Fig. 4.7 One-dimensional piezoelectric rod with a square cross section.

We take strain S and electric field E as independent variables and we use Hamilton's variational approach (see monographs [Ikeda (1990); Craig and Kurdila (2006)] for more details). We apply the Lagrangian function $\mathcal{L}$ (Lagrangian density $\mathcal{L}_d$ [J/m^3]). Owing to Hamilton's principle, one gets:

$$\int_{t_0}^{t_1} \delta\mathcal{L}dt = 0, \qquad \mathcal{L}dt = 0, \tag{4.32}$$

where t denotes time and

$$\mathcal{L} = \int_V \mathcal{L}_d dV = \int_V (T - \kappa_d)\, dV. \tag{4.33}$$

The kinetic energy

$$T = \frac{1}{2}\rho\left(\frac{dW}{dt}\right)^2. \tag{4.34}$$

In order to derive the electric enthalpy density $\mathcal{L}_d$, we use the rod in the uniaxial stress state (only T_3 is generated). Since S and E are used as input vector functions, Eq. (4.7) allows to find from conditions $T_1 = T_2 = 0$ that $S_1 = S_2$, and then

$$S_1 = S_2 = -\frac{k_{13}^E}{k_{11}^E + k_{12}}S_3 + \frac{e_{31}}{k_{11}^E + k_{12}^E}E_3. \tag{4.35}$$

The remaining specific potential and kinetic energies of rod are given by the following formulas:

$$\mathcal{L}_d - T = -\left(\frac{1}{2}k_{11}^E S_1^2 + \frac{1}{2}k_{12}^E S_2^2 + \frac{1}{2}k_{33}^E S_3^2 - e_{33}E_3 S_3 - \frac{1}{2}\varepsilon_{33}^s E_3^2\right) =$$

$$-\left[\frac{1}{2}\left(k_{11}^E + k_{12}^E\right)\left(-\frac{k_{13}^E}{k_{11}^E + k_{12}^E}S_3 + \frac{e_{31}}{k_{11}^E + k_{12}^E}E_3\right)^2 + \frac{1}{2}k_{13}^E S_3^2 - \frac{1}{2}\varepsilon_{33}^s E_3^2\right] =$$

$$-\left[\frac{1}{2}\left(\frac{(k_{13}^E)^2}{k_{11}^E + k_{12}^E} + k_{33}^S\right)S_3^2 - \left(e_{33} + \frac{k_{13}^E e_{31}}{(k_{11}^E + k_{12}^E)^2}\right)S_3 E_3 +$$

$$+\frac{1}{2}\left(\frac{e_{31}^2}{(k_{11}^E + k_{12}^E)^2} - \varepsilon_{33}^S\right)E_3^2\right] = -\left(\frac{1}{2}\tilde{k}_{13}^E S_3^2 - \tilde{e}_{33}S_3 E_3 - \frac{1}{2}\varepsilon_{33}^S E_3^2\right). \tag{4.36}$$

In the piezoelectric materials, one deals with both electric and mechanical fields, which are coupled. The electric field is defined by the vectors D (electric field flux density) and E (electric field intensity), whereas the second order tensor (matrices) T and S refer to strain and stress fields, respectively. Any point x from a piezoelectric continuum is characterized by its displacement $u = u(x_1, x_2, x_3)$ and corresponding potential $\Phi = \Phi(x_1, x_2, x_3)$. The piezoelectric 3D cell state is governed by the dynamic state of the cells mass derived from the impulse-type equations (i), then through stress-strain relations (ii), and finally, the electric phenomena are governed by Maxwell's equations (iii).

(i) Dynamics

Impulse-type equations follow:

$$\frac{\partial T_{11}}{\partial x_1} + \frac{\partial T_{12}}{\partial x_2} + \frac{\partial T_{13}}{\partial x_3} + F_1 = \rho \frac{d}{dt}\left(\dot{u}_1\right),$$

$$\frac{\partial T_{21}}{\partial x_1} + \frac{\partial T_{22}}{\partial x_2} + \frac{\partial T_{23}}{\partial x_3} + F_2 = \rho \frac{d}{dt}\left(\dot{u}_2\right), \tag{4.37}$$

$$\frac{\partial T_{31}}{\partial x_1} + \frac{\partial T_{32}}{\partial x_2} + \frac{\partial T_{33}}{\partial x_3} + F_3 = \rho \frac{d}{dt}\left(\dot{u}_3\right),$$

where $u = (u_1, u_2, u_3)$, F_i are the volume forces and ρ denotes the piezoelectric material density.

In the Einstein summation convention, equations take the following simple form

$$T_{ij,j} + F_i = \rho \frac{d}{dt}\left(\dot{u}_i\right), \quad i,j = 1,2,3, \tag{4.38}$$

and $T_{ij,j}$ denotes differentiation of T_{ij} regarding the coordinate j, where $j = 1,2,3$ or $j = x_1, x_2, x_3$.

(ii) Stress-strain linear relations:

$$s_{11} = \frac{\partial u_1}{\partial x_1},$$

$$s_{21} = s_{12} = \frac{1}{2}\left(\frac{\partial u_1}{\partial x_2} + \frac{\partial u_2}{\partial x_1}\right),$$

$$s_{31} = s_{13} = \frac{1}{2}\left(\frac{\partial u_1}{\partial x_3} + \frac{\partial u_3}{\partial x_1}\right),$$

$$s_{22} = \frac{\partial u_2}{\partial x_2}, \tag{4.39}$$

$$s_{32} = s_{23} = \frac{1}{2}\left(\frac{\partial u_2}{\partial x_3} + \frac{\partial u_3}{\partial x_2}\right),$$

$$s_{33} = \frac{\partial u_3}{\partial x_3}.$$

Equation (4.39) can be cast in the following compact form

$$s_{ij} = \frac{1}{2}\left(u_{i,j} + u_{j,i}\right). \tag{4.40}$$

(iii) Maxwell's equations.

The influence of the possibly occurred magnetic field is neglected and we further consider only the electrostatic behavior. Maxwell's equations have the following form:

$$\frac{\partial D_1}{\partial x_1} + \frac{\partial D_2}{\partial x_2} + \frac{\partial D_3}{\partial x_3} = F_\rho,$$

$$E_1 = -\frac{\partial \varphi}{\partial x_1},$$

$$E_2 = -\frac{\partial \varphi}{\partial x_2}, \tag{4.41}$$

$$E_3 = -\frac{\partial \varphi}{\partial x_3},$$

where F_ρ denotes the density of a free electric load (for an electrically isolated dielectric we have $F_\rho = 0$). Equation (4.39) can be presented in the following abbreviated form

$$D_{i,i} = F\rho, \quad E_i = -\varphi_i, \quad i = 1, 2, 3. \tag{4.42}$$

(iv) Hamilton's principle for mechatronical continua.

In order to derive relations between electric and mechanical fields, one may apply the classical energy approach, using the principle of mechanical and electric energy conservation as well as Hamilton's principle for piezoelectric continua.

Let us take a piezoelectric 3D object with the volume U and the surface Ω. On the surface Ω, both mechanical and electrical boundary conditions can be formulated (we follow here considerations presented by [von Wagner (2003)]).

In the case of mechanical boundary conditions, the surface $\Omega = \Omega_f + \Omega_u$, where Ω_u corresponds to the boundary conditions associated with displacement

$$u_i = \bar{u}_i \quad \text{on} \quad \Omega_u, \tag{4.43}$$

and Ω_f corresponds to the surface part loaded by mechanical forces $\bar{f}_i$ of the form

$$T_{ij} n_j = \bar{f}_i \quad \text{on} \quad \Omega_f. \tag{4.44}$$

Similarly, in the case of electric boundary conditions, $\Omega = \Omega_\sigma + \Omega_\Phi$, where Ω_σ corresponds to the boundary conditions associated with the surface change $\bar{\sigma}$ of the form

$$-D_i n_i = \bar{\sigma} \quad \text{on} \quad \Omega_\sigma, \tag{4.45}$$

and Ω_Φ corresponds to the surface potential

$$\varphi = \bar{\varphi} \quad \text{on} \quad \Omega_\varphi, \tag{4.46}$$

where $n_i(n_j)$ denotes the normal vector.

In what follows, we use Eqs. (4.37), (4.42) and Eqs. (4.44), (4.45), but in the following adapted form:

$$T_{ij,j} + F_i - \rho \frac{d}{dt}(\dot{u}_i) = 0, \tag{4.47}$$

$$T_{ij} n_j - \bar{f}_i = 0, \tag{4.48}$$

$$D_i n_i + \bar{\sigma} = 0. \tag{4.49}$$

Equations (4.47) and (4.48) are multiplied by δu_i, whereas Eqs. (4.42) (of the form $D_{i,i} = 0$) and (4.49) are multiplied by $\delta\varphi$, and hence the weak integral form of the mechatronic problem is formulated in the following way

$$-\int_V T_{ij,j} + F_i - \rho \frac{d}{dt}(\dot{u}_i)\, \delta u_i dV - \int_{\Omega_f} (T_{ij} n_j - \bar{f}_i) \delta u_i d\Omega -$$
$$-\int_V D_{k,k} \delta\varphi dV + \int_{\Omega_\sigma} (D_i n_i + \bar{\sigma})\, \delta\varphi d\Omega = 0. \tag{4.50}$$

By using Eq. (4.40) in (4.50) and carrying the integration of Eq. (4.50) by parts, the following form of equations is yielded

$$\int\limits_{V} (T_{ij}\delta s_{ij} - D_k\delta E_k)dV - \int\limits_{V} \left[F_i - \rho\frac{d}{dt}\left(\dot{u}_i\right) \right] \delta u_i dV +$$
$$+ \int\limits_{\Omega_\sigma} \bar{\sigma}d\varphi d\Omega - \int\limits_{\Omega_f} \bar{f}_i\delta u_i d\Omega = 0. \tag{4.51}$$

The total differential

$$d\mathcal{H} = T_{ij}ds_{ij} - D_k dE_k, \tag{4.52}$$

and s_{ij} and E_k are treated as independent input variables (they can be also represented by u_i and φ). Now, taking the total variation $\delta\mathcal{H}$ represented also by Eq. (4.52), where instead of d we use δ and, since the integral form of F_i action is zero, Eq. (4.51) is cast to the following form

$$\int\limits_{V} \delta\mathcal{H}dV + \int\limits_{V} \rho\frac{d}{dt}\left(\dot{u}_i\right)\delta u_i dV + \int\limits_{\Omega} \left(\bar{\sigma}\delta\varphi - \bar{f}_i\delta u_i\right) d\Omega = 0. \tag{4.53}$$

In addition, we integrate Eq. (4.53) once more to obtain

$$\int\limits_{t_0}^{t}\int\limits_{V} \left(\delta\mathcal{H} - \frac{1}{2}\rho\delta\left(\dot{u}_i^2\right)\right) dVdt + \int\limits_{t_0}^{t}\int\limits_{\Omega} \left(\bar{\sigma}\delta\varphi - \bar{f}_i\delta u_i\right) d\Omega dt = 0, \tag{4.54}$$

and $\delta u_i(t_0) = \delta u_i(t) = 0$ and $\delta\varphi(t_0) = \delta\varphi(t) = 0$.

Hamilton's principle applied to the piezoelectric continuum has the form

$$\delta\int\limits_{t_0}^{t} \mathcal{L}(u_i, \varphi)dt + \int\limits_{t_0}^{t} \delta W(u_i, \varphi)dt = 0, \tag{4.55}$$

where virtual work

$$\delta W = \int\limits_{\Omega} \left(\bar{f}_i\delta u_i - \bar{\sigma}\delta\varphi\right)d\Omega, \tag{4.56}$$

and the Lagrangian

$$\mathcal{L} = \int\limits_{V} \left(\frac{1}{2}\rho\dot{u}_i^2 - \mathcal{H}\right) dV. \tag{4.57}$$

It should be emphasized that the formulas (4.56) and (4.57) are derived using the earlier introduced assumptions regarding independence of S and E.

Now, we apply the following Legendre's transformation

$$\mathcal{U} = \mathcal{H} + E_k D_k. \tag{4.58}$$

Then,

$$d\mathcal{U} = T_{ij}ds_{ij} + E_k dD_k, \tag{4.59}$$

and hence the following formulas hold:

$$\mathcal{G} = \mathcal{H} - s_{ij}T_{ij}, \tag{4.60}$$

$$d\mathcal{G} = -s_{ij}dT_{ij} - D_k dE_k, \tag{4.61}$$

for T_{ij} and E_k treated as independent variables and:

$$\mathcal{G} = \mathcal{H} - s_{ij}T_{ij} + E_k D_k, \tag{4.62}$$

$$d\mathcal{G} = -s_{ij}dT_{ij} + E_k dD_k, \tag{4.63}$$

where now s_{ij} and D_k are independent of each other.

If the mechanical s_{kl} and electric E_k fields are considered as independent input values, then the output in the form of mechanical T_{ij} and electrical D_i quantities is expressed through the following linear constitutive equations:

$$dT_{ij} = \frac{\partial T_{ij}}{\partial s_{kl}}ds_{kl} + \frac{\partial T_{ij}}{\partial E_k}dE_k, \tag{4.64}$$

$$dD_i = \frac{\partial D_i}{\partial s_{jk}}ds_{jk} + \frac{\partial D_i}{\partial E_j}dE_j. \tag{4.65}$$

Equations (4.64), (4.65) can be inversed, i.e., T_{ij} and D_i can be treated as the input values, whereas s_{kl} and D_i are output quantities. However, this is valid only for purely elastic material, where piezoelectric and dielectric phenomena, as well as internal and electric field damping effect, are omitted.

Since $\mathcal{H} = \mathcal{H}(s, E)$, hence

$$d\mathcal{H} = \frac{\partial \mathcal{H}}{\partial s_{ij}}ds_{ij} + \frac{\partial \mathcal{H}}{\partial E_i}dE_i, \tag{4.66}$$

and taking into account Eq. (4.52), one gets

$$\begin{aligned} T_{ij} &= \frac{\partial \mathcal{H}}{\partial s_{ij}}, \\ D_i &= -\frac{\partial \mathcal{H}}{\partial E_i}. \end{aligned} \tag{4.67}$$

The linear constitutive equations (4.64), (4.65) are associated with the following mechatronic enthalpy

$$\mathcal{H} = \frac{1}{2}c_{ijkl}^{E}s_{ij}s_{kl} - e_{ijk}E_i s_{jk} - \frac{1}{2}\varepsilon_{ij}^{S}E_i E_j, \tag{4.68}$$

where $c_{ijkl}^{E}, e_{ijk}, \varepsilon_{ij}^{S}$ are mechanical (elastic) piezoelectric and dielectric constants, respectively, being measured for either constant E or s (superscripts) values. Equations (4.67) and (4.68) yield:

$$T_{ij} = c_{ijkl}^{E}s_{kl} - e_{kij}E_k, \tag{4.69}$$

$$D_i = e_{kij}s_{kl} - \varepsilon_{ik}^{S}E_k. \tag{4.70}$$

If we now take S and E as inputs, then the internal energy of the piezoelectric continuum takes the following form

$$\mathcal{U} = \mathcal{H} + E_k D_k = \frac{1}{2} s^E_{ijkl} T_{ij} T_{kl} + d_{ijk} T_{jk} E_i + \frac{1}{2} \varepsilon^T_{ij} E_i E_j, \qquad (4.71)$$

where $s^E_{ijkl}, d_{ijk}, \varepsilon^T_{ij}$ are mechanical, piezoelectric and dielectric constants, and superscripts E and T indicate constant fields E and T, respectively.

This allows to derive the following linear constitutive equations corresponding to the enthalpy $\mathcal{U}$:

$$S_{ij} = \frac{\partial \mathcal{U}}{\partial T_{ij}} = s^E_{ijkl} T_{kl} + d_{kij} E_k, \qquad (4.72)$$

$$D_i = \frac{\partial \mathcal{U}}{\partial E_i} = d_{ikl} T_{kl} + \varepsilon^T_{ik} E_k. \qquad (4.73)$$

In general, a piezoelectric element (PE) exhibits a variety of nonlinear effects, as shown in Fig. 4.8.

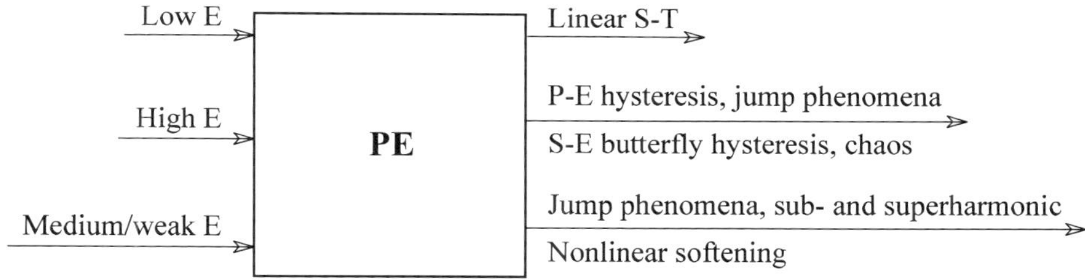

Fig. 4.8   Inputs and outputs of a PE (piezoelectric element).

Low electric field E input results in getting a linear stress-strain PE state. An increase in E results in the well-known jump phenomena, bifurcations yielding subharmonic and/or super-harmonic vibrations, dependence of the amplitude vibration on the applied frequency. Further increase in E yields P-E *hysteresis* and S-E *butterfly hysteresis* phenomena. Therefore, owing to numerous experimental results, it is clear that a derivation of constitutive equations of the piezoelectric continuum requires introduction of a nonlinear *enthalpy density function*. Therefore, in what follows, we derive the constitutive nonlinear equations of a piezoelectric continuum following the works [Maugin (1988); von Wagner and Hagedorn (2002)]. Since the stress and strain tensors are symmetric, the following form of the *electric enthalpy density* function has been proposed in reference [Samal *et al.* (2006)]

$$\mathcal{H} = \frac{1}{2} S^T C S - E^T d C S - \frac{1}{2} E^T v_0 E + \varepsilon \left[\frac{1}{3} S^T C_1 S^2 + \frac{1}{4} \left(S^2\right)^T C_{21} S^2 + \frac{1}{4} S^T C_{22} S^3 \right.$$

$$- \frac{1}{2} E^T \gamma_{11} S^2 - \frac{1}{2} \left(E^2\right)^T \gamma_{12} S - \frac{1}{3} E^T v_1 E^2 - \frac{1}{3} E^T \gamma_{21} S^3 - \frac{1}{2} \left(E^2\right)^T \gamma_{22} S^2$$

$$\left. - \frac{1}{3} \left(E^3\right)^T \gamma_{23} S - \frac{1}{4} E^T v_{21} E^3 - \frac{1}{2} \left(E^2\right)^T v_{22} E^2 \right].$$

$$(4.74)$$

Note that for $\varepsilon = 1$, we have a nonlinear enthalpy form, whereas for $\varepsilon = 0$, we have the widely used linear enthalpy density function. For $1 \ll \varepsilon$, we have a weak nonlinear enthalpy density function representation, and hence in many cases, analytical approximate methods can be used to solve the governing equations.

In the above formula, superscripts E (constant electric field) and s (constant strain) are omitted for clarity, C is the linear elasticity matrix, d is the linear piezoelectricity matrix, ε is the dielectric coefficient matrix, C_1 is quadratic, whereas C_{21} and C_{22} are cubic elasticity matrices, respectively; γ_{11}, γ_{12} are quadratic piezoelectric matrices, whereas γ_{21}, γ_{22} and γ_{23} are cubic piezoelectric matrices; $v_0 = \varepsilon - dcd^T, v_1, v_{21}$ and v_{22} are linear, quadratic and cubic dielectric matrices (constant s), respectively. Superscripts 2, 3 indicate the nonlinearity type, for instance $s^3 = \left(s_1^3, s_2^3, s_3^3, s_4^3, s_5^3, s_6^3\right)$. Observe that $T = \frac{\partial \mathcal{H}}{\partial S}$, $D = -\frac{\partial \mathcal{H}}{\partial E}$, and taking into account the linear part of Eq. (4.43) ($\varepsilon = 0$), one may see that T (stress matrix) and D (elastic displacement vector) are coupled with S (strain matrix) and E (electric field vector), where $s_{ij} = \left(u_{i,j} + u_{j,i}\right)/2, E_i = -\varphi_{,i}$.

In Figure 4.9, both mechanical and electric forces applied to a piezoelectric 3D continuum object are presented.

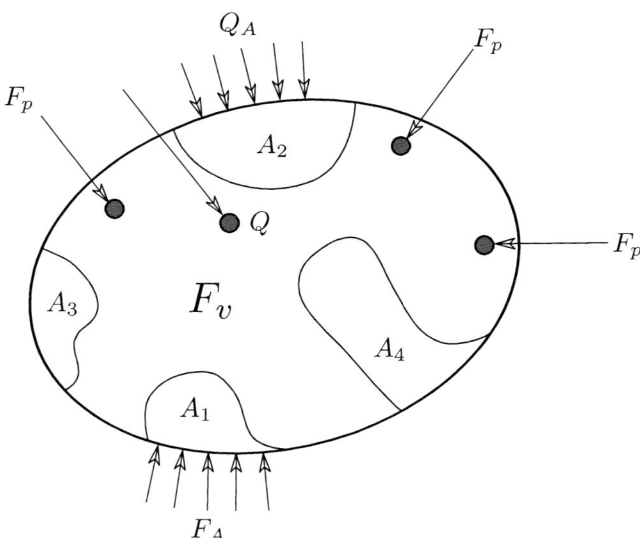

Fig. 4.9   Mechanical forces F and electric charges Q associated with the displacement δu, the electrical potential $\delta\varphi$ and the corresponding boundary condition $\delta u = 0$ ($\delta\varphi = 0$) on the surface $A_3(A_u)$. F_V – body force vector, F_A – surface force vector, F_p – point load vector, Q_A – surface charge.

For a 3D piezoelectric material of the mass density ρ under the action of the generalized displacement u and vector velocity $\dot{u}$ fields, the Lagrangian $\mathcal{L} = \frac{1}{2}\rho\dot{u}^T\dot{u} - \mathcal{H}$, where $\mathcal{H}$ is defined by Eq. (4.74) and the virtual work $\mathcal{W}$ is used to define

Hamilton's principle of the following form

$$\delta \int_{t_0}^{t} \left(\int_{V} \mathcal{L} dV \right) dt + \int_{t_0}^{t} \delta \mathcal{W} dt = 0, \tag{4.75}$$

where t_0 and t_1 are time instants $(t_1 > t_0)$. Taking into account the notation used in Fig. 4.9, the variation of the virtual work is cast to the form

$$\delta \mathcal{W} = \int_{V} u^T F_V dV + \int_{A_1} \delta u^T F_A dA + \delta u^T F_p - \int_{A_2} \delta \varphi Q dA_2 - \delta \varphi Q + \delta \mathcal{W}_D, \tag{4.76}$$

where $\delta \mathcal{W}_D$ is the virtual work done by damping forces. Substituting Eq. (4.73) into (4.74) and then into Eq. (4.79), and using (4.76), the following formula (see [Samal *et al.* (2006)]) is yielded

$$- \int_{t_0}^{t} \int_{V} \rho \delta u^T \ddot{u} dV dt - \int_{t_0}^{t} \int_{V} \delta S^T \left[CS - C^T dE - S_d \gamma_{11}^T E^2 - \frac{1}{2} \gamma_{12}^T E^2 \right.$$

$$+ S_d C_{21} S_d S + \frac{1}{4} C_{22} S^3 + \frac{3}{4} S_d C_{22}^T S + \frac{1}{3} C_1 S^2 + \frac{2}{3} S_d C_1^T S - S_d^2 \gamma_{21}^T E \tag{4.77}$$

$$\left. - S_d \gamma_{22}^T E^2 - \frac{1}{3} \gamma_{23}^T E^3 \right] dV dt - \int_{t_0}^{t} \int_{V} \delta E^T \left[dCS + u_0 E + \frac{1}{2} \gamma_{11} S^2 \right.$$

$$+ E_d \gamma_{12} S + \frac{1}{3} \gamma_{13} S^2 + \frac{2}{3} E_d \gamma_{13}^T E + \frac{1}{3} \gamma_{21} S^3 + E_d \gamma_{22} S^2$$

$$\left. + E_d^2 \gamma_{23} S + \frac{1}{4} \nu_1 E^3 + \frac{3}{4} E_d^2 \nu_1^T E + E_d \nu_2 E_d E \right] dV dt + \int_{t_0}^{t} \int_{V} \delta u^T F_V dV dt$$

$$+ \int_{t_0}^{t} \int_{A_1} \delta u^T F_A dA dt + \int_{t_0}^{t} \delta u^T F_p dt + \int_{t_0}^{t} \int_{A_2} \delta \varphi Q_x dA dt - \int_{t_0}^{t} \delta \varphi Q dt - \int_{t_0}^{t} \delta \mathcal{W}_D dt = 0,$$

where the matrix with the subscript "*d*" refers to a diagonal matrix with the main diagonal terms being the terms of the vector belonging to it. In the linear case, the virtual work $\delta \mathcal{W}_D$ includes dissipation regarding a viscous-type damping $\dot{s}$ as well as both linear piezoelectric γ_{od} and dielectric ν_{od} damping of the form [Ikeda (1990); von Wagner (2004)]:

$$\delta \mathcal{W}_D = \delta \int_{V} \left(S^T c_d^{(o)} \dot{S} - S^T \gamma_{od} \dot{E} - \dot{S}^T \gamma_{od} E - E^T \nu_{od} \dot{E} \right) dV. \tag{4.78}$$

Assuming that variations occurred in the integral Hamilton's equations from Eq. (4.77) are independent, the corresponding nonlinear form of the constitutive equations for a 3D piezoelectric continuum can be derived.

In what follows, we derive the *nonlinear enthalpy density function* for a 1D piezoelectric beam in the case of the d_{31}-effect (see [von Wagner *et al.* (2001); von

Wagner and Hagedorn (2002)]). The linear constitutive equations take the following form:

$$T_{x_1x_1} = E_c s_{x_1x_1} - d_{31}E_cE_3,$$
$$D_3 = d_{31}E_c s_{x_1x_1} + \left(\varepsilon_{33}^T - d_{31}^2 E_c\right) E_3, \tag{4.79}$$

where x_1 is the direction of the beam length, E_3 is the polarization direction, E_c is the elastic piezoceramic beam modulus orthogonal to F_3, d_{31} and ε_{33}^T correspond to the d_{31} piezoelectric effect and dielectric constant for constant stress T, respectively. The following nonlinear relations are applied:

$$E_c = E_c^{(0)} + E_c^{(1)}s_{x_1x_1} + E_c^{(2)}s_{x_1x_1}^2,$$
$$d_{31} = d_{31}^{(0)} + d_{31}^{(1)}s_{x_1x_1} + d_{31}^{(2)}s_{x_1x_1}^2, \tag{4.80}$$

and we take:

$$T_{x_1x_1} = E_c^{(0)}s_{x_1x_1} + E_c^{(1)}s_{x_1x_1}^2 + E_c^{(2)}s_{x_1x_1}^3 - \gamma_0 E_3 - \gamma_1 s_{x_1x_1}E_3 - \gamma_2 s_{x_1x_1}^2 E_3,$$
$$D_3 = \gamma_0 s_{x_1x_1} + \frac{1}{2}\gamma_2 s_{x_1x_1}^2 + \frac{1}{3}\gamma_2 s_{x_1x_1}^3 + \nu_0 E_3, \tag{4.81}$$

where:

$$\nu_0 = \varepsilon_{33}^T - \left(d_{33}^{(0)}\right)^2 E_c^0, \gamma_0 = E_c^{(0)}d_{31}^{(0)}, \gamma_1 = E_c^{(0)}d_{31}^{(1)} + E_c^{(1)}d_{31}^{(0)},$$
$$\gamma_2 = E_c^{(0)}d_{31}^{(2)} + E_c^{(2)}d_{31}^{(0)} + E_c^{(1)}d_{31}^{(1)}. \tag{4.82}$$

Taking into account the compatibility equation:

$$T_{x_1x_1} = \frac{\partial \mathcal{H}}{\partial s_{x_1x_1}},$$
$$D_3 = -\frac{\partial \mathcal{H}}{\partial E_3}, \tag{4.83}$$

and Eqs. (4.79)–(4.81), the following nonlinear enthalpy function is derived

$$\mathcal{H} = \frac{1}{2}E_c^{(0)}s_{x_1x_1}^2 + \frac{1}{3}E_c^{(1)}s_{x_1x_1}^3 + \frac{1}{4}E_c^{(2)}s_{x_1x_1}^4$$
$$+ \gamma_{00}s_{x_1x_1}E_3 - \frac{1}{2}\gamma_1 s_{x_1x_1}^2 E_3 - \frac{1}{3}\gamma_2 s_{x_1x_1}^3 E_3 - \frac{1}{2}\nu_0 E_3^2. \tag{4.84}$$

One may check that $\mathcal{H}$ satisfies necessary and sufficient condition of its existence of the following form

$$\frac{\partial^2 \mathcal{H}}{\partial s_{x_1x_1}\partial E_3} = \frac{\partial T_{x_1x_1}}{\partial E_3} = -\frac{\partial D_3}{\partial s_{x_1x_1}} = \frac{\partial^2 \mathcal{H}}{\partial E_3 \partial s_{x_1x_1}}. \tag{4.85}$$

Chapter 5

Modeling of Mechanical Fluid Systems

In Tables 5.1 and 5.2, the most common cross sections of hydraulic and pneumatic flow elements, formulas for flow rates, and critical Reynolds number Re defining applicability range of these elements are presented.

Table 5.1   Losses coefficients for the flow of oil and air through throttle elements.

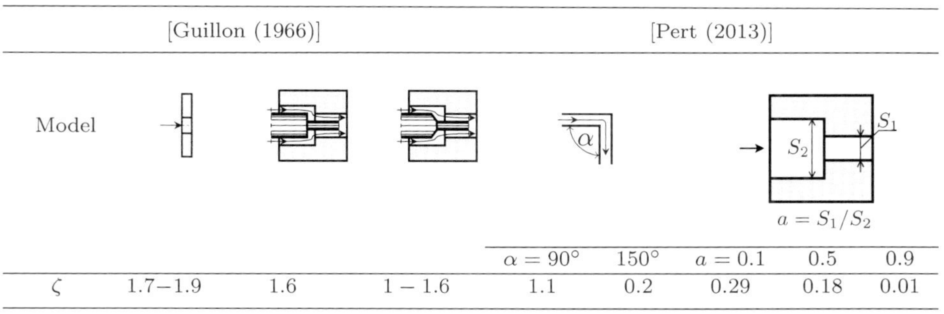

	[Guillon (1966)]					[Pert (2013)]		
Model								$a = S_1/S_2$
				$\alpha = 90°$	$150°$	$a = 0.1$	0.5	0.9
ζ	$1.7-1.9$	1.6	$1-1.6$	1.1	0.2	0.29	0.18	0.01

The loss coefficient ζ for various fluid mechanical mechatronic devices is given in Tab. 5.1. Coefficients λ, expressed by formulas a-d in the sixth row of Tab. 5.2, can be estimated also from the Nikuradse diagram [Nakayama and Boucher (2000)] presented in Fig. 5.2.

The model of fluid friction (laminar), i.e., the Newton formula for the fluid friction F_T, shown in Fig. 5.1, is a function of dynamic viscosity η, surface of friction S_T and velocity gradient v/h.

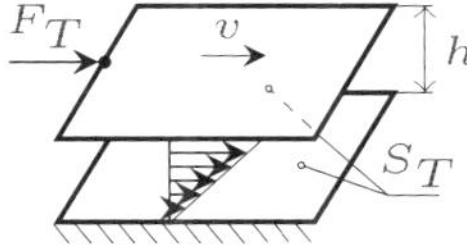

Fig. 5.1   Newton's fluid friction force F_T as a function of dynamic viscosity η, the surface of friction S_T and velocity gradient v/h, where $F_T = \eta S_T \frac{v}{h}$.

Mathematical description elaborated during analysis of a technological

Table 5.2 Volumetric flow rates of oil and air through throttle elements in normal conditions.

Element	Q – oil	Q_N – air	Re
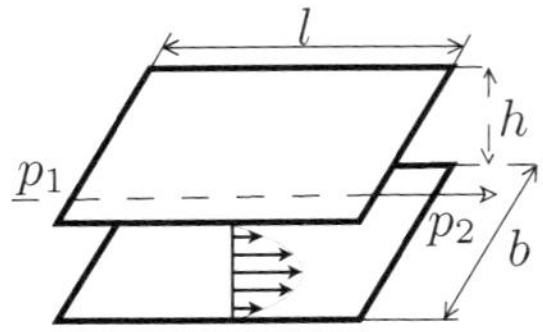	$\frac{bh^3}{12\eta l}(p_1 - p_2)$	$\frac{bh^3}{24\eta l}\frac{(p_1^2 - p_2^2)}{p_N}$	≤ 1100
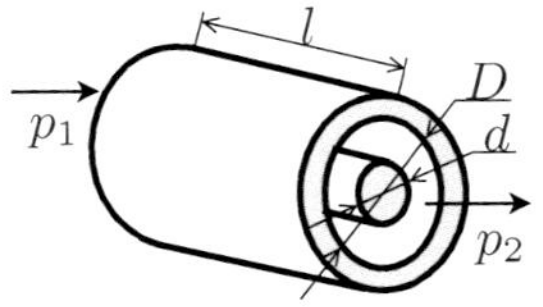	$\frac{\pi D h^3}{12\eta l}(p_1 - p_2)$ $h = (D-d)/2$	$\frac{\pi D h^3}{24\eta l}\frac{(p_1^2 - p_2^2)}{p_N}$ $h = (D-d)/2$	≤ 1100
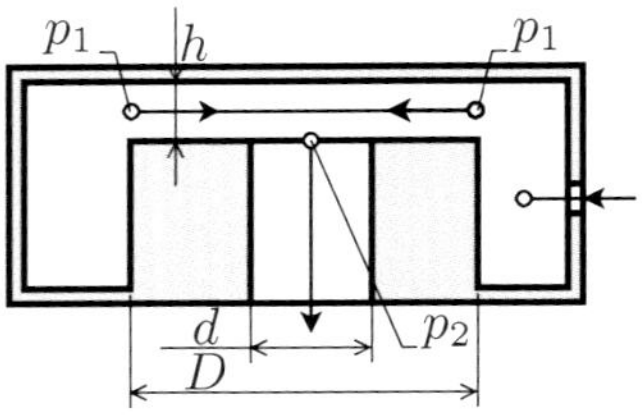	$\frac{\pi h^3}{6\eta \ln \frac{D}{d}}(p_1 - p_2)$	$\frac{\pi h^3}{12\eta \ln \frac{D}{d}}\frac{(p_1^2 - p_2^2)}{p_N}$	≤ 1100
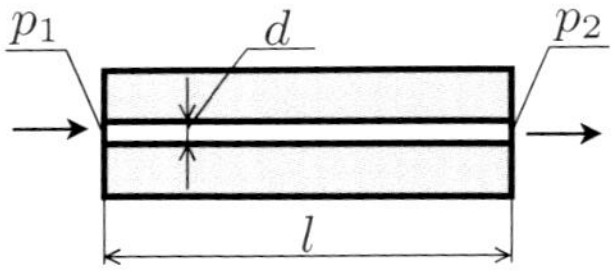 $l/d > 20$	$\frac{\pi d^4}{128\eta l}(p_1 - p_2)$	$\frac{\pi d^4}{256\eta l}\frac{(p_1^2 - p_2^2)}{p_N}$	≤ 2300
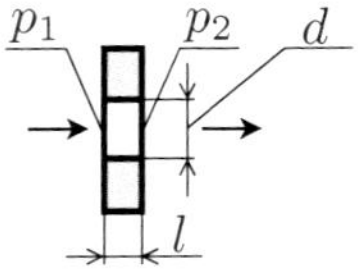 $l/d < 1$	$S\sqrt{\frac{2}{\zeta\rho}}\sqrt{p_1 - p_2}$	$S\sqrt{\frac{1}{\zeta\rho_N p_N}}\sqrt{p_1^2 - p_2^2}$	Tab. 5.1
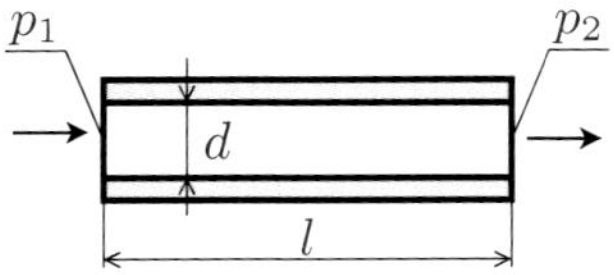 $\eta = \lambda l/d$	a) Darcy friction factor $\lambda = 64/Re$ for laminar flow		≤ 2300
	b) $\lambda = 0.316/\sqrt[4]{Re}$ according to Blasius correlation		≥ 2300 ≤ 50000
	c) $\lambda = 0.0096 + 4\sqrt{\frac{s}{r}} + 1.2\sqrt{\frac{2}{Re}}$ according to Mises correlation		≥ 2300 $\leq 10^6$
	d) λ from Nikuradse diagram		Fig. 5.2

Reynolds number $Re = vd/\nu$ or $Re = vh/\nu$, $\nu = \eta/\rho$, η – dynamic viscosity, ξ – losses coefficient, hydraulic diameter $d_H = 4S/l_o$

solution and some dynamic responses of the investigated object comes from idealization of a physically existing dynamical system.

In order to develop a mathematical model of a dynamical system, one needs to describe system's physical counterpart by mathematical equations that are valid in the fields that the system is devoted to.

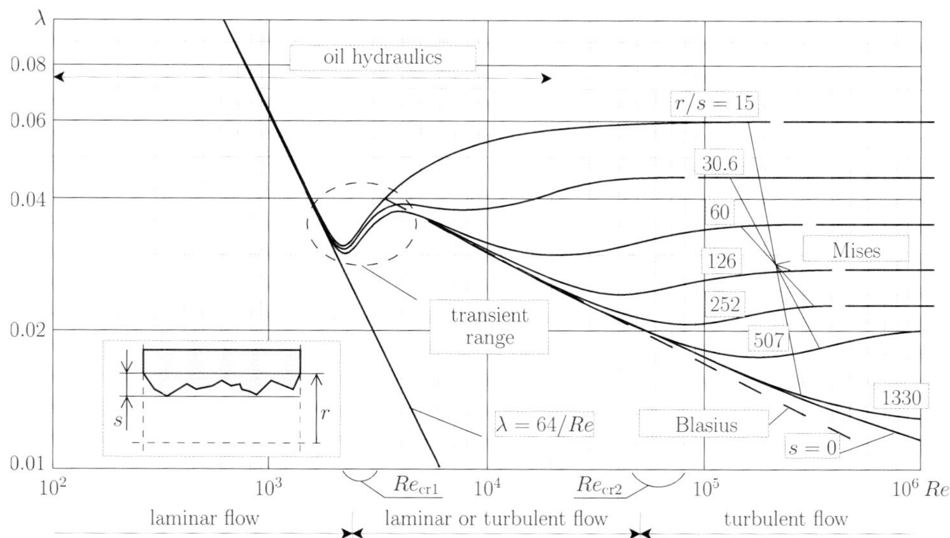

Fig. 5.2   Nikuradse diagram [Nakayama and Boucher (2000)].

A simulation model of the investigated physical object introduces an analytical description being a part of the mathematical model of the object.

The simulation models considered in the book are created in two programming environments, i.e., `LabVIEW` with an additional toolbox *Control Design and Simulation* and `Scilab` with `Xcos` supporting preparation and running of numerical simulations. Construction of a simulation model is divided into several basic steps:

1. Setting numerical values of model variables and numerical simulation.
2. Choosing from libraries and inserting the necessary basic blocks.
3. Connecting the selected blocks with the use of lines ended with arrows indicating the direction of signal flow.

Basic blocks in `Xcos` model many available structures and mathematical functions applied in elaboration of mathematical models.

Exploration of a simulation model is based on the numerical solution of equations describing the analyzed object. Before running the simulation, we initiate parameters, the method of numerical integration, insert the desired function of external forcing, set numerical values, and define parameters of elementary blocks. Results of the simulation are produced in the form of responses of the modeled object to the external forcing and are monitored on virtual oscilloscopes. Analysis of the results of exploration allows us to detect and correct errors committed in both the design of the modeled object as well as in the selection of model parameters.

Table 5.3   Assessment of the relative influence of terms in Bernoulli equation in the context of regulation and driving fluid systems.

$E_c = p$	$E_p = \rho g h$	$E_k = \rho v^2/2$	$E_s =$ $E_c + E_p + E_k$
p $E_c/p = 1$	$\rho g h$ $E_p/p = \rho g h/p$	$\rho v^2/2$ $E_k/p = \rho v^2/(2p)$	Elements E_s/p
Hydraulic systems			
$p = 2.5 \cdot 10^6$ 1	$p = 2.5 \cdot 10^6,\ \rho = 900$ $h = 3$ $1.1 \cdot 10^{-2}$	$p = 2.5 \cdot 10^6,\ \rho = 900$ $v = 8$ $1.2 \cdot 10^{-2}$	Data for hydr. syst. $E_s = 1.023$
Pneumatic systems			
$p = 0.7 \cdot 10^6$ 1	$p = 0.7 \cdot 10^6$ $\rho_p = 8.61,\ h = 3$ $0.036 \cdot 10^{-2}$	$p = 0.7 \cdot 10^6$ $\rho_p = 8.61,\ v_p = 20$ $0.25 \cdot 10^{-2}$	Data for pneum. syst. $E_s = 1.00286$

p – pressure, ρ – density of oil, ρ_p – density of air at pressure p,
h – difference in flow levels, v – mean velocity of oil in the line,
v_p – mean velocity of air at pressure p

5.1   The Balance of Fluid Flow

Due to low density of air (in pneumatics) and relatively high pressure of the fluid used (in hydraulic systems), inertial forces are omitted in the description of phenomena occurring in oil and air. Therefore, to describe these phenomena, the principle of mass conservation can be used. Confirmation of the validity of this assumption is found after analysis of terms of the Bernoulli equation. The results of this analysis are summarized in Tab. 5.3. In conclusion, these phenomena can be described by the principle of mass conservation [Guillon (1966); Lewandowski (1996, 1971)].

In Figure 5.3a, volume V surrounded by a total control surface S is described analytically by the principle of mass conservation

$$\frac{dm_p}{dt} = \iint\limits_{s} \rho\, v_N\, dS\,, \tag{5.1}$$

where: m_p is a mass of fluid in the volume V, ρ – density, v_N – vector of velocity normal to the element of the control surface dS (see Fig. 5.4).

It results from the analysis of fluid mechanical systems, both presented earlier and those analyzed in the next chapters of this monographs, that the design of such systems is modular. Various methods of design developed through years have caused that almost all elements of these systems following the technical characteristics are offered by specialized manufacturing and trading companies. There is also far-reaching unification and standardization in this field. Sold in the trade are, for instance, pumps, motors, filters, mufflers, manifolds, flow controllers, relays,

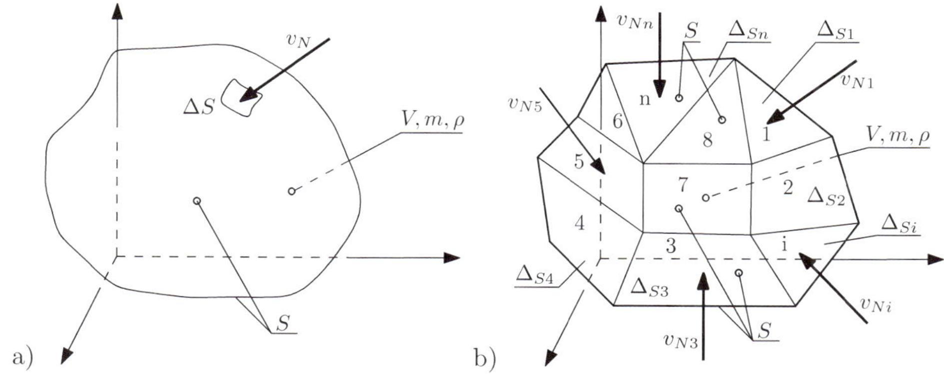

Fig. 5.3 Volumetric models of a fluid in the continuous (a) and discrete system (b).

and many other special components. Hence, the conclusion is that a successful discretization in fluid mechanical systems is possible. Discretization can embrace a lot of phenomena, such as flow, leakages, elasticity of housing and control systems. Therefore, it is possible to pass from the continuous model described in the fluid dynamics (see Fig. 5.3) to a more convenient analytical description of the discrete model.

Passing from the continuous model shown in Fig. 5.3a to Fig. 5.3b via discretization, the formula (5.1) takes the form

$$\frac{dm_p}{dt} = \sum_{i=1}^{n} \rho\, v_{Ni}\, \Delta S_i\,.\tag{5.2}$$

By introducing the elementary mass flow rate Q_{mi} given in the form

$$Q_{mi} = \rho v_{Ni}\Delta S_i\,,\tag{5.3}$$

we get

$$\frac{dm_p}{dt} = \sum_{i=1}^{n} Q_{mi}\,.\tag{5.4}$$

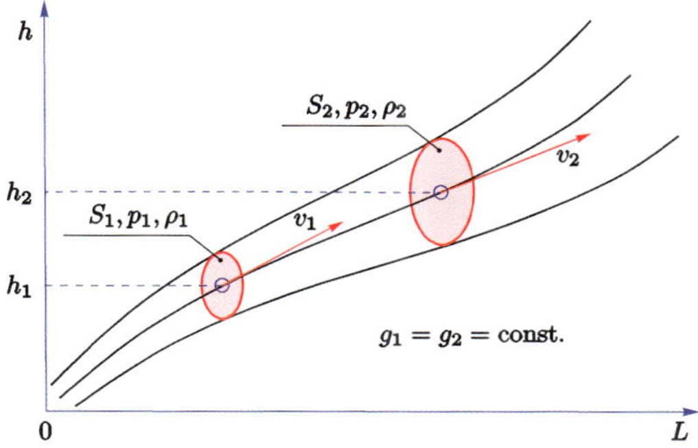

Fig. 5.4 Illustration of divergence of the control surface S (S_1 and S_2 appear subsequently) and the vector of normal velocity v_N (v_1 and v_2 appear subsequently too).

Assuming the density field m_p of a homogeneous mass, one writes

$$Q_{mi} = \rho Q_i \quad \text{and} \quad m_p = \rho V, \tag{5.5}$$

where:

   Q_i – elementary volumetric flow rate,
   V – volume of a fluid surrounded by a control surface S.

Substituting Eq. (5.5) to (5.4), one finds

$$\sum_{i=1}^{n} Q_i = \frac{1}{\rho} \frac{d(\rho V)}{dt}. \tag{5.6}$$

Differentiating the product ρV in Eq. (5.6) with respect to time, we get the equation of flow balance and divergence of volume of the fluid surrounded by the control surface S (see Fig. 5.4) as follows

$$\sum_{i=1}^{n} Q_i = \frac{dV}{dt} + \frac{V}{\rho} \frac{d\rho}{dt}. \tag{5.7}$$

Equation (5.7) describes relations for fluids. Thus, it is valid either for liquids or gases. Particular considerations devoted to liquids and gases require a separate analysis with respect to their principally different properties. The procedure in the case of hydromechanical systems is explained below.

Oil is the most commonly used fluid in hydromechanical systems. According to the study in [Guillon (1966)], a relative increase in the density of oil is given by the formula

$$\frac{d\rho}{\rho} = \frac{dp}{E}, \tag{5.8}$$

in which: p is a pressure, E – oil bulk modulus of elasticity.

After substituting Eq. (5.8) in (5.7), and using a dot above the symbols denoting a derivative with respect to time, the following equation is derived

$$\sum_{i=1}^{n} Q_i = \dot{V} + \frac{V}{E} \dot{p}, \tag{5.9a}$$

$$V = V_0 \pm Sx + k_0 p. \tag{5.9b}$$

In Eq. (5.9b), the symbol "$\pm$" is to be replaced by "$+$" if with the displacement in the direction $+x$ the volume V of a fluid increases or by "$-$" if with the displacement in the same direction the volume V decreases. Moreover, one introduces also notations: S – area of the hydraulic piston, x – displacement of the rod, V_0 – total initial volume of fluid in the cylinder and lines, k_0 – coefficient of housing elasticity.

Expansion and physical interpretation of Eq. (5.9a) will be stated on the physical model of the hydromechanical system shown in Fig. 5.5.

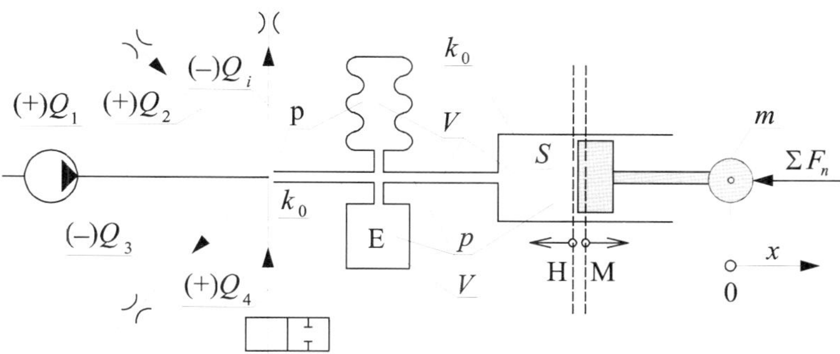

Fig. 5.5   Physical model of a hydromechanical system (H – region of hydraulics, M – region of mechanics).

The first derivative of volume on the right-hand side of Eq. (5.9a) is found by differentiation of Eq. (5.9b) with respect to time

$$\dot{V} = \pm S\dot{x} + k_0\dot{p}. \tag{5.10}$$

The first term of the sum on the right-hand side of Eq. (5.10) expresses a time variation of volume that resulted from a displacement of the piston of the surface of the area S with velocity $\dot{x}$. Similarly, the second term expresses time variation of a volume filled with oil that results from the deformation of the control surface of housing including some surfaces of other elements such as lines, measuring instruments, cylinders, accumulators. Expression $k_0\dot{p}$ is an approximation arising after idealization by taking the assumption that volume of the enumerated elements changes proportionally to the changes of oil volume.

Combining Eqs. (5.9b) and (5.10) with (5.9a), we get

$$\sum_{i=1}^{n} Q_i = \pm S\dot{x} + \left(k_0 + k_0\frac{p}{E} + \frac{V_0}{E} \pm \frac{S}{E}x \right)\dot{p}. \tag{5.11}$$

Pressure in hydraulic systems usually ranges from 1 to 25 MPa and oil bulk modulus of elasticity E is greater than 1000 MPa, therefore $p/E \ll 1$. Due to these values, in practical applications, one assumes that $k_0(1 + p/E) \approx k_0$, so Eq. (5.11) is transformed to the form

$$\sum_{i=1}^{n} Q_i = \pm S\dot{x} + \left(k_0 + \frac{V_0}{E} \pm \frac{S}{E}x \right)\dot{p}. \tag{5.12}$$

In the case of rotary hydraulic motors (see Fig. 5.6), one assumes that volumes of chambers are constant. Then, Eq. (5.12) takes the form

$$\sum_{i=1}^{n} Q_i = \pm qn + \left(k_0 + \frac{V_0}{E} \right)\dot{p}, \tag{5.13}$$

where: q is a volumetric efficiency of the vane motor $q \cong \pi(D+d)be$, n – rotational velocity, e – eccentricity of rotor, b – width of rotor's vanes.

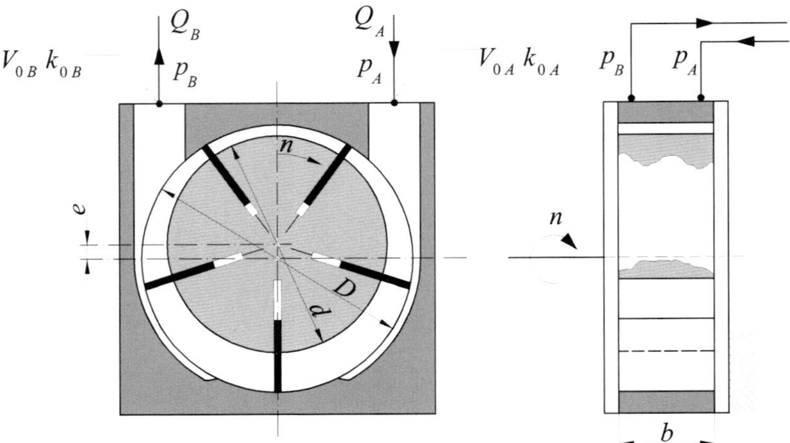

Fig. 5.6   Hydraulic vane motor (V_{0A}, V_{0B} – volume of liquid in both chambers; k_{0A}, k_{0B} – elasticity of housing of both chambers of the motor; Q_A, Q_B – flow rates of fluid through inlets and outlets).

Pneumomechanical systems with air as the fluid work practically in constant temperature. Therefore, transformations occurring there are isothermal. In this case, the Clapeyron equation holds

$$pV = p_N V_N . \qquad (5.14)$$

For the time-dependent functions V and V_N, differentiating Eq. (5.14) with respect to time and introducing these definitions: $\dot{V} = Q$, $\dot{V}_N = Q_N$, one gets

$$Qp = Q_N p_N . \qquad (5.15)$$

Dividing both sides of Eq. (5.15) by m, and taking into assumption that $V/m = 1/\rho$ and $V_N/m = 1/\rho_N$, the proportionality occurs

$$\frac{p}{\rho} = \frac{p_N}{\rho_N} , \qquad (5.16)$$

where p_N, ρ_N denote normal values of pressure and density, respectively.

Putting Q, p/ρ, $\dot{p}/\dot{\rho}$ calculated from Eqs. (5.15), (5.16) and the time derivative of both sides of Eq. (5.15) into the Eq. (5.7), we find

$$\sum_{i=1}^{n} Q_{Ni} = \frac{p}{p_N} \dot{V} + \frac{V}{p_N} \dot{p} . \qquad (5.17)$$

In view of high compressibility of air, the elasticity of housing in pneumome-chanical systems (e.g. in Fig. 5.7) is neglected ($k_0 = 0$). On the basis of that assumption and Eqs. (5.9b), (5.10) and (5.16), the flow equation (an equation of flow balance) is obtained

$$\sum_{i=1}^{n} Q_{Ni} = \pm \frac{S}{p_N}\dot{x}\,p + \left(\frac{V_0}{p_N} \pm \frac{S}{p_N}x\right)\dot{p}\,. \qquad (5.18)$$

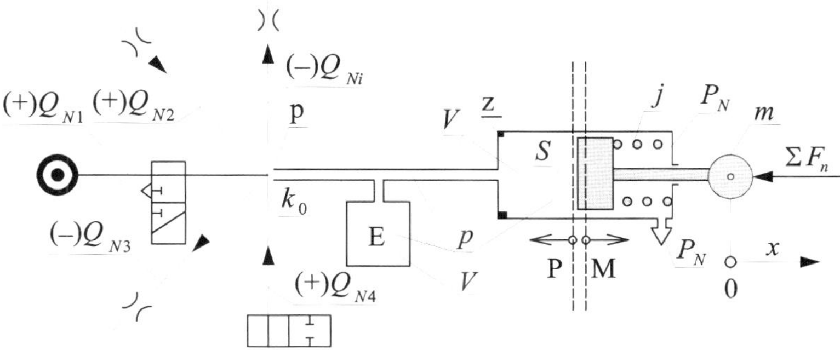

Fig. 5.7 Physical model of a pneumomechanical system (P – region of pneumatics, M – region of mechanics, z – stop, j – spring).

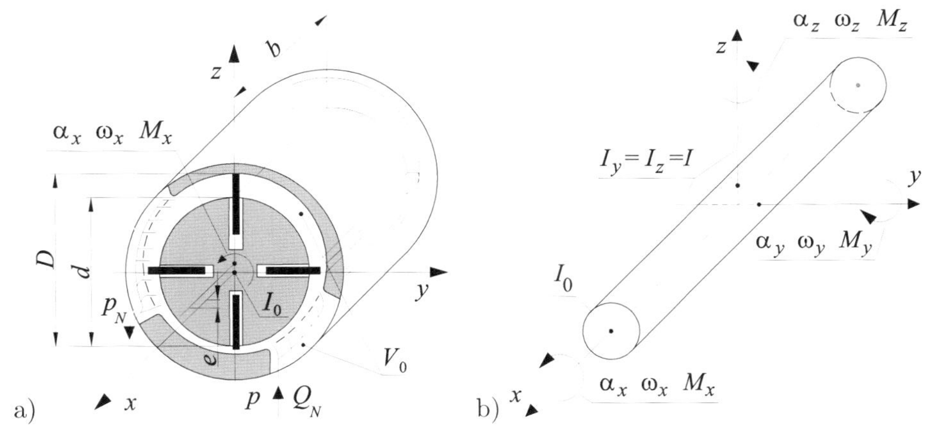

Fig. 5.8 Flow model (a) and the mechanical model of a flow impeller (b) of the pneumatic motor of rotational motion.

In the case of rotary pneumatic motor shown in Fig. 5.8a, proceeding as while deriving the Eq. (5.5), as well as introducing some parameters of the motor, the flow equation is transformed to the form

$$\sum_{i=1}^{n} Q_{Ni} = \frac{q}{p_N}np + \frac{V_0}{p_N}\dot{p}\,. \qquad (5.19)$$

5.2 Description of the Dynamics of a Mechanical System

When describing the dynamics of motion of a mechanical system [Awrejcewicz (2012a); Schmid (2002); Zierep (1978)], it is required to use Newton's second law. According to the law, the general equation of the dynamics of motion takes the form

$$m\ddot{x} = \sum F(x, \dot{x}, \dots) + \sum F(p) + \sum [F + F(t)], \qquad (5.20)$$

where m is the mass of the moving part of the mechanical system, and:

$$
\begin{aligned}
F(x, \dot{x}, \dots) &- \text{ forces dependent on the displacements and their derivatives;} \\
F(p) &- \text{ pressure forces on active surfaces of the hydraulic actuator;} \\
F + F(t) &- \text{ external forces among others dependent on time.}
\end{aligned}
$$

Description of the dynamics of rotation is obtained using the angular momentum equation

$$\frac{d\bar{K}}{dt} = \sum \bar{M}. \qquad (5.21)$$

In the case of idealization of the rotor model shown in Fig. 5.8a, the angular momentum $\bar{K} \cong I_0\bar{\omega}$ at relatively small angular displacements α_z, α_y and the condition ω_z, $\omega_y \ll \omega_x$. Using the mentioned assumption, the equation of rotational motion with regard to gyroscopic forces is found [Awrejcewicz (2012a)]:

$$
\begin{aligned}
I\dot{\omega}_y + I_0\omega_x\omega_z &= \sum M_y, \\
I\dot{\omega}_z - I_0\omega_x\omega_y &= \sum M_z, \qquad (5.22)\\
I_0\dot{\omega}_x &= \sum M_x,
\end{aligned}
$$

where: I, I_0 are the moments of inertia, ω_x, ω_y, ω_z – angular velocities, M_x, M_y, M_z – torques with respect to axes x, y, z.

For instance, a full system of equations (5.22) is applied in a description of high-speed ($90\text{-}180\cdot10^3$ rev/min) aerostatic ball-bearing spindles [Lewandowski (1996)]. Typically, in hydraulic and pneumatic motors working at relatively small rotational velocities, the third equation of the system (5.22) is used.

The physical model in Fig. 5.8a is described by a general equation:

$$
\begin{aligned}
I_0\dot{\omega} &= \sum M_x \\
&= \sum M(\alpha, \omega, \dot{\omega}, \dots) + \sum M(p) + \sum [M + M(t)]. \qquad (5.23)
\end{aligned}
$$

Notations of components of Eq. (5.23) are analogous to the notations used in Eq. (5.20).

A sum of torques M_n and M_p coming from any pressure forces acting in the hydraulic (see Fig. 5.6) and pneumatic motor (see Fig. 5.8), respectively, is calculated from the formulas:

$$M_n = \frac{q}{2\pi}(p_A - p_B), \quad M_p = \frac{q}{2\pi}(p - p_N). \qquad (5.24)$$

5.3 Modeling of an Open Hydromechanical System of Linear Displacement

Mathematical model

The physical model depicted in Fig. 5.9 and the equation of the flow balance (5.12), which is valid in the places denoted by digits 1 and 2, can be used to derive the following equations:

$$Q - Q_A - k_1 p_A = 0\,, \tag{5.25a}$$

$$Q_A - k_2(p_A - p_B) = S\dot{x} + \left(\frac{V_0}{E'} + \frac{S}{E'}x\right)\dot{p}_A\,. \tag{5.25b}$$

In equation (5.25) and Fig. 5.9, the following notations were used: Q – pump inflow without leaking, Q_A – the inflow sourcing the cylinder, k_1 – flow leakage on the input, k_2 – flow leakage on the piston, p_A – pressure on input lines and in the cylinder, p_B – pressure on the output, V_0 – initial volume, E' – oil bulk modulus of elasticity with regard to free air in the hydraulic system and elasticity of housing [Schmid (2002)].

The definitions under Fig. 5.9 supplement the above notations.

By transforming Eq. (5.25) and putting relations for Q and p_B, we get the equation of the oil flow balance in the considered system

$$\left(\frac{V_0}{E'} + \frac{S}{E'}x\right)\dot{p}_A = Q_0\mathbf{1}(t) - S\dot{x} - (k_1 + k_2)p_A + k_2 p_0\mathbf{1}(t)\,. \tag{5.26}$$

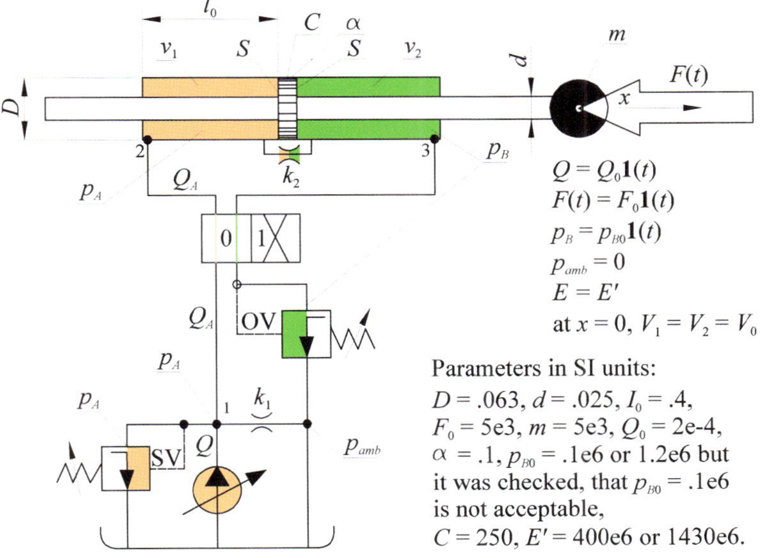

Fig. 5.9 Physical model of a hydromechanical system of linear displacement (OV – overflow vale, SV – safety valve).

Function $\mathbf{1}(t)$ in Eq. (5.26) is a unit step, which is characterized by an initial time, initial value (0 or 1) and final value (1 or 0, respectively). Equation (5.26) is the first component of the considered system's mathematical model. The second component of the model is found from the successively transformed Eq. (5.23) describing dynamics of the hydromechanical system in Fig. 5.9. The original form of this equation follows

$$m\ddot{x} = Sp_A - \alpha S|p_A - p_B|\operatorname{sgn}\dot{x} - C\dot{x} - Sp_B - F(t), \qquad (5.27)$$

where: α is the coefficient of resistances of the contact friction of seals, C – coefficient of total resistance of fluid friction.

The final form of mathematical model's second component can be estimated by substitution of definitions for p_B and $F(t)$ (see the caption to Fig. 5.9). After arranging Eq. (5.27), we get

$$m\ddot{x} = S\left[p_A - \mathbf{1}(t)p_0\right] - \alpha S\left|p_A - p_0\mathbf{1}(t)\right|\operatorname{sgn}\dot{x} - F_0\mathbf{1}(t) - C\dot{x}. \qquad (5.28)$$

The full mathematical model of the analyzed hydromechanical system is constituted of Eqs. (5.26) and (5.28).

Simulation model of a hydromechanical system.

The basis of the simulation model presented in Fig. 5.10 is the physical model (Fig. 5.9) expressed by the mathematical model (5.12) and (5.15).

In the introduced model, some lines of displacement, its derivative and input pressure, being subsequently monitored on oscilloscopes $x(t)$, $\dot{x}(t)$, $p_A(t)$, can be distinguished. In addition, selected excitations such as ideal outflow $Q(t)$ of the pump, loading $F(t)$ of the hydraulic cylinder and the output pressure $p_B(t)$ are monitored. It is possible to make more connections with oscilloscopes at arbitrary nodes of the diagram, what is especially useful in testing and observation of time histories of transient signals.

Figure 5.11 presents the results of numerical simulation of a model of the hydromechanical system depicted in Fig. 5.8, obtained in `Scilab`.

Time histories 5.11a and b represent changes in the excitation signals $Q(t)$ and $F(t)$. Analyzing the time history of $p_A^{(2)}$ (red line) in Fig. 5.11c, it is seen that its value decreases to an unreal negative value $-0.8\text{e}6$ after a step of pressure p_A. In this case, the result of simulation is encumbered with errors. It is not recommended for any hydraulic systems to work at pressures $p_A < 0$, because of emission of gases dissolved in oil as well as of air intake caused by vacuum leakages. It is therefore necessary to use other design or change the parameters of the analyzed system. In the considered example, the positive effect is obtained by increasing p_{B0} of the pressure p_B in the outflow from $0.1\text{e}6$ to $1.2\text{e}6$ (see Fig. 5.11b).

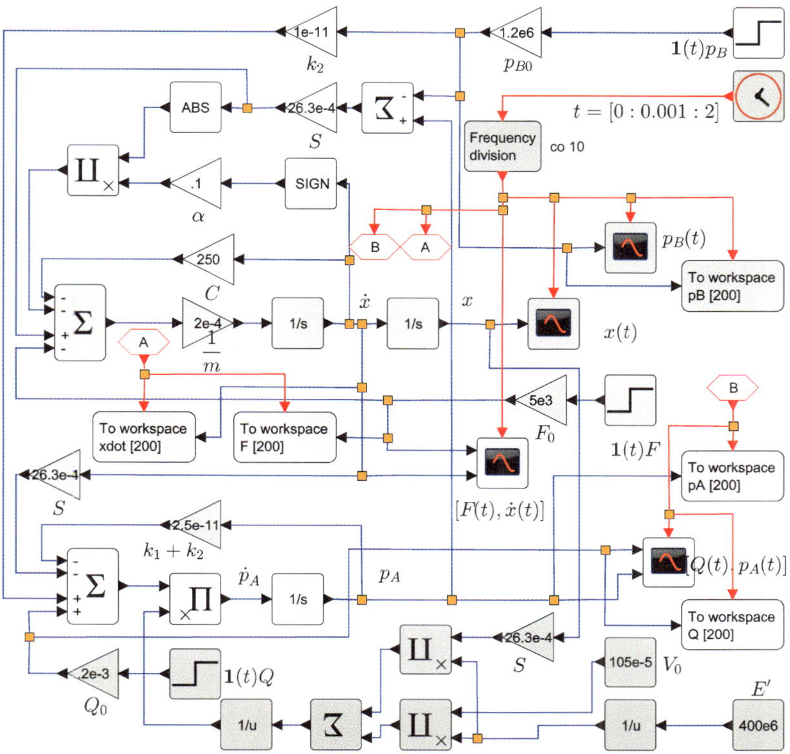

Fig. 5.10   Simulation model of the hydromechanical system described by the physical model shown in Fig. 5.9.

5.4   Modeling of a Hydraulic Electromechanical Servomechanism of Rotational Motion

By using a physical model of the servomechanism (given in Fig. 5.12), the mathematical model described by equations (5.29)–(5.31) is yielded, where: I – balance of flows, II – balance of torques, III – automatic regulation.

Flow equations in lines A, B and points 1-4:

$$
\begin{aligned}
&\text{I A1} && -Q_A + kx\sqrt{p_z - p_A} = 0\,, \\[4pt]
&\text{I A2} && qn + \frac{V_0}{E}\dot{p}_A = Q_A - Q_A(1 - \eta_v)\,, \\[4pt]
&\text{I A1, 2} && \frac{V_0}{E}\dot{p}_A = \eta_v kx\sqrt{p_z - p_A} - qn\,, \\[4pt]
&\text{I B3} && -Q_B = -qn + \frac{V_0}{E}\dot{p}_B\,, \\[4pt]
&\text{I B4} && Q_B - kx\sqrt{p_B} = 0\,, \\[4pt]
&\text{I B3, 4} && \frac{V_0}{E}\dot{p}_B = -kx\sqrt{p_B} + qn\,.
\end{aligned}
\tag{5.29}
$$

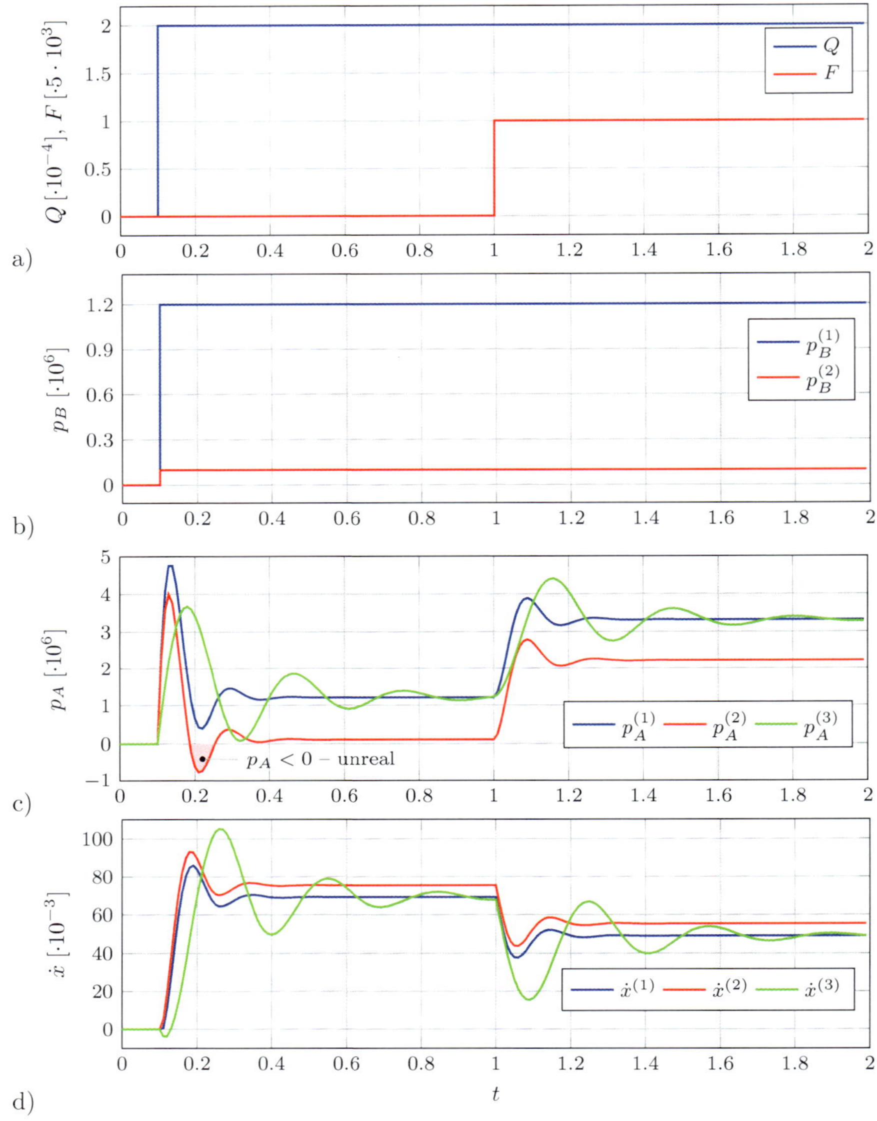

Fig. 5.11   Time histories of hydromechanical system's responses (Fig. 5.9) after the numerical solution of the simulation model 5.10: (a) changes in pump's outflow $Q(t)$ and loading $F(t)$; (b) changes in pressure $p_B(t)$ on the output in the case: 1) $p_{B0} = 0.1\,\text{e}6$, 2) $p_{B0} = 1.2\,\text{e}6$; (c) changes in input pressure $p_A(\text{t})$; (d) changes of piston rod's velocity $\dot{x}(t)$. Superscripts (1-3) on the right of the symbols of variables shown in sub-figures c and d denote: (1) $E' = 1.43\,\text{e}9$, $\varepsilon_N = 5\text{e}{-}4$, $p_{B0} = 1.2\,\text{e}6$, (2) $E' = 1.43\,\text{e}9$, $\varepsilon_N = 5\text{e}{-}4$, $p_{B0} = 0.1\,\text{e}6$, (3) $E' = 0.4\,\text{e}9$, $\varepsilon_N = 5\text{e}{-}3$, $p_{B0} = 1.2\,\text{e}6$.

Torque equation

$$\text{II} \qquad 2\pi I_x \dot{n} = \frac{q}{2\pi}(p_A - p_B) - Cn - M \, . \tag{5.30}$$

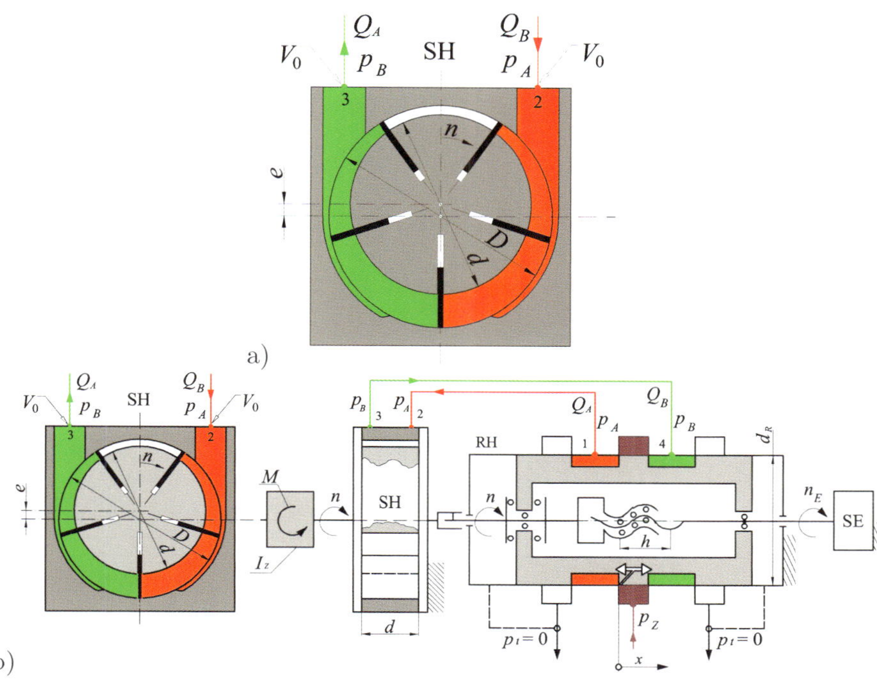

Fig. 5.12 Physical model of a servomechanism of rotational motion: (a) cross sectional view of the hydraulic motor, (b) longitudinal cross section of the hydraulic divider.

Equation of automatic regulation

$$\text{III} \qquad \dot{x} = (n_E - n)s. \tag{5.31}$$

In equations (5.29)–(5.30), the following notations are applied: η_m – coefficient of local losses, Q – flow rate, η_v – volumetric efficiency of the motor, V_0 – volume of oil in lines A and B, E – oil bulk modulus of elasticity, C – coefficient of circumferential viscous friction, q – motor capacity, $k = \pi d_r \sqrt{2/(\rho \xi_m)}$ – capacity constant of the throttle valve.

In Figure 5.12, the following notations are introduced: SE – electric motor's set-point of input rotational velocity n_E, SH – controllable hydraulic motor reaching the output velocity n (a coordinate of rotational velocity), RH – hydraulic divider, d_R – diameter of the piston, h – thread pitch of the lead screw, x – coordinate of the throttling edge displacement of the piston (width of the throttle gaps in the hydraulic divider), D – diameter of a stator (stationary part of the electric motor), d – diameter of the rotor, b – width of rotor's vanes, e – eccentricity of the rotor, I_x – reduced moment of inertia with respect to the axis of rotation, M – torque loading the hydraulic motor; p_z – input pressure; p_A, p_B pressures in lines A and B; p_T – ambient pressure; 1-4 – points of mathematical description.

A simulation model obtained by means of equations I A1, I B3, I B4, I B3 given in Eq. (5.29) is proposed in Fig. 5.14.

The time histories depicted in Fig. 5.13 present trajectories of the discrete series: $x(t)$, $\dot{x}(t)$ – displacement and velocity of the piston of the hydraulic divider; $p_Z(t)$ – pressure of inflow, $p_A(t)$, $p_B(t)$ – pressures in lines A and B; $n(t)$ – velocity of the rotor in the hydraulic motor; $n_E(t)$ – electric motor's set-point velocity; $M(t)$ – torque loading the motor; $P(t)$ – power of the hydraulic motor.

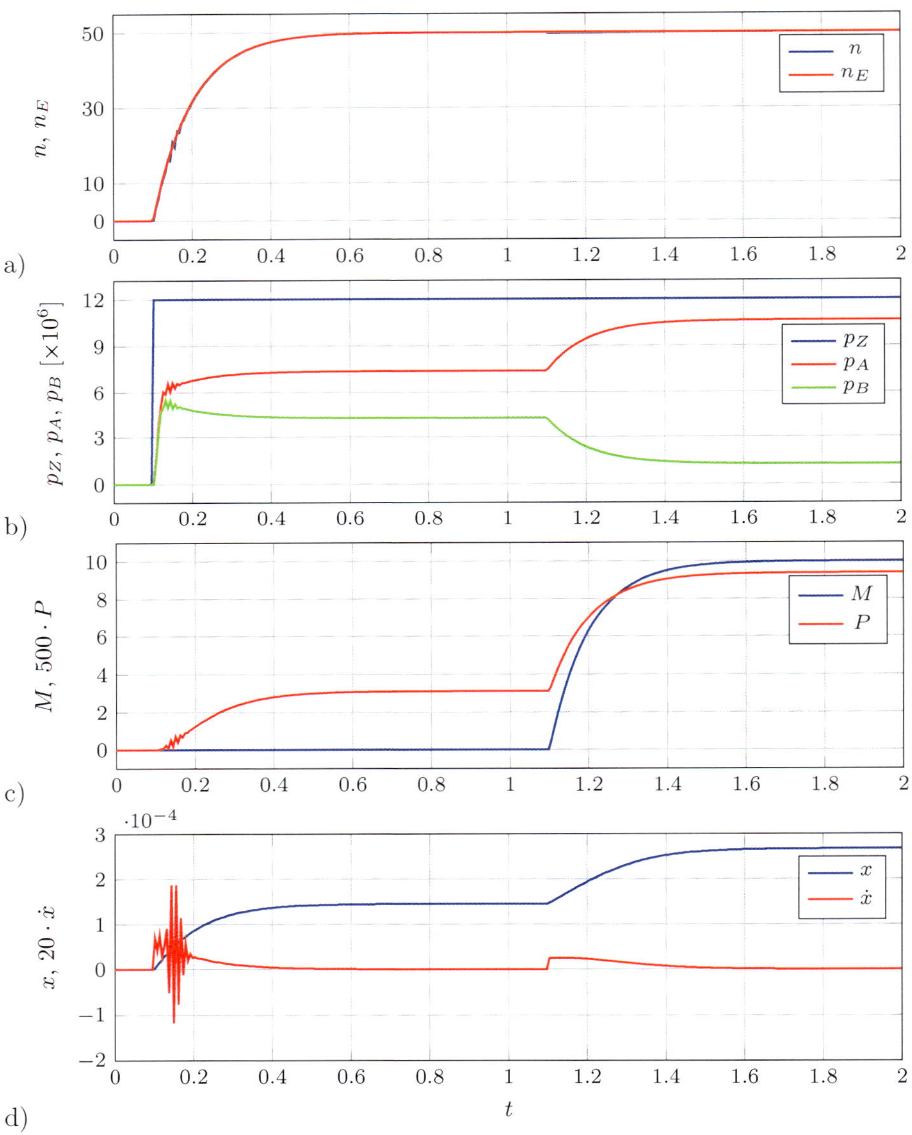

Fig. 5.13   Time histories resulting from numerical solution of the simulation model of the analyzed servomechanism of rotational motion.

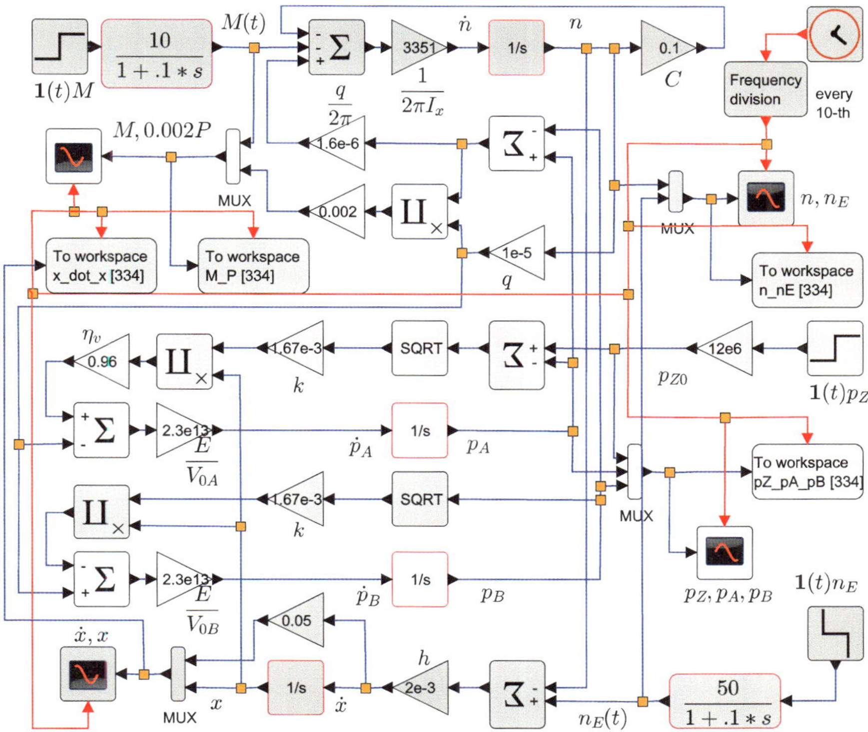

Fig. 5.14   Simulation model of the servomechanism of rotational motion for the parameters: $n = 50$, $Q_{max} = 1\,\text{e–}3$, $p_z = 12\,\text{e}6$, $q_m = 2\,\text{e–}5$, $D = 39.54\,\text{e–}3$, $d = 35\,\text{e–}3$, $b = 40\,\text{e–}3$, $e = 2.27\,\text{e–}3$, $d_T = 16\,\text{e–}3$, $m = 0.31$, $I_x = 47.5\,\text{e–}6$, $V_0 = V_{0A} = V_{0B} = 61.3\,\text{e–}6$, $E = 1400\,\text{e}6$, $k = 1.67\,\text{e–}3$, $\mu = 7\,\text{e–}3$, $C = 0.1$, $M = 10$, $1/(2\pi I_x) = 3351$, $q/(2\pi) = 3.18\,\text{e–}6$, $E/V_0 = 2.3\,\text{e}13$.

In Figure 5.13, just after a step change (0.1 second of the simulation) in the supply pressure p_z, an irregular behavior of all components of the servomechanism system takes place in the time interval [0.1,0.2]. The temporal high amplitude variations of velocity $\dot{x}$ of movement of the throttling edge of the piston (see Fig. 5.12) are the most significant. After supplying the hydraulic system, the load $M(t)$, turned at 1.1 second of the simulation, makes inertially smooth increase in pressures p_A and p_B in lines A and B, respectively to 10.63 MPa and 1.26 MPa. To conclude, the system characterized by the set of parameters written in Fig. 5.14 caption is well designed and can be a starting point to realize the real prototype.

5.5   A Proportional Valve in the Drive and Control of a Hydromechanical System

A physical model of the drive and control of the hydromechanical system of a hydraulic cylinder with a proportional valve is shown in Fig. 5.15 [Lewandowski and Awrejcewicz (2012)].

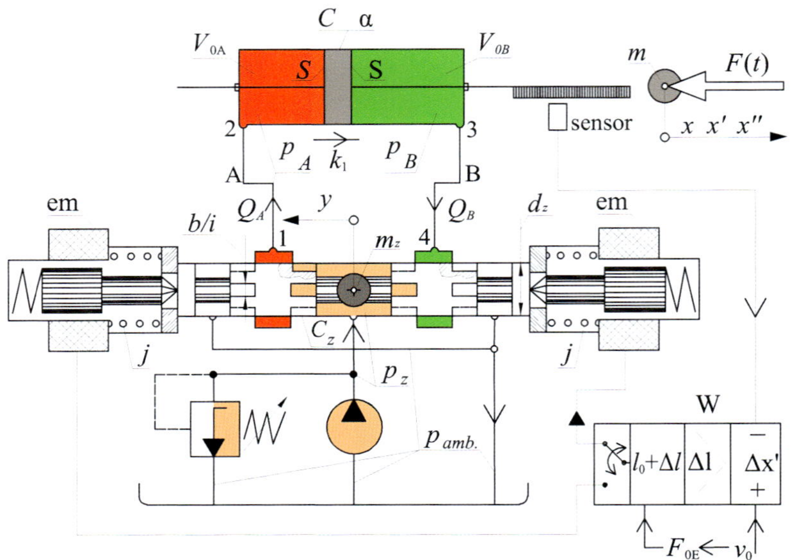

Fig. 5.15   Proportional valve in the drive and control of the hydromechanical system.

Below we present a derivation of the mathematical model of a hydromechanical
system controlled with the use of a proportional valve.

Equations of flows in lines A, B and points (1-4) have the form:

$$\text{I A1} \qquad -Q_A + kby\sqrt{p_z - p_A} = 0, \quad \text{and} \quad Q_A = kby\sqrt{p_z - p_A}\,,$$

$$\text{I A2} \qquad Q_A - k_1(p_A - p_B) = S\dot{x} + \left(\frac{V_{0A}}{E} + \frac{S}{E}x\right)\dot{p}_A\,,$$

$$\text{I A} \qquad \left(\frac{V_{0A}}{E} + \frac{S}{E}x\right)\dot{p}_A = -S\dot{x} + kby\sqrt{p_z - p_A} - k_1(p_A - p_B)\,,$$

$$\text{I B3} \qquad -Q_B + k_1(p_A - p_B) = -S\dot{x} + \left(\frac{V_{0A}}{E} - \frac{S}{E}x\right)\dot{p}_B\,,$$

$$\tag{5.32}$$

$$\text{I B4} \qquad Q_B - kby\sqrt{p_B} = 0, \quad \text{and} \quad Q_B = kby\sqrt{p_B}\,,$$

$$\text{I B} \qquad \left(\frac{V_{0B}}{E} - \frac{S}{E}x\right)\dot{p}_B = +S\dot{x} - kby\sqrt{p_B} + k_1(p_A - p_B)\,.$$

The equation of load in the hydraulic cylinder:

$$\text{II} \qquad m\ddot{x} = S(p_A - p_B) - \alpha S|p_A - p_B|\operatorname{sgn}\dot{x} - c\dot{x} - F(t)\,. \tag{5.33}$$

The equation of load in the proportional valve acting on the piston:

$$\text{III} \qquad m_z\ddot{y} = -C_z\dot{y} - jy + F_E\,. \tag{5.34}$$

The equation of automatic regulation:

$$\text{IV} \qquad F_E = F + RW(v_0 - \dot{x})\,. \tag{5.35}$$

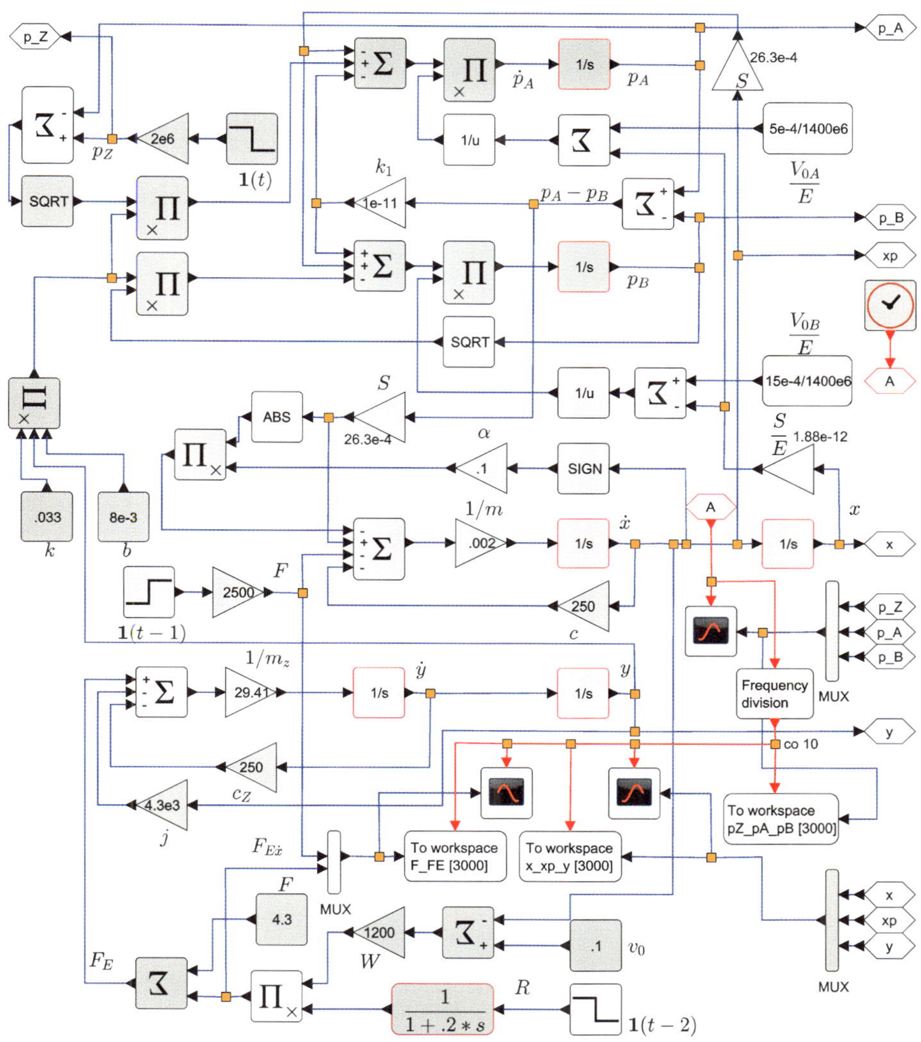

Fig. 5.16   Simulation model of the analyzed hydromechanical system.

Equations (5.32)–(5.35) and Fig. 5.15 uncovers the following variables and parameters: Q – flow rate, m – reduced mass of the piston, m_z – mass of valve's piston, b – total width of throttling slot in the valve, y – coordinate of spool displacement of the valve, j – elasticity of springs of the valve, F_E – force exerted by the electromagnet, F_{E0} – electromagnet force at a manual setting of piston's velocity, v_0 – reference velocity, C_z – a coefficient of viscous friction in the proportional valve, k_1 – capacity of leakages of the hydraulic cylinder, S – active surface of piston, α – coefficient of motion resistance of the piston, R – automatic speed control switch (0 or 1), W – gain coefficient, $k = \sqrt{2/(\rho \xi_m)}$ – constant of throttle valve's capacity.

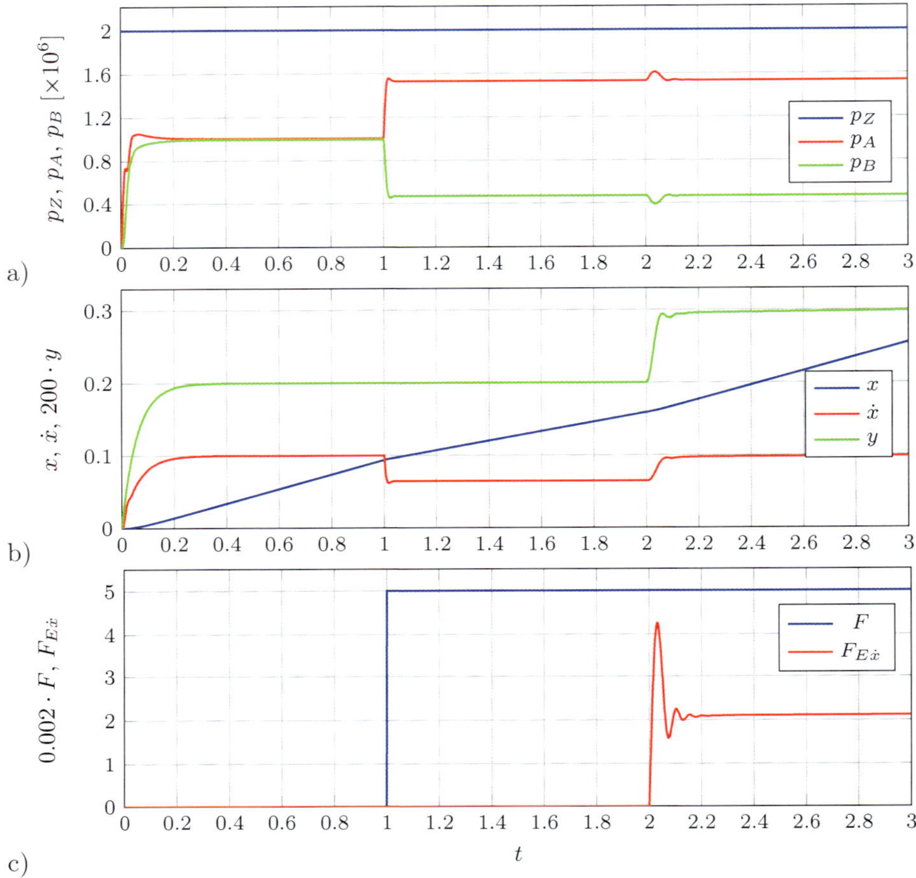

Fig. 5.17 Results of numerical solution of the analyzed simulation model of the hydromechanical system with proportional valve. The time characteristics in sub-figures (a-c) represent: $y(t)$ – displacement of the spool of the proportional valve, $p_Z(t)$ – inflow pressure, $p_A(t)$ and $p_B(t)$ – pressures in lines A and B on both sides of hydraulic cylinder's piston, $\dot{x}(t)$ and $x(t)$ – velocity and displacement of loading, $F(t)$ – loading force, $F_{E\dot{x}}(t)$ – force acting on the solenoid (electromagnet), which is a function of the difference $v_0 - \dot{x}(t)$ between the reference and measured velocity.

A simulation model of the hydromechanical driving system controlled by a proportional valve is shown in Fig. 5.16.

In Figure 5.17, some time histories of state variables of the investigated system are presented. The results of numerical simulation confirm that the designed hydraulic system carries the load (see function $F(t)$ turned on at $t = 1\,\mathrm{s}$) almost without oscillations.

5.6 Physical and Mathematical Model of the Pneumatic Hydromechanical System

The necessary condition for the creation of a simulation model is the knowledge of a theoretical description of the investigated system. In the case of a feeding mechanism of a lathe, the basic laws described ealier can be applied. On the basis of the laws of fluid and classical mechanics [Awrejcewicz (2014)], a physical model of the system is proposed.

The physical model of a system is derived from an idealization of a real object that is based on object's data sheet, its scheme of functioning and some conditions of operation. The mentioned idealization must provide analogy (adequacy) between the derived physical model and the modeled real object.

While attempting to model lathe's feeding mechanism, some theoretical background described in [Lerner and Trigg (2005)] was taken into consideration in our study as well as the following assumptions have been made:

- the mass of the driven carriage is concentrated in one point;
- leakages in the hydraulic system are neglected;
- pressure losses in lines are neglected;
- flows in throttle valves are sub-critical and turbulent;
- dry and viscous friction exist simultaneously in piston's contact and in the throttle valves.

A scheme of the physical model of the cutter feeding mechanism, including geometrical and exploitation parameters, is shown in Fig. 5.18. The mathematical description of the proposed physical model is elaborated in accordance to the following scheme.

In the analyzed system, the balance of air flows related to normal conditions is described by Eq. (5.18).

The balance of flows in line A is described by the equation

$$k_{dA}\sqrt{p_s^2 - p_A^2} = \frac{S - S_d}{p_N}\dot{x}p_A + \left(\frac{V_0}{p_N} + \frac{S - S_d}{p_N}x\right)\dot{p}_A \,, \qquad (5.36)$$

the balance of flows in line B follows

$$-k_{dB}\sqrt{p_B^2 - p_N^2} = -\frac{S - S_d}{p_N}\dot{x}p_B + \left[\frac{L_0(S - S_d)}{p_N} - \frac{S - S_d}{p_N}x\right]\dot{p}_B \,, \qquad (5.37)$$

while the balance of oil flow in the hydraulic cylinder of the damper, according to Eq. (5.11), is expressed by the formula

$$-k_t\sqrt{p_t - p_s} = -S_t\dot{x} \,. \qquad (5.38)$$

With respect to high elasticity of housing, which is a cylinder in this case, it is assumed that $k_0 = 0$. In equations (5.36)–(5.38):

k_{dA}, k_{dB} – capacities of throttling valves of the pneumatic drive are given by the formulas:

$$k_{dA} = k_A \cdot \mathbf{1}(t)_0^\infty, \quad k_{dB} = k_B \cdot \mathbf{1}(t)_0^\infty ; \tag{5.39}$$

k_t – capacity of the throttling valve of the damping system;

p_S – input pressure of the pneumatic system.

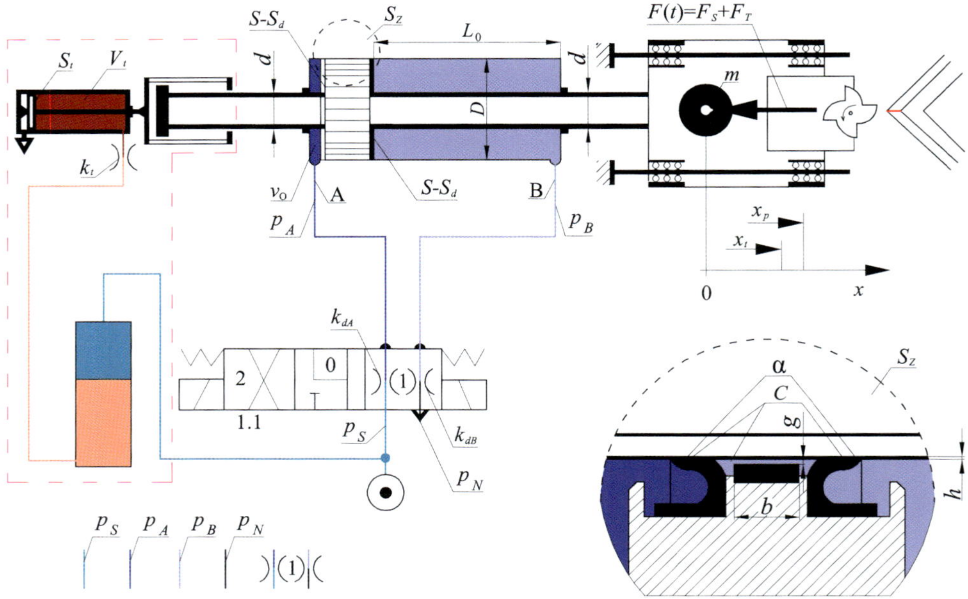

Fig. 5.18   A physical model of the pneumatic hydromechanical system. Parameters: $D = 50\mathrm{e}{-3}$, $d = 25\mathrm{e}{-3}$, $h = g = 1.5\mathrm{e}{-6}$, $m = 20$, $L_0 = 0.5$, $p_N = 0.1\mathrm{e}6$, $p_s = 0.7\mathrm{e}6$, $V_0 = 9.8\mathrm{e}{-6}$, $k_A = 0.2\,\mathrm{e}{-9}$, $k_B = 0.05\mathrm{e}{-12}$, $I_D = I_4 = 5\mathrm{e}{-3}$, $b_D = b_4 = 10\mathrm{e}{-3}$, $P_0 = 5$, $\eta = 16\mathrm{e}{-3}$, $S = 1.96\mathrm{e}{-3}$, $S_d = 0.49\mathrm{e}{-3}$, $C = 50$, $S_t = 1730\mathrm{e}{-7}$, $k_1 = 2889\mathrm{e}{-12}$, $C_t = 620\mathrm{e}3$, $x_t = 0.12$, $x_p = 0.122$, period of step function $T = 53\mathrm{e}{-4}$, $\nu = 187$ Hz, duty cycle $W = 30\,\%$ of periodic step function.

Balance equations of flows in lines A and B of the pneumatic drive are found using Eqs. (5.36) and (5.37). Substituting capacity functions k_{dA} and k_{dB}, we get:

$$\left(\frac{V_0}{p_N} + \frac{S - S_d}{p_N}x\right)\dot{p}_A = -\frac{S - S_d}{p_N}\dot{x}p_A + k_A\mathbf{1}(t)\sqrt{p_s^2 - p_A^2}, \tag{5.40}$$

$$\left[\frac{L_0(S - S_d)}{p_N} - \frac{S - S_d}{p_N}x\right]\dot{p}_B = +\frac{S - S_d}{p_N}\dot{x}p_B - k_B\mathbf{1}(t)\sqrt{p_B^2 - p_N^2}. \tag{5.41}$$

In the case of the damper, the description of which states the equation of flows (5.12), and the equation of forces acting on it, i.e.,

$$F_T - S_T(p_t - p_S) = 0, \tag{5.42}$$

we obtain a formula for the damping force

$$F_T = C_T \dot{x}^2 + p_S S_T \,, \tag{5.43}$$

where $C_T = S_T^3/k_T^2$.

Equations (5.42) and (5.43) come from the principle of mass conservation and they describe the first component of the analyzed system's mathematical model. In the description of the dynamics of the pneumatic hydromechanical system, one applies Eq. (5.20), which after a rearrangement takes the form:

$$\begin{aligned} m\ddot{x} = &- C\dot{x} + (S - S_d)(p_A - p_B) + \\ &- \operatorname{sgn}\dot{x}\,[S|p_A - p_B| + S_d(p_A + p_B - 2p_N)]\,\alpha - F(t)\,, \end{aligned} \tag{5.44}$$

in which:

$$\begin{aligned} F(t) = F_S + F_T = &\underbrace{0.5[\operatorname{sgn}(x - x_p) + 1]F_0 I(T, W)}_{F_S} + \\ &+ \underbrace{0.5[\operatorname{sgn}(x - x_t) + 1]\left(C_t\dot{x}^2 + S_t p_S\right)}_{F_T}\,. \end{aligned} \tag{5.45}$$

The second component of the mathematical model is found by putting relations (5.39) and (5.45) into the Eq. (5.44), getting the second order differential equation as follows:

$$\begin{aligned} m\ddot{x} = &- C\dot{x} + (S - S_d)(p_A - p_B) + \\ &- \operatorname{sgn}\dot{x}\,[S|p_A - p_B| + S_d(p_A + p_B - 2p_N)]\,\alpha + \\ &- 0.5[\operatorname{sgn}(x - x_p) + 1]F_0 I(T, W) + \\ &- 0.5[\operatorname{sgn}(x - x_t) + 1]\left(C_t\dot{x}^2 + S_t p_S\right)\,, \end{aligned} \tag{5.46}$$

where: x_t – switching threshold of the damper, x_p – a starting point threshold of machining.

Construction and exploration of the simulation model.

After taking into account external forcing, the transformed theoretical model states a kind of mathematical model, which is created by some nonlinear equations having no strict analytical solution.

A simulation model of the pneumatic hydromechanical system shown in Fig. 5.18 is presented in Fig. 5.19.

The simulation model was obtained by means of Eqs. (5.40), (5.41), (5.43) and (5.46) stating the mathematical description of the investigated system. Initially, equations (5.40) and (5.41) were used to determine the lines of air pressures p_A, p_B and their derivatives $\dot{p}_A$, $\dot{p}_B$ originating from summing junctions. Pressures p_A, p_B, p_S and state variables x, $\dot{x}$ are monitored on three oscilloscopes.

Function $F_S(t)$ with the amplitude $F_{S0} = 20\,\mathrm{N}$, period $T_S = 53 \cdot 10^{-4}\,\mathrm{s}$ and duty cycle $d_S = 30\,\%$ simulates some periodic forcing exerted by the cutting forces propagated by the cutter. The two stops switching the damper during machining

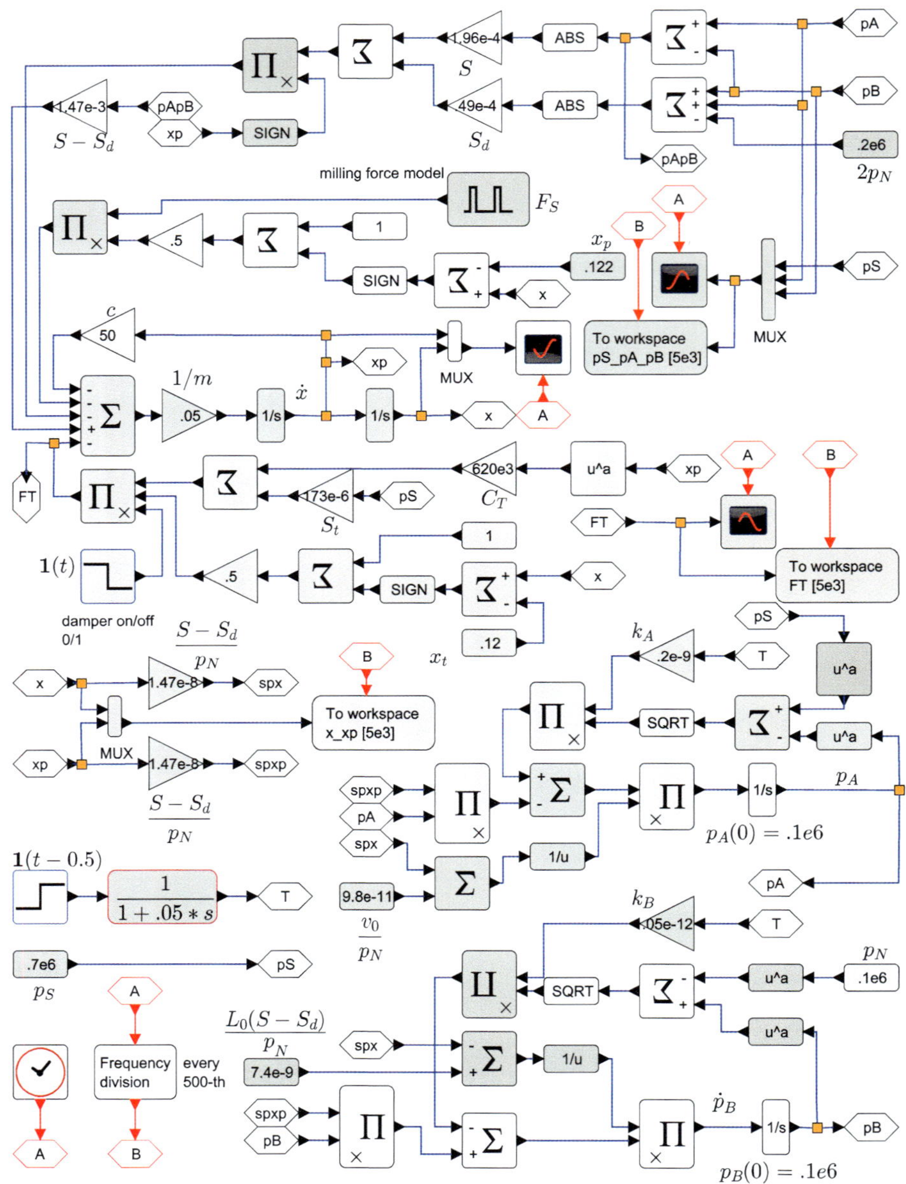

Fig. 5.19   A simulation model of the pneumatic hydromechanical system shown in Fig. 5.18.

are simulated by function $0.5[\,\mathrm{sgn}\,(x_t - x) + 1\,]$ and $0.5[\,\mathrm{sgn}\,(x_p - x) + 1\,]$. Results of exploration of the simulation model shown in Fig. 5.19 are pictured in Fig. 5.20.

The time histories presented in Fig. 5.20a and c illustrate trajectories of pressures p_S, p_A and p_B, displacement x and velocity $\dot{x}$ of the system with (solid line) and without the hydraulic damper (dashed line). The time history presented in

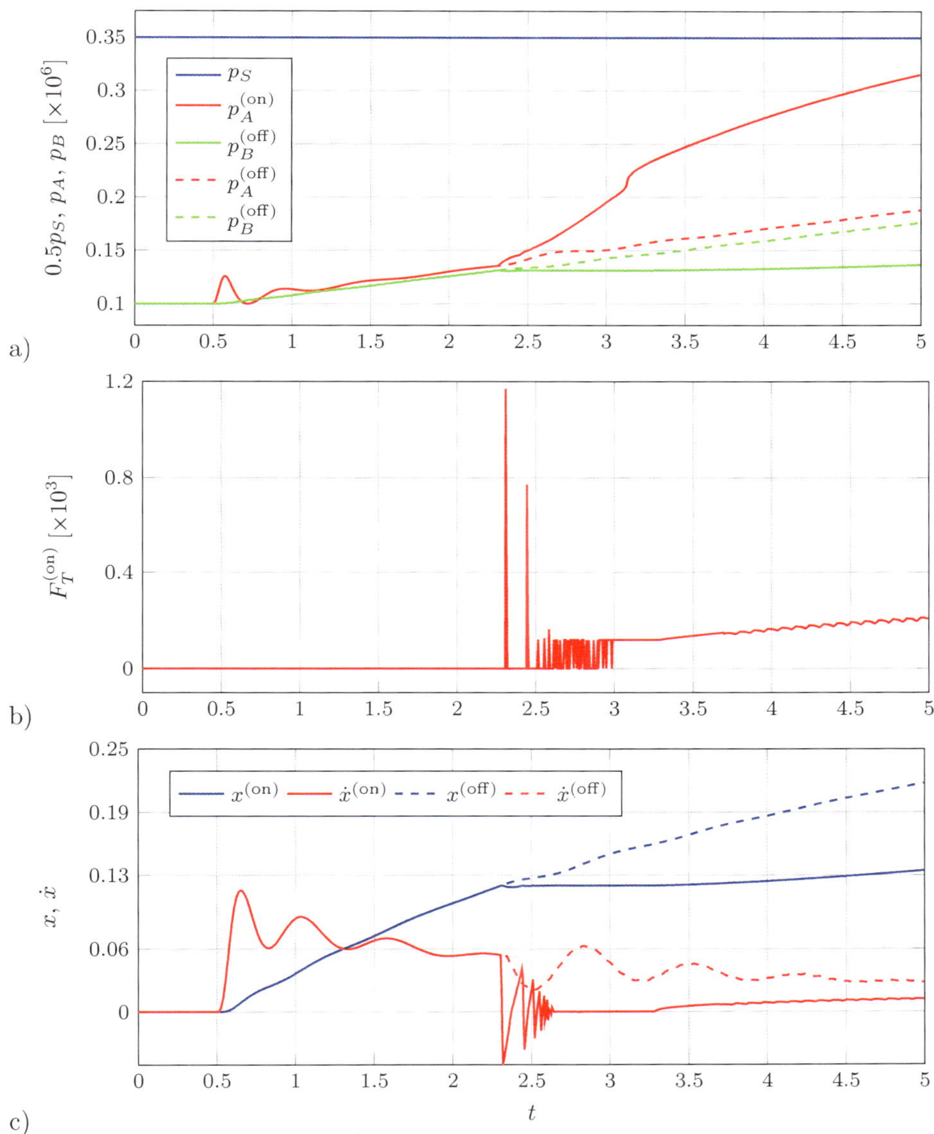

Fig. 5.20 Results of numerical simulation of the pneumatic hydromechanical system.

Fig. 5.20b illustrates changes in the damping force $F_T(t)$ caused by sudden turn off of the hydraulic damper. Denotations *on* and *off* in superscripts of symbols relate to the responses obtained for the damper being turned on and off, respectively. Information about the simulation parameters is given in the caption to Fig. 5.18 (a physical model) and of the simulation model in Fig. 5.19. In Fig. 5.19b, only a trajectory of $F_T(t)$ at the working damper is shown, because the function takes zero value when the damper is turned off.

Comparison of results of numerical simulation in scope of modification of crucial parameters of the machining process, like the distance and velocity of motion of the tool with and without the damper, indicates a definite improvement of working conditions of the feeding mechanism. A selection of sufficiently small and stable feed providing the high quality of the machining process became possible.

Analysis of results allows to detect and correct any mistakes that could be made at the initial stage of modeling. The structure and parameters of the virtual model are usually subjected to many corrections until the best analogy between the model and its real counterpart is achieved.

The cognitive nature is connected with some important and positive features of the presented method of modeling. An observation of selected nodes of the simulation diagram is possible during ongoing simulation as well as investigation of system's reaction on changes of model parameters and construction of the physical model.

Experimentally verified models of real objects can be used for utilitarian purposes and in the design processes. The main advantage of conducting numerical simulations in the industry is the ability to reduce the number of tests on real objects and the number of specially constructed models designated for various testing experiments.

Chapter 6

Modeling of Electrohydraulic Servomechanisms

Electrohydraulic servomechanisms are a perfect example of mechatronic devices which, by their design and processing of signals, create a link between various fields of science and engineering such as solid mechanics, fluid mechanics, electronics, electrical engineering and computer science (numerical programming, CAD). The issues related to control and optimization of dynamic processes and optimization of shapes are not less important. To ensure smooth operation of these devices, a good understanding of the relationships between theoretical and experimental aspects that merge these fields of science is highly required.

The aforementioned systems have many advantages, but require an extensive knowledge in the design and prototyping. Any improperly designed mechatronic system may be subject to frequent breakdowns, and even its functioning may be be hazardous to its operators. Therefore, there is a need of numerical virtualization of the designed system using software dedicated to numerical simulation and analysis of discrete dynamical models. Another benefit is the opportunity to examine the validity of the designed control system. This stage of designing a real device allows to detect weaknesses in the proposed engineering solution and to reduce costs of the prototype production. One of such tools is the `LabVIEW` programming environment made by National Instruments (NI), which allows for a quick and easy creation of simulation blocks (virtual instruments) modeling equations of dynamics describing the investigated mechatronic system, and much more.

A *servomechanism*, also referred to as a *servo*, is a fundamental and the most important part (a closed subsystem) of today's industrial automation.

According to the first definition, a *servomechanism* is a closed loop control system with an error-sensing negative feedback in which the output signal is a physical quantity subject to adjustment (regulation).

According to the second definition, a *servomechanism* is a control system which measures its own output and forces the system generating that output to quickly and accurately follow the reference signal. The control system guarantees its resistance to external forcing (excitation) with a large stability margin.

The servo can control various physical quantities, e.g., force, torque, linear or angular position, temperature, electric voltage, amperage or pressure.

There are three basic types of a servomechanical system:

- electromechanical,
- electropneumatic,
- electrohydraulic.

The main criteria taken into account when choosing the appropriate system include performance, costs of purchase and maintenance, dimensions, weight, conditions of work in the target environment (e.g., resistance to vibrations, shock occurrence or temperature variations).

Electromechanical servos are mechatronic systems in which electronics cooperates with mechanical actuators [Rydberg (2008)]. These systems can include many kinds of controlled machines or mechanisms with the negative feedback line. They fulfill the tasks of control of electric motors, mechanical gears, and others. Electromechanical systems provide a very efficient control in complex technical applications of mechatronic devices and systems. Those of them that operate at low loads are cheap, but this aspect becomes less beneficial when they operate under heavy loads and with a requirement of fast dynamic responses.

Electropneumatic servos work mainly in low loads regimes, where a fast and repeatable motion of actuators between subsequent positions is required. Systems of such type are often noisy. Mostly, they are required to achieve precision in extreme positions. The advantages include mainly low cost and high positioning speed required to perform a fast movement of light elements.

Electrohydraulic servos combine the advantages of hydraulic and electronic systems. They have a high power to weight ratio (moment of inertia) [Rabie (2009)]. Their positive properties are extended on a very good thermal conductivity and self-lubrication [Sadeghieh *et al.* (2012)], which is the desired result of the presence of oil – the working liquid. Application of these systems is particularly advantageous in applications requiring high precision of motion mapped in accordance to a reference function.

The disadvantage of electrohydraulic servos is the variability of oil parameters affecting the repeatability of the duty cycle. These parameters are: density, viscosity or oil bulk modulus of elasticity, depending on its composition, and the operating temperature [Sadeghieh *et al.* (2012)]. Bulk modulus of elasticity depends on many factors, e.g., on pressure, temperature and undissolved air volume. An amount of undissolved air has the greatest influence on the hydraulic oil bulk modulus of elasticity due to high air volume compressibility compared to the oil volume compressibility.

A number of nonlinearities resulting from friction of movable surfaces, leakages and hysteresis of dynamic states, which lead to many uncertainties of unknown parameters and some unknown errors of modeling, state a significant difficulty in the accurate and robust control of these devices. In addition, often used electric power supply is more accessible in industrial plants and prototype testing stands than any hydraulic source of power. Costs of hydraulic systems are high due to the need of manufacturing various components with high accuracy associated with the expensive and advanced manufacturing processes. Proper oil filtration should be assured, since oil must be clear [Rabie (2009)].

Electrohydraulic servos are widely used in industries, such as oil production, control of satellite antennas, fatigue tests, manufacturing techniques, positioning of rocket launchers, generation of vibrations, control of aircraft's components, industrial robotics, agricultural accessories and devices, conveyors, cranes, control of machining processes (e.g., in automatic drives of drilling heads), and other [Younkin (2002)].

Hydraulic servo valves are used to control the flow of fuel in internal combustion engines. For example, the hydromechanical unit can be actuated to deliver a fuel mixture at suitable pressure to a specific location of the combustion system.

Other uses of electrohydraulic servos relate to control parameters of the cable, support landing of helicopters on ships operating in difficult weather conditions [Zhu (2009)]. Many hydraulic control systems are often used in aircraft. The assistance systems that facilitate piloting and control of flying machine's parameters in the air [Fowler and O'Connor (1976); Viswanath and Nagarajan (2002)] can be given as an example.

Guidance of missiles uses servos to control the position of the hydraulic cylinders. The electric signals from the control unit control the operation of valves in a way to automatically adapt the guidance and stabilize the track of the projectile [Holtrop (1983)].

High-pressure hydraulic servo cylinders (called also hydraulic servo actuators) are applied to stabilize the starting platform and arm the rocket launcher passing into a state of combat readiness [Sreekumar and Ramchandani (2005)]. Similar systems are installed on a turret armed combat vehicles [Brandstadter and Taylor (1974)].

Hydraulic servos are present in the systems of active reduction of vibrations of vehicles' chassis. Each column of the chassis is controlled independently to achieve the best driving comfort or the best tracking ability of the vehicle.

Three basic types of the electrohydraulic servomechanisms can be distinguished:

- position servos (linear and angular),
- velocity servos (linear and angular),
- force (momentum) servo.

Position servo. Figure 6.1a introduces an idealized schematic diagram of a linear position servo belonging to a group of basic closed loop control systems of automatic regulation.

Position of the piston of the hydraulic cylinder (Fig. 6.1a), or even of the load being attached to it is measured by a position encoder (a transducer of the physical quantity – linear displacement). The encoder produces an electric voltage signal on the basis of which the output signal u_z – information about actual position (state) of the body of mass M is obtained. The amplifier of the servo, acting also as a comparator, compares the reference signal u_s with the signal u_z in the negative feedback line. The resulting error signal is directed to the amplifier acting as the basic element of the proportional gain K. The signal of electric current present at the amplifier output subsequently controls a state of the servo valve.

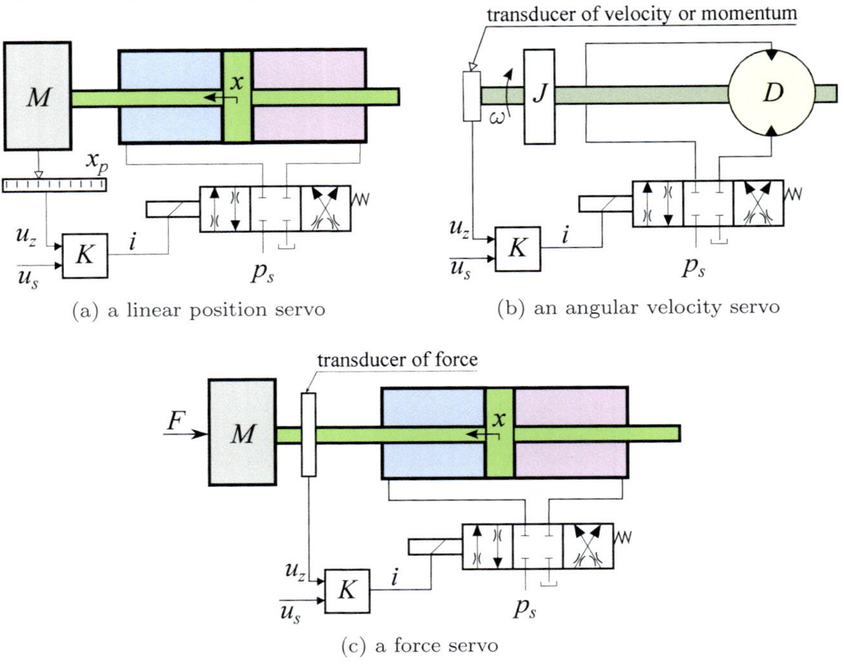

Fig. 6.1   Idealized models of servos of various physical quantities.

Velocity servo. Another example of the servomechanism is a velocity servo. In their design, velocity servos are very similar to position servos, with the difference that the transducer provides information about the movement velocity, but the controller may have different characteristics. The velocity servo is used with an amplifier gaining the integral action realized in the electronic circuit. The integration is essential to minimize static errors and to ensure the stability of system's dynamic response. The design of the velocity servo of rotational motion is schematically shown in Fig. 6.1b.

Force (momentum) servo. A force servo is similar to the position servo, but its transducer measures force or a moment of force. The voltage signal of a measured quantity is directed backwards to the amplifier. There is also a servo implementation, wherein the sensor of the voltage signal coming from pressure of external load closes the feedback loop. Implementation of this type is then close to the true servomechanism of force, where, in addition, the friction force in the actuator is taken into account. An exemplary force servo is presented in Fig. 6.1c.

6.1 Simplified Model of a Servo With a Proportional Valve

An experimental setup with an electrohydraulic servo modeled in [Milič *et al.* (2010)] is schematically presented in Fig. 6.2. The simplification is made by the derivation of Eq. (6.7) of the dynamics, described by a vector state $\bar{x}$ according to definition (6.6), and by neglecting friction between the piston and the cylinder.

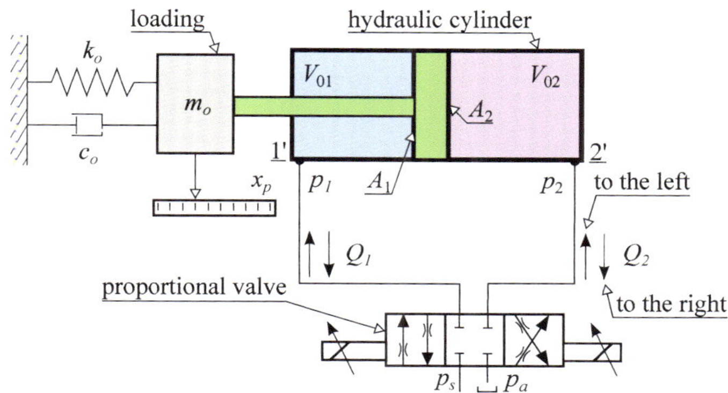

Fig. 6.2 Simplified model of a servo with a proportional valve.

The system depicted in Fig. 6.2 is composed of a gear pump, a proportional valve and a hydraulic cylinder with a load (another body of a mass greater than the piston mass). Analyzing the schematic diagram, the second order linear ordinary differential equations, describing dynamics of proportional valve's spool, can be derived:

$$\ddot{y} + 2\zeta\omega\dot{y} + \omega^2 y = k_p\omega^2 U, \tag{6.1}$$

where k_p is the proportional gain of the valve, ω – natural frequency of vibrations of the spool, ζ – damping ratio of the proportional valve, y – position of spool, U – control input voltage.

Based on the equations of continuity of flow through valve's orifices in ports 1' and 2' (see Fig. 6.2), one writes:

$$Q_1 = \begin{cases} C_d\sigma y\sqrt{2(p_s - p_1)/\rho}, & y \geq 0, \\ C_d\sigma y\sqrt{2(p_1 - p_a)/\rho}, & y < 0, \end{cases} \tag{6.2}$$

$$Q_2 = \begin{cases} C_d\sigma y\sqrt{2(p_2 - p_a)/\rho}, & y \geq 0, \\ C_d\sigma y\sqrt{2(p_s - p_2)/\rho}, & y < 0, \end{cases} \tag{6.3}$$

where Q_1, Q_2 are the volumetric flows through the orifices, p_1, p_2 – pressures in the left and right cylinder chambers, p_s – supply pressure, p_a – pressure of liquid in the tank, C_d – valve coefficient of discharge, σ – valve orifice area gradient.

Neglecting the internal and external leakages, the dynamics of the hydraulic pressure behavior in the compressible fluid volumes is given by the equations:

$$\dot{p}_1 = \frac{\beta}{V_{01} + A_1 x_p} \left(Q_1 - A_1 \dot{x}_p\right), \tag{6.4a}$$

$$\dot{p}_2 = \frac{\beta}{V_{02} - A_2 x_p} \left(-Q_1 + A_2 \dot{x}_p\right), \tag{6.4b}$$

where β is the fluid bulk modulus of elasticity, A_1, A_2 – annulus areas of the piston and the rod side of the cylinder, $V_{0i} = A_i l/2$ $(i = 1, 2)$ – the cylinder half-volumes on both sides of the piston.

The equation of dynamics of the mechanical part of the system is given by

$$m\ddot{x}_p = p_1 A_1 - p_2 A_2 - c_o \dot{x}_p - k_o x_p - F_L, \tag{6.5}$$

where m is the total mass of the piston and the load – a body attached to the hydraulic cylinder, c_o and k_o are the viscous damping coefficient of the servo actuator and the load stiffness, respectively, F_L – unknown external disturbance of the force coming from the load attached to the hydraulic cylinder. In a further simplification, any friction proportional to the hydraulic pressure has been omitted.

Taking into account Eqs. (6.1)–(6.5), the vector of state variables is defined by

$$\bar{x} = [x_1, x_2, x_3, x_4, x_5, x_6]^T = [y, \dot{y}, p_1, p_2, x_p, \dot{x}_p]^T. \tag{6.6}$$

The above assumptions allow to write a nonlinear model of dynamics of the servo from Fig. 6.2, expressed by the system of differential equations [Milič *et al.* (2010)]:

$$\dot{x}_1 = x_2, \tag{6.7a}$$

$$\dot{x}_2 = -\omega^2 x_1 - 2\zeta\omega x_2 + k_p \omega^2 U, \tag{6.7b}$$

$$\dot{x}_3 = \frac{\beta}{V_{01} + A_1 x_5} \left(C_d \sigma x_1 \sqrt{2\Delta p_P/\rho} - A_1 x_6\right), \tag{6.7c}$$

$$\dot{x}_4 = \frac{\beta}{V_{02} - A_2 x_5} \left(-C_d \sigma x_1 \sqrt{2\Delta p_R/\rho} + A_2 x_6\right), \tag{6.7d}$$

$$\dot{x}_5 = x_6, \tag{6.7e}$$

$$\dot{x}_6 = \frac{1}{m} \left(A_1 x_3 - A_2 x_4 - k_o x_5 - c_o x_6 - F_L\right). \tag{6.7f}$$

In Eq. (6.7)c-d, Δp_P and Δp_R are given by:

$$\Delta p_P = \begin{cases} |p_s - x_3|, & x_1 \geq 0, \\ |x_3 - p_a|, & x_1 < 0, \end{cases} \tag{6.8a}$$

$$\Delta p_R = \begin{cases} |x_4 - p_a|, & x_1 \geq 0, \\ |p_s - x_4|, & x_1 < 0. \end{cases} \tag{6.8b}$$

6.2 Torque Motor

Electromechanical transducers (e.g., torque motors) belong to the group of basic elements of electrohydraulic servomechanisms. They are used to transform an electric signal into a usable displacement of mechanical elements [Rabie (2009)].

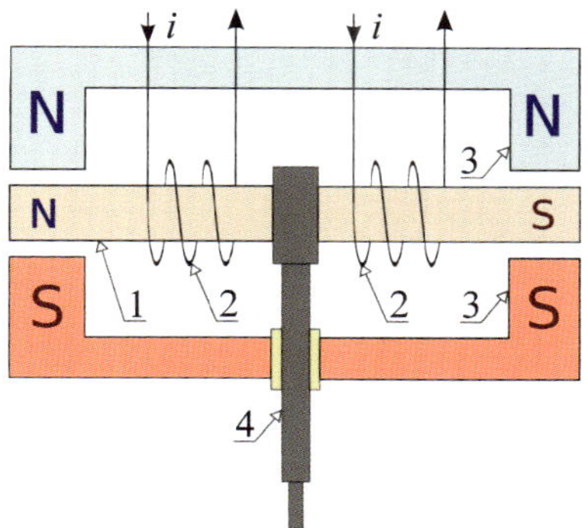

Fig. 6.3   Schematic diagram of a torque motor.

The typical torque motor, schematically shown in Fig. 6.3, is composed of an armature 1 with coils 2 and permanent magnets 3 placed at a distance to both poles of the armature being fixed to a flapper with feedback spring 4. Appearance of electric current in the coils gains the magnetic flow generated by the magnets. Depending on the control signal, one of the ends of the armature is attracted to the neighboring permanent magnet stronger. Angular motion of the armature is transferred onto the feedback spring that moves a spool valve.

Four air gaps create the dominant reluctance in the magnetic circuit of the torque motor. It results from the negligible reluctance of poles that can be omitted. With respect to subsystems' symmetry, the air gaps placed on the opposite sides of the diagonal have the same width. Reluctance R_1 and R_2 on both sides of the armature can be described by these formulas:

$$R_1 = \frac{x_0 - x_a}{\mu_0 A}, \quad R_2 = \frac{x_0 + x_a}{\mu_0 A}, \quad x_a = \vartheta \frac{L}{2}. \tag{6.9}$$

In Eq. (6.9), the following parameters are assumed: x_0 – width of air gap in the neutral position of the armature [m], x_a – displacement of the armature end [m], μ_0 – magnetic permeability of the air gaps [V s/(A m)], A – surface of the air gap in the direction perpendicular to the magnetic field between poles [m^2], ϑ – angle of rotation of the armature [rad].

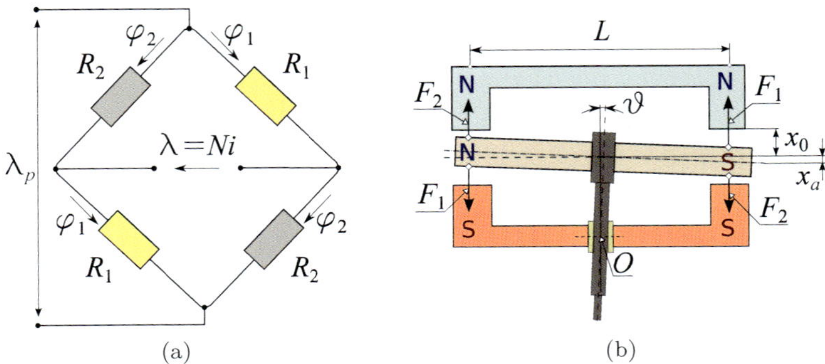

Fig. 6.4   Analogy between the electric circuit (a) and an approximate distribution of the magnetic field (b). Coils are neglected.

The magnetic circuit of the torque motor from Fig. 6.4b is symmetric, therefore the magnetic fluxes on both sides of the diagonals are identical. The observation of correspondence between the electric (a) and magnetic field (b) allows to refer an electrical resistance to the magnetic reluctance. As a result of the existing analogy, a distribution of forces of the magnetic field propagated in the torque motor is approximated by the electric circuit in Fig. 6.4a.

Equating the magnetomotive forces in electric circuit's loops yields:

$$-\lambda + R_1\varphi_1 - R_2\varphi_2 = 0, \quad -\lambda_p + R_1\varphi_1 + R_2\varphi_2 = 0, \tag{6.10}$$

where magnetic fluxes are expressed with:

$$\varphi_1 = \frac{\lambda_p + \lambda}{2R_1}, \quad \varphi_2 = \frac{\lambda_p - \lambda}{2R_2}, \tag{6.11}$$

and $\lambda = iN$ is the magnetomotive force produced by the coil current i [A], λ_p – magnetomotive force of the permanent magnet [A], N – a number of coil windings of the permanent magnet.

Substituting Eq. (6.9) to (6.11), the equations for magnetic fluxes in the air gaps are found:

$$\varphi_1 = \frac{(\lambda_p + iN)\mu_0 A}{2(x_0 - x_a)}, \quad \varphi_2 = \frac{(\lambda_p - iN)\mu_0 A}{2(x_0 + x_a)}. \tag{6.12}$$

Forces F_1 and F_2 acting on ends of the armature produce a moment of forces T against the point O, which is calculated as follows:

$$F_1 = \frac{\varphi_1^2}{2\mu_0 A}, \tag{6.13a}$$

$$F_2 = \frac{\varphi_2^2}{2\mu_0 A}, \tag{6.13b}$$

$$F = F_1 - F_2 = \frac{\varphi_1^2 - \varphi_2^2}{2\mu_0 A}, \tag{6.13c}$$

$$T = FL = \frac{L}{2\mu_0 A}\left[\frac{(\lambda_p + iN)^2\mu_0^2 A^2}{4(x_0 - x_a)^2} - \frac{(\lambda_p - iN)^2\mu_0^2 A^2}{4(x_0 + x_a)^2}\right], \tag{6.13d}$$

then

$$T = \frac{\mu_0 AL}{8(x_0^2 - x_a^2)^2} \left[(\lambda_p + iN)^2 (x_0 + x_a)^2 - (\lambda_p - iN)^2 (x_0 - x_a)^2 \right]. \qquad (6.14)$$

The magnetomotive force λ generated by the current in windings of coils is relatively small in comparison to the magnetomotive force λ_p generated by the permanent magnet. Analogously, the displacement x_a of the armature's ends is significantly smaller than x_0 – width of the air gap. According to the relations, the influence of terms λ^2 and x_a^2 on the moment T with regard to the more significant influence of terms λ_p^2 and x_0^2 is negligible.

Continuation of derivation of the formula for the moment of forces acting on the armature takes the form

$$T = \frac{\mu_0 AL}{8x_0^4} \left[4x_0 x_a \lambda_p^2 + 4iN\lambda_p x_0^2 \right] = \frac{\lambda_p^2 \mu_0 AL}{2x_0^3} x_a + \frac{N\lambda_p \mu_0 AL}{2x_0^2} i. \qquad (6.15)$$

Equation (6.15) is now transformed to the sum of linear terms

$$T = K_x x_a + K_i i \quad \text{or} \quad T = K_e \vartheta + K_i i, \qquad (6.16)$$

where:

$$\vartheta = \frac{2x_a}{L}, \quad K_x = \frac{\lambda_p^2 \mu_0 AL}{2x_0^3}, \quad K_i = \frac{N\lambda_p \mu_0 AL}{2x_0^2}, \quad K_e = \frac{\lambda_p^2 \mu_0 AL^2}{4x_0^3}.$$

In fact, the term $K_e \vartheta$ (servo valve parameter) is very small in relation to $K_i i$, thus the Eq. (6.16) can be simplified to the form

$$T = K_i i. \qquad (6.17)$$

The above mathematical description of the torque motor will be used in next Sec. 6.4.

6.3 Piezoelectric Plate Transducer

The servo analyzed in [Piefort (2001)] is schematically shown in Fig. 6.5. The principle of operation of such a servo bases on a continuous deformation of an elastic plate transducer subjected to the bending force applied in the point of attachment of the linear displacement vector x_m. The piezoelectric plate $\underline{1}$ functions as a flapper placed between nozzles $\underline{2}$. Restricted motion of the plate enforces the position of the spool $\underline{3}$. Its displacement is measured by the position transducer $\underline{4}$. The throttle valves $\underline{5}$ are also marked.

The following bending force acts on the free end of transducer's plate

$$F(t) = -k_m x_m(t) + d \cdot k_m U(t), \qquad (6.18)$$

which is the effect of the difference of pressures of fluids flowing through the nozzles $\underline{2}$.

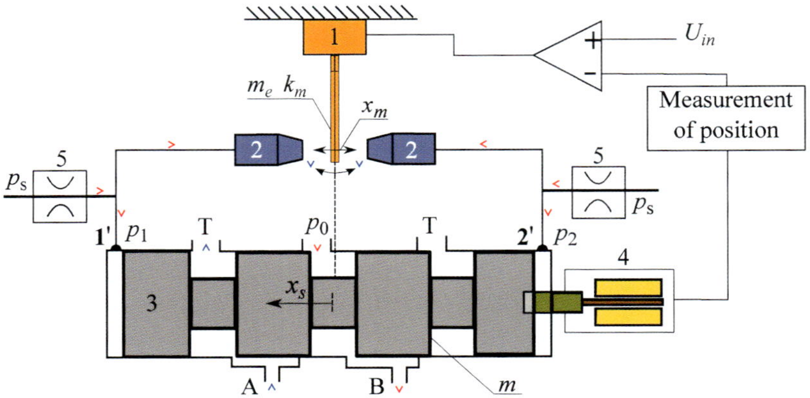

Fig. 6.5   A servo with the piezoelectric plate transducer.

The equation of motion of the piezoelectric element between the nozzles is written with the use of the second order differential equation

$$m_e \ddot{x}_m(t) + c_t \dot{x}_m(t) = F(t) + F_s(t),$$

where, after taking Eq. (6.18),

$$m_e \ddot{x}_m(t) + c_t \dot{x}_m(t) + k_m x_m(t) = F_s(t) + d \cdot k_m U(t). \tag{6.19}$$

State variables and parameters of the piezoelectric plate transducer used in Eqs. (6.18) and (6.19) are as follows: x_m – linear displacement of the end of the piezoelectric element [m], m_e – effective mass of the piezoelectric element [kg], F_s – force of external loading (e.g., from the spool in Fig. 6.6) [N], k_m – coefficient of stiffness [N/m], c_t – coefficient of motion resistance [N·s/m], d – parameter of shear deformation of the piezoelectric crystal [m/V], U – electric voltage on the plates of the piezoelectric element [V].

While introducing a simplified electric model of the transducer, one assumes that it is built as a resistor-capacitor serial connection. At this assumption, the differential equation describing the dynamics of the electric voltage on plates of the piezoelectric element is given in the form

$$RC\dot{U}(t) + U(t) = U_z(t), \tag{6.20}$$

where U_z is supply voltage, U – voltage on the piezoelectric plates, R – resistance to the currents flowing in the electric circuit and plates of the piezoelectric element, C – electrical capacitance of plates of the piezoelectric element.

With respect to the sum of terms of the right-hand side of Eq. (6.19)b, a two-dimensional vector of forcing can be extracted. Therefore, one writes two transfer functions between input F_s and output x_m and between input U and the same output. If one assumes that F_s does not exist, then, using Eqs. (6.19) and (6.20), the linear dynamical system of the analyzed piezoelectric plate transducer is given in the operator form

$$G(s) = \frac{x_m(s)}{U_z(s)} = \frac{K_z \cdot \omega_0^2}{(RCs + 1)(s^2 + 2\zeta\omega_0 s + \omega_0^2)}, \tag{6.21}$$

in which: $\omega_0 = \sqrt{k_m/m_e}$ – natural frequency, $\zeta = \dfrac{c_t}{2\sqrt{m_e k_m}}$ – damping ratio, $K_z = d \cdot k_m$ – proportional gain.

If some influence of friction and any hydrodynamic forces is neglected as well, then the equation of the spool motion is given as follows

$$m_s\ddot{x}_s + c_s\dot{x}_s = \xi A_s x_m, \tag{6.22}$$

where x_s is a displacement of the spool, m_s – mass of the spool and the moved working fluid, c_s – damping parameter of the spool motion, A_s – area of the spool face from the side of ports $\underline{1}$' and $\underline{2}$', ξ – scaling coefficient $[\text{N/m}^3]$.

The differential equation (6.22) is linear and has an equivalent operator form

$$x_s(s) = \frac{\xi A_s}{s(m_s s + c_s)} x_m(s) = \frac{\xi_s}{s(T_s s + 1)} x_m(s), \tag{6.23}$$

where $T_s = m_s/c_s$ and $\xi_s = \xi A_s/c_s$.

6.4 Control System of Load Positioning Using a Hydraulic Servo Valve

An exemplary physical model of servo valve's control system has been introduced in [Rabie (2009)]. A scheme of a similar system is shown in Fig. 6.6. It has served for realization of a `LabVIEW` simulation model [Bialkowski (2014)].

Dynamics of motion of the armature is described by the moment balance equation

$$J\ddot{\vartheta} + f_\vartheta\dot{\vartheta} + K_T\vartheta + T_P + T_F + T_L = T, \tag{6.24}$$

in which T is given by Eq. (6.17), and also

$$T_p = \frac{\pi}{4}d_f^2(P_2 - P_1)L_f, \tag{6.25a}$$

$$T_F = F_s L_s = K_s(L_s\vartheta + x)L_s, \tag{6.25b}$$

$$T_L = \begin{cases} 0 & |x_f| < x_i, \\ R_s\dot{\vartheta} - (|x_f| - x_i)K_{\text{Lf}}L_f \, \text{sgn}\,(\vartheta) & |x_f| > x_i, \end{cases} \tag{6.25c}$$

$$x_f = L_f\vartheta. \tag{6.25d}$$

Equation (6.25)b defines the backward moment, and Eq. (6.25)c, which is valid in the interval of variability of x_f, defines extreme positions of the flapper. The simulation model of the torque motor (at the top of Fig. 6.6) is shown on a scheme in Fig. 6.7.

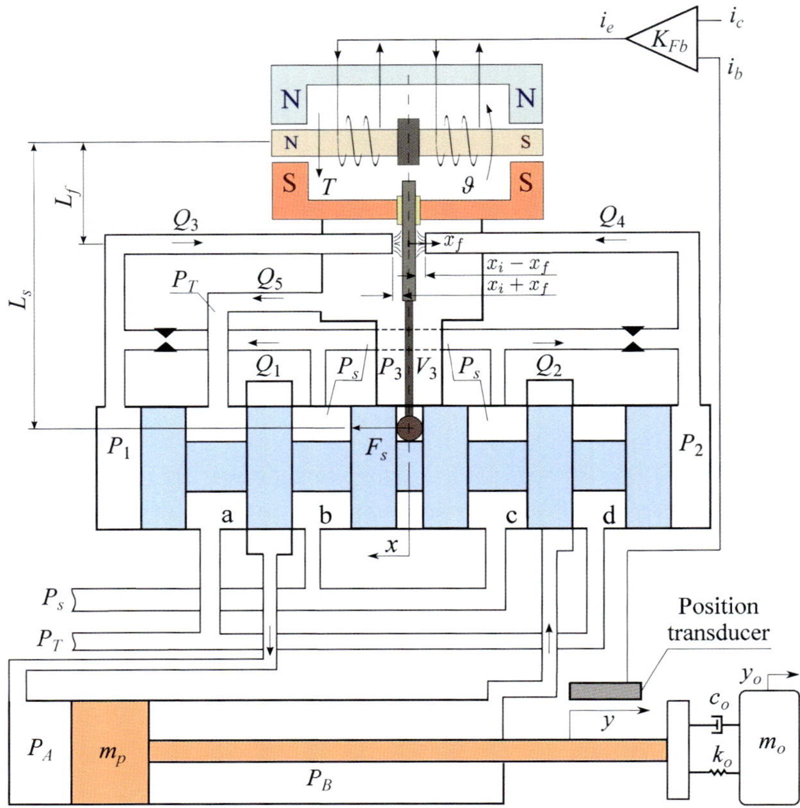

Fig. 6.6   An exemplary physical model of a servomechanism. Numerical solution of dynamics of the system behavior is given in Sec. 6.5.

Definitions of state variables and system parameters appeared in Eq. (6.24) and (6.25): T_p – torque caused by the difference of pressures on both sides of the flapper [N·m], T_F – torque of force F_s caused by the spool motion [N·m], T_L – torque generated by the flapper [N·m], ϑ – angle of rotation of the armature [rad], x_i – limit position of the flapper [m], x_f – displacement of end of the flapper [m], x – displacement of the spool [m], P_i – pressure on both sides of the spool ($i = 1, 2$) [Pa], K_T – stiffness coefficient of feedback spring of the torque motor [N·m/rad], K_{Lf} – stiffness coefficient of the flapper seat [N/m], K_s – stiffness coefficient of the feedback spring [N/m], J – moment of inertia of the system armature-flapper [kg·m^2], F_s – spool force exerted on the end of the piezoelectric element [N], L_f – length of the flapper measured from the armature's axis to the axis of nozzles [m], L_s – total length of torque motor's feedback spring and the flapper [m], f_ϑ – damping ratio of motion of the armature [N·s·m/rad], R_s – damping ratio of the flapper seat [N·s·m/rad], d_f – hydraulic diameters of nozzles [m].

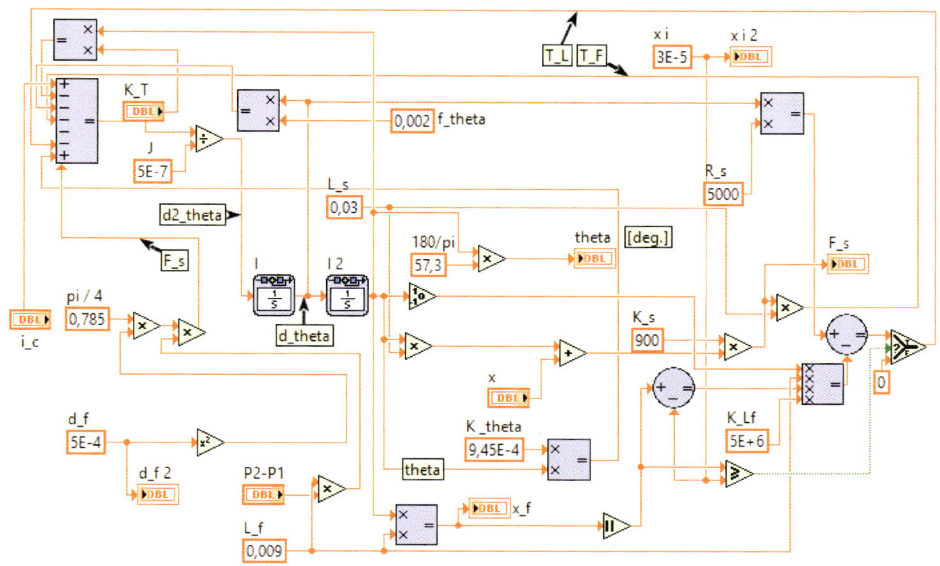

Fig. 6.7   Simulation model created in **LabVIEW**, solving torque motor's dynamics.

A mathematical description of remaining components of the modeled system of the servomechanism from Fig. 6.6 is presented below in points a-h:

a) flow rates through the spool valve (see Fig. 6.9):

$$Q_a = C_d A_a(x)\sqrt{2(P_A - P_T)/\rho}, \tag{6.26a}$$

$$Q_b = C_d A_b(x)\sqrt{2(P_s - P_A)/\rho}, \tag{6.26b}$$

$$Q_c = C_d A_c(x)\sqrt{2(P_s - P_B)/\rho}, \tag{6.26c}$$

$$Q_d = C_d A_d(x)\sqrt{2(P_B - P_T)/\rho}; \tag{6.26d}$$

b) flow rates in lines connected to the valve (see Fig. 6.8):

$$Q_1 = C_{12}\sqrt{P_s - P_1}, \tag{6.27a}$$

$$Q_2 = C_{12}\sqrt{P_s - P_2}, \tag{6.27b}$$

$$Q_3 = C_{34}(x_i + x_f)\sqrt{P_1 - P_3}, \tag{6.27c}$$

$$Q_4 = C_{34}(x_i - x_f)\sqrt{P_2 - P_3}, \tag{6.27d}$$

$$Q_5 = C_5\sqrt{P_3 - P_T}, \tag{6.27e}$$

where: $C_{12} = C_d A_0\sqrt{2/\rho}$, $C_{34} = C_d \pi d_f\sqrt{2/\rho}$, $C_5 = C_d A_5\sqrt{2/\rho}$;

c) equations of flow continuity in the chambers around the flapper (see Fig. 6.10):

$$Q_1 - Q_3 + A_s\dot{x} = \frac{V_0 - A_s x}{B}\dot{P}_1, \tag{6.28a}$$

$$Q_2 - Q_4 - A_s\dot{x} = \frac{V_0 + A_s x}{B}\dot{P}_2, \tag{6.28b}$$

$$Q_3 + Q_4 - Q_5 = \frac{V_3}{B}\dot{P}_3; \tag{6.28c}$$

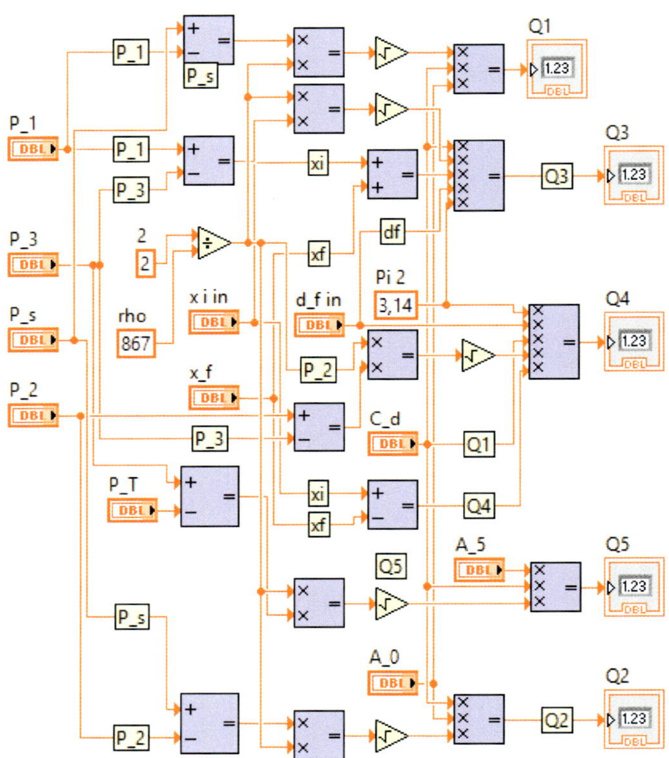

Fig. 6.8   Simulation model created in `LabVIEW`, computing the flow rates $Q_{1,\dots,5}$.

d) the equation of the dynamics of the spool motion (see Fig. 6.10)

$$A_s(P_2 - P_1) = m_s\ddot{x} + f_s\dot{x} + F_s; \tag{6.29}$$

e) the equation of the dynamics of the piston motion (see Fig. 6.11)

$$A_p(P_A - P_B) = m_p\ddot{y} + f_p\dot{y} + K_b y; \tag{6.30}$$

f) equations of flow continuity in chambers of the cylinder (see Fig. 6.12):

$$Q_b - Q_a - A_p\dot{y} - (P_A - P_B)/R_i = (V_c + A_p y)/B \cdot \dot{P}_A, \tag{6.31a}$$

$$Q_c - Q_d - A_p\dot{y} + (P_A - P_B)/R_i = (V_c - A_p y)/B \cdot \dot{P}_B; \tag{6.31b}$$

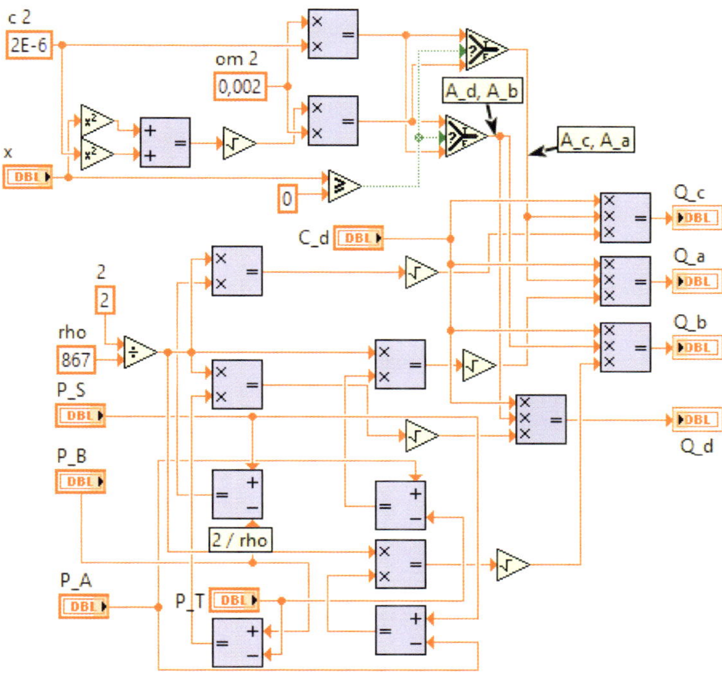

Fig. 6.9   Simulation model created in **LabVIEW**, computing flow rates through the spool valve.

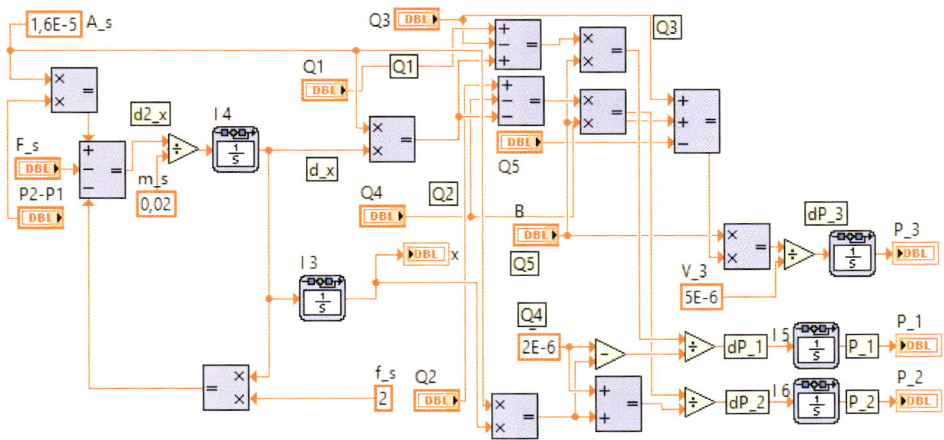

Fig. 6.10   Simulation model created in **LabVIEW**, solving the equations of flow continuity and the dynamics of the spool motion.

g) limitations for the valve:

$$\left.\begin{aligned} A_a = A_c &= \beta r_s \\ A_b = A_d &= \beta\sqrt{(x^2 + r_s^2)} \end{aligned}\right\} \qquad \text{if } x \geq 0, \tag{6.32}$$

$$\left.\begin{aligned} A_a = A_c = \beta\sqrt{(x^2 + r_s^2)} \\ A_b = A_d = \beta r_s \end{aligned}\right\} \quad \text{if } x < 0; \tag{6.33}$$

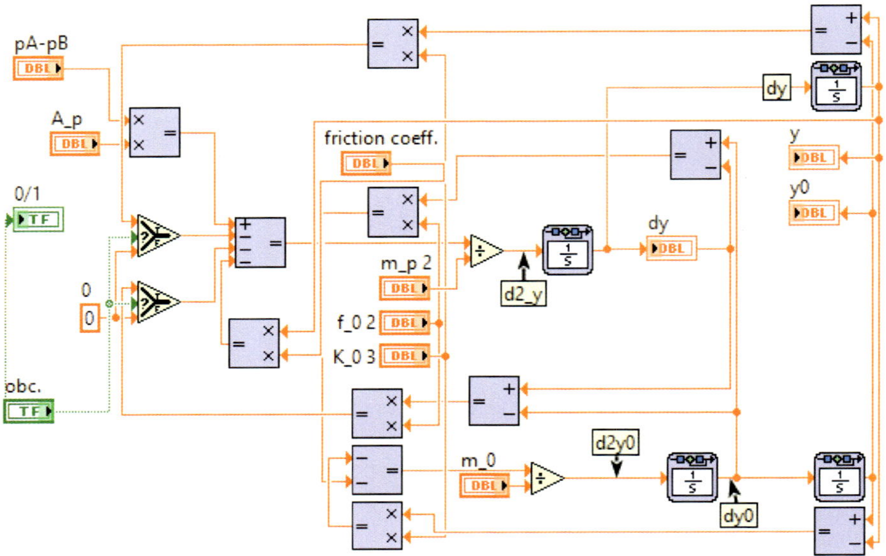

Fig. 6.11   Simulation model created in `LabVIEW`, solving dynamics of loaded piston.

h) electric feedback (see Fig. 6.13)

$$i_e = i_c - i_b, \quad i_b = K_{\text{Fb}}y. \tag{6.34}$$

As it is seen, a body that loads the hydraulic cylinder has been introduced. Viscous damping coefficient c_o of the servo load and its stiffness k_o is properly selected to imitate elasticity of real mounting.

Differential equations of the dynamics of the elastically connected load and hydraulic cylinder's rod clevis (for instance, like in Fig. 6.2) describe changes in displacements y_o and y, respectively:

$$m_o\ddot{y}_o = c_0(\dot{y} - \dot{y}_o) + k_o(y - y_o), \tag{6.35a}$$

$$m_p\ddot{y} = c_o(\dot{y}_o - \dot{y}) + k_o(y_o - y) + A_p(P_A - P_B). \tag{6.35b}$$

Small elasticity of the load and the hydraulic cylinder is guaranteed by large values of parameters k_o and c_o. Equations (6.35), realized in the simulation model, are shown in Fig. 6.11.

State variables and system parameters used in Eqs. (6.24)–(6.35) are as follows: i_c – input current (reference value) [A], i_b – current on output of the position transducer [A], i_e – error current [A], y – piston displacement [m], K_i – torque constant dependent on the current in the armature coils [N · m/A], K_e – gain constant of the

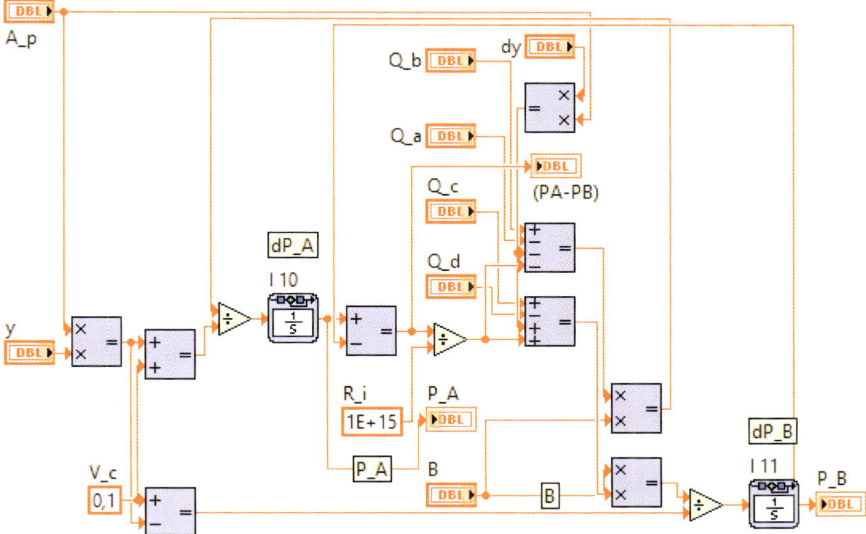

Fig. 6.12   Simulation model created in LabVIEW, solving equations of flow continuity in chambers of the cylinder.

relation angle-moment of the armature $[\text{N} \cdot \text{m/rad}]$, K_{Fb} – proportional gain of the controller $[\text{A/m}]$, d_5 – diameter of the return line $[\text{m}]$, d_s – diameter of the spool $[\text{m}]$, ρ – oil density $[\text{kg/m}^3]$, B – oil bulk modulus of elasticity $[\text{Pa}]$, V_3 – initial volume of oil in the return chamber $[\text{m}^3]$, m_s – spool mass $[\text{kg}]$, m_p – mass of the piston and the rod with clevis $[\text{kg}]$, β – width of spool's port $[\text{m}]$, r_s – radial clearance of the spool $[\text{m}]$, f_s – kinetic friction coefficient on the contact surface of the spool and the housing of the proportional valve $[\text{N} \cdot \text{s/m}]$, f_p – kinetic friction coefficient on the contact surfaces of the piston and the cylinder $[\text{N} \cdot \text{s/m}]$, P_s – supply pressure $[\text{Pa}]$, P_T – ambient pressure $[\text{Pa}]$, A_p – area of the piston $[\text{m}^2]$, V_c – initial volume of oil in cylinder's chamber $[\text{m}^3]$, R_i – resistance to external leakages $[\text{Pa} \cdot \text{s/m}^3]$, k_o – stiffness of connection of piston's clevis and load $[\text{N/m}]$, c_o – viscous damping coefficient of the servo load $[\text{N} \cdot \text{s/m}]$.

6.5   Numerical Simulations of the Dynamics of an Electrohydraulic Servo Subjected to Dynamic Loading

The numerical simulations of the dynamics of the electrohydraulic servo presented in Fig. 6.6 were performed in LabVIEW.

The abovementioned programming environment gives access to many libraries called *virtual instruments*. It supports hardware based on GPIO input/output that is applied in many industrial measurement, control and monitoring systems. The *G programming language*, introduced by NI, uses a graphical user interface (GUI) having a form of a block diagram. Routes for data transfer between the functional blocks representing the `LabVIEW` procedures are marked with lines, the color, thickness and pattern of which define the type of transferred data (scalar, vector, matrix). A simulation diagram of the dynamics of the analyzed servo, created in `LabVIEW`, is shown in Fig. 6.13.

By using `LabVIEW`, a numerical solution to algebraic and differential equations can be found. The environment is equipped with many components for fast and clear presentation of structural and numerical data on various displays and scalable time graphs placed in control panel. The control panel presented in Fig. 6.14 allows to observe the solution and modify parameters of the numerical simulation also during running simulation program.

A reference value of the control current i_c can be adjusted in the control panel using a slider. Proper setting of this input causes backward or forward displacement of the piston with regard to its zero initial position. With respect to the application of the automatic control system with a feedback loop, the piston position follows the reference value i_c by means of the proportional regulation of the displacement of torque motor's end.

By investigating the step and linear input responses of the analyzed dynamical system, the efficiency of positioning of a body of the mass m_o attached to hydraulic cylinder's rod can be evaluated. Results of exploration of system's simulation model (see diagram 6.13) at external loading $m_o = \{50, 150, 300\}$ kg of the piston in a response to the step (from zero) and timely linear change in the control current $i_c = \{40, 70, 100\}$ mA are illustrated in Fig. 6.15-6.19. The ranges of the axis of ordinates of the presented time histories of θ, x, y, P_A and P_B have been fixed in the same range for better visualization of common relations between the variables, differences in the shape of oscillations as well as visualization of discrepancy between some steady-state values reached at the end of each numerical experiment.

During elaboration of the servo's model described in Sec. 6.4, dynamical and design parameters were defined, which take the following values in the simulation:

$$
\begin{array}{lll}
K_i = 0.556\,\text{N} \cdot \text{m/A} & f_\vartheta = 0.002\,\text{N} \cdot \text{m} \cdot \text{s/rad} & K_T = 1000\,\text{N} \cdot \text{m/rad} \\
J = 5\text{e}7\,\text{kg} \cdot \text{m}^2 & K_e = 9.45\text{e–}4\,\text{N} \cdot \text{m/rad} & K_{\text{Fb}} = 3\,\text{A/m} \\
L_f = 9\text{e–}3\,\text{m} & L_s = 30\text{e–}3\,\text{m} & x_i = 30\text{e–}6\,\text{m} \\
d_f = 0.5\text{e–}3\,\text{m} & R_s = 5\text{e}3\,\text{N} \cdot \text{s} \cdot \text{m/rad} & K_{Lf} = 5\text{e}6\,\text{N/m} \\
d_5 = 0.6\text{e–}3\,\text{m} & f_p = 0.11\,\text{N} \cdot \text{s/m} & \rho = 867\,\text{kg/m}^3 \\
B = 1.5\text{e}9\,\text{Pa} & d_s = 4.6\text{e–}3\,\text{m} & V_3 = 5\text{e–}9\,\text{m}^3 \\
m_s = 0.02\,\text{kg} & \beta = 2\text{e–}3\,\text{m} & r_s = 2\text{e–}6\,\text{m} \\
f_s = 2\,\text{N} \cdot \text{s/m} & K_s = 900\,\text{N/m} & P_s = 25\text{e}6\,\text{Pa} \\
P_T = 0\,\text{Pa} & A_p = 12.5\text{e–}4\,\text{m}^2 & V_c = 500\text{e–}6\,\text{m}^3 \\
R_i = 1\text{e}14\,\text{Pa} \cdot \text{s/m}^3 & m_p = 10\,\text{kg} & c_o = 10\text{e}3\,\text{N} \cdot \text{s/m} \\
k_o = 100\text{e}3\,\text{N/m}. & &
\end{array}
$$

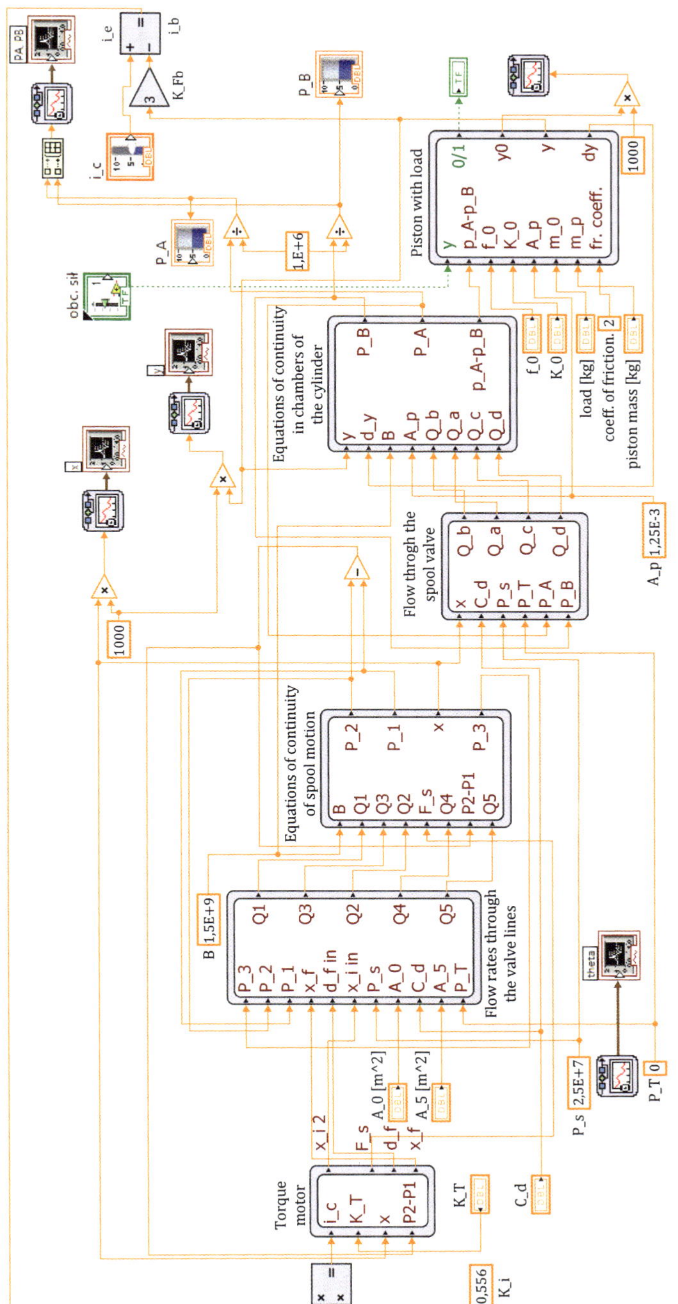

Fig. 6.13   General simulation model solving dynamics of the investigated servomechanism.

The conducted experimental tests, relying on placing the piston with load not greater than $150\,\mathrm{kg}$ in the position y, exhibit satisfactorily stable responses of the piston reflected in fairly small oscillations of selected state variables of the analyzed system, i.e., a) the angle θ of the feedback spring of the torque motor, b) the displacement x of the spool of the proportional valve, c) pressures P_A and P_B in the cylinder's chambers on both sides of the piston.

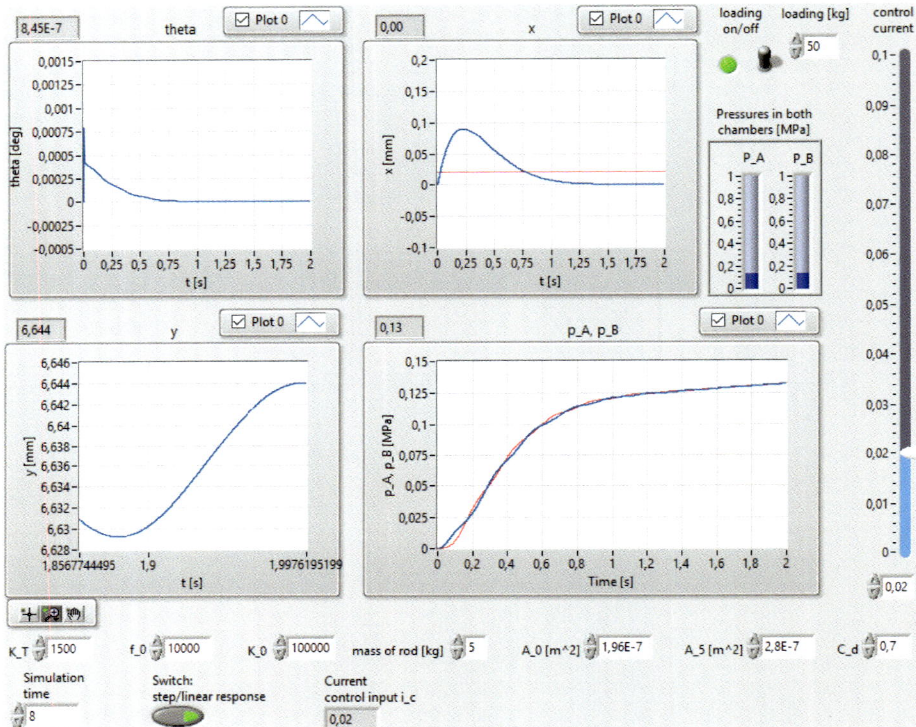

Fig. 6.14   Front panel created in `LabVIEW`, used for monitornig of the servo valve.

Attachment of the mass $m_o = 300\,\mathrm{kg}$ makes some very unstable response of the servo, manifested by oscillations of all observable state variables, visible. In particular, an attempt of forcing the body to the position $y \approx 34\,\mathrm{mm}$ ($i_c = 100$ mA) causes appearance of slowly damped oscillations of $y(t)$ around a steady-state value (see red line in Fig. 6.17c).

The time histories presented in Figs. 6.15d–6.19d illustrate variations of pressures in the cylinder on both sides of the piston. It is clearly seen that at a step-shaped function of excitation and at some loading of the plant realized by an attached body of the mass $m_o = 300\,\mathrm{kg}$, the control system having some predetermined parameters does not allow for a quick and stable positioning of the system to the reference input values.

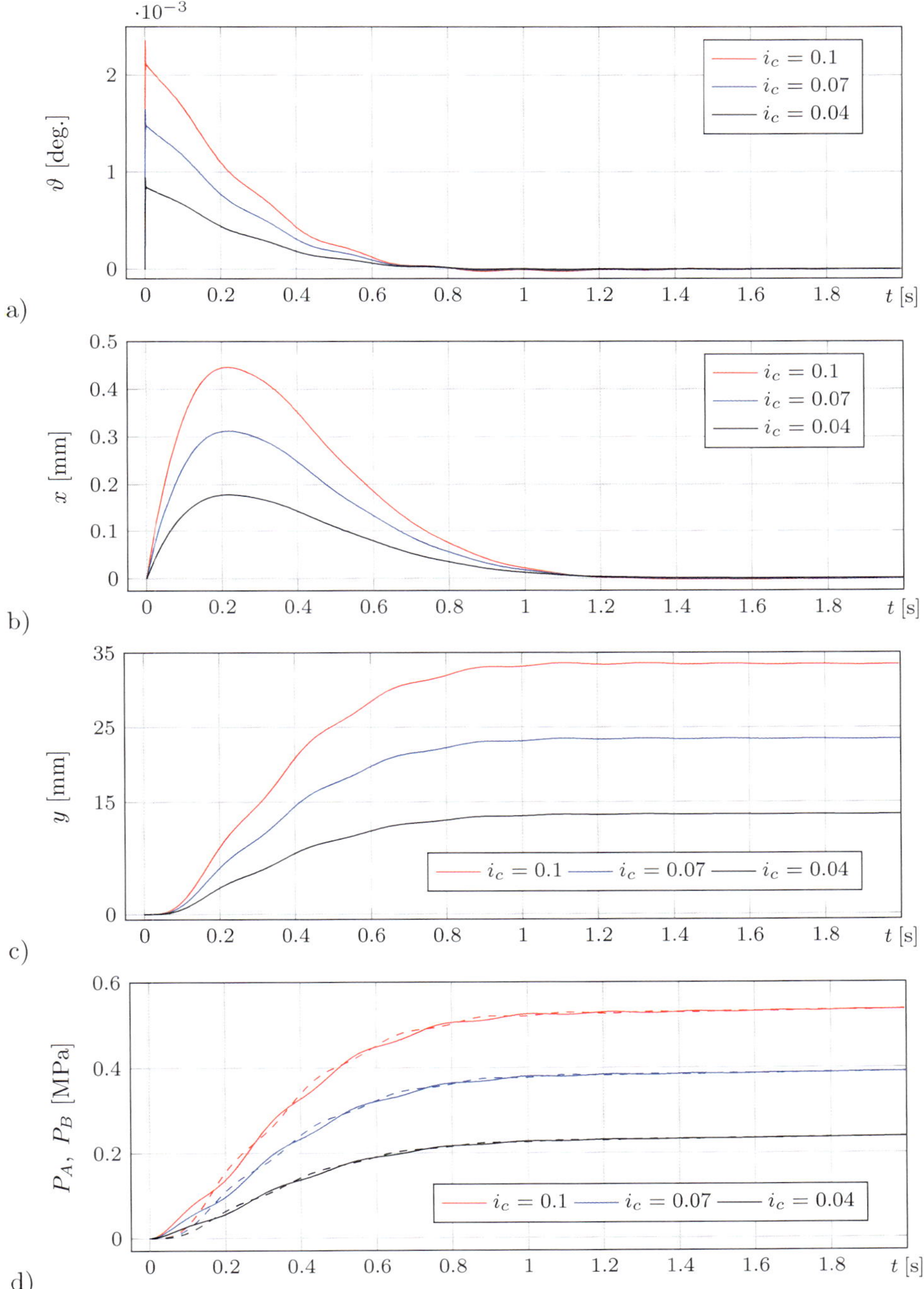

Fig. 6.15   Results of exploration of the simulation model, given in Fig. 6.13, as an effect of the system response to a step change in the reference current at loading mass $m_o = 50\,\text{kg}$. In figure d, variations of pressure P_A are marked with solid lines while variations of pressure P_B by dashed lines.

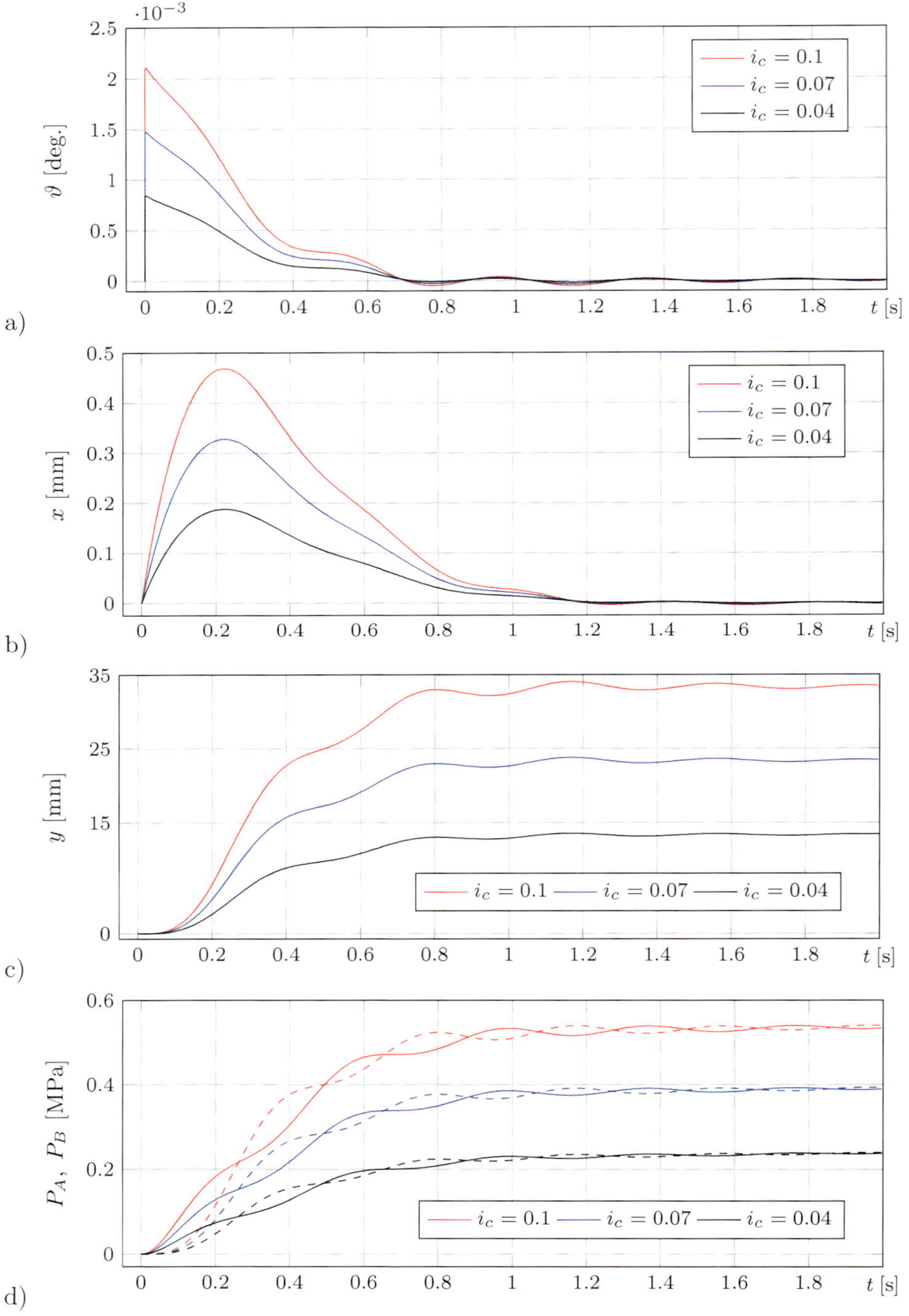

a)

b)

c)

d)

Fig. 6.16   Results of exploration of the simulation model, given in Fig. 6.13, as an effect of the system response to a step change in the reference current at loading mass $m_o = 150\,\mathrm{kg}$.

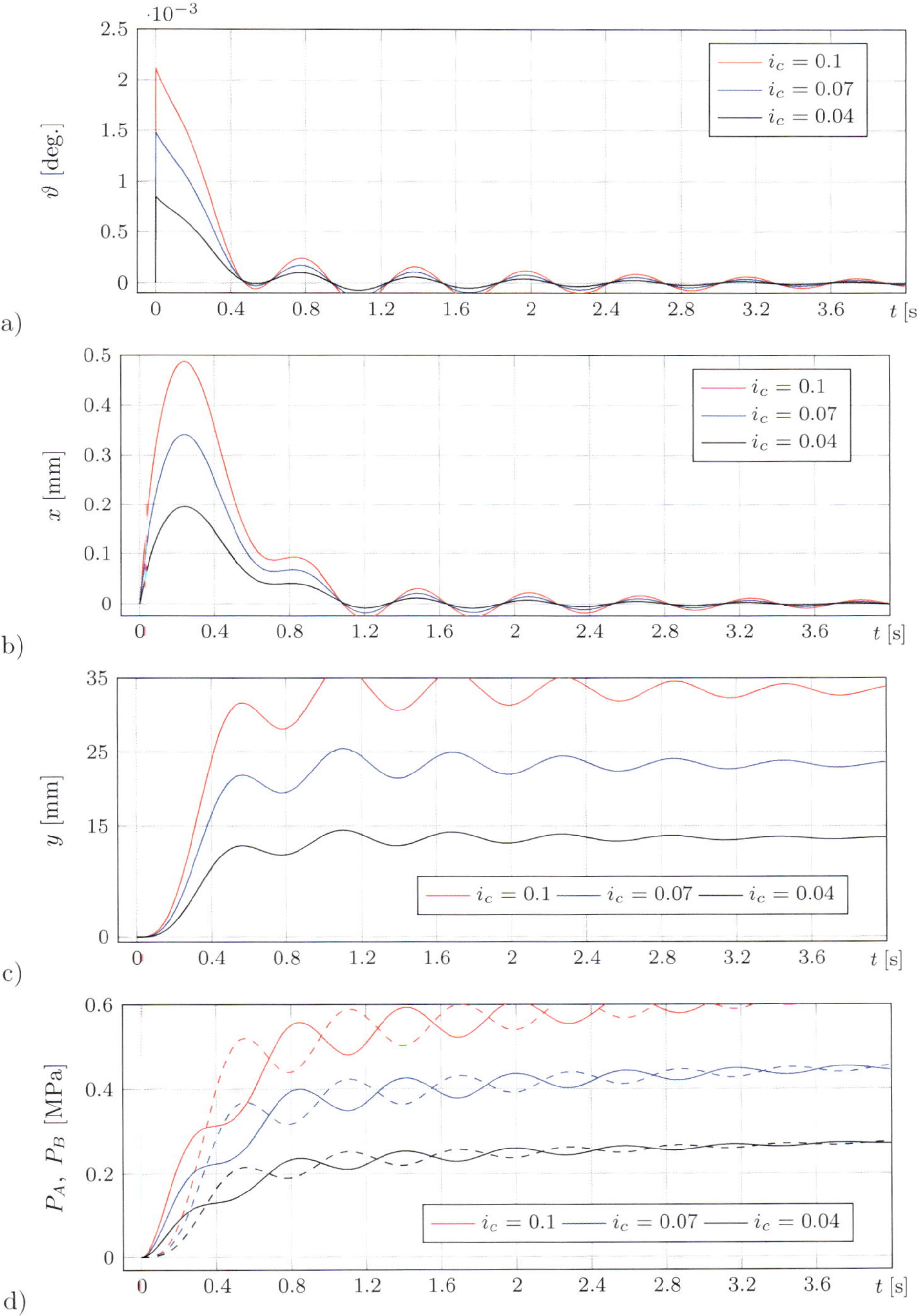

Fig. 6.17   Results of exploration of the simulation model, given in Fig. 6.13, as an effect of the system response to a step change in the reference current at loading mass $m_o = 300$ kg.

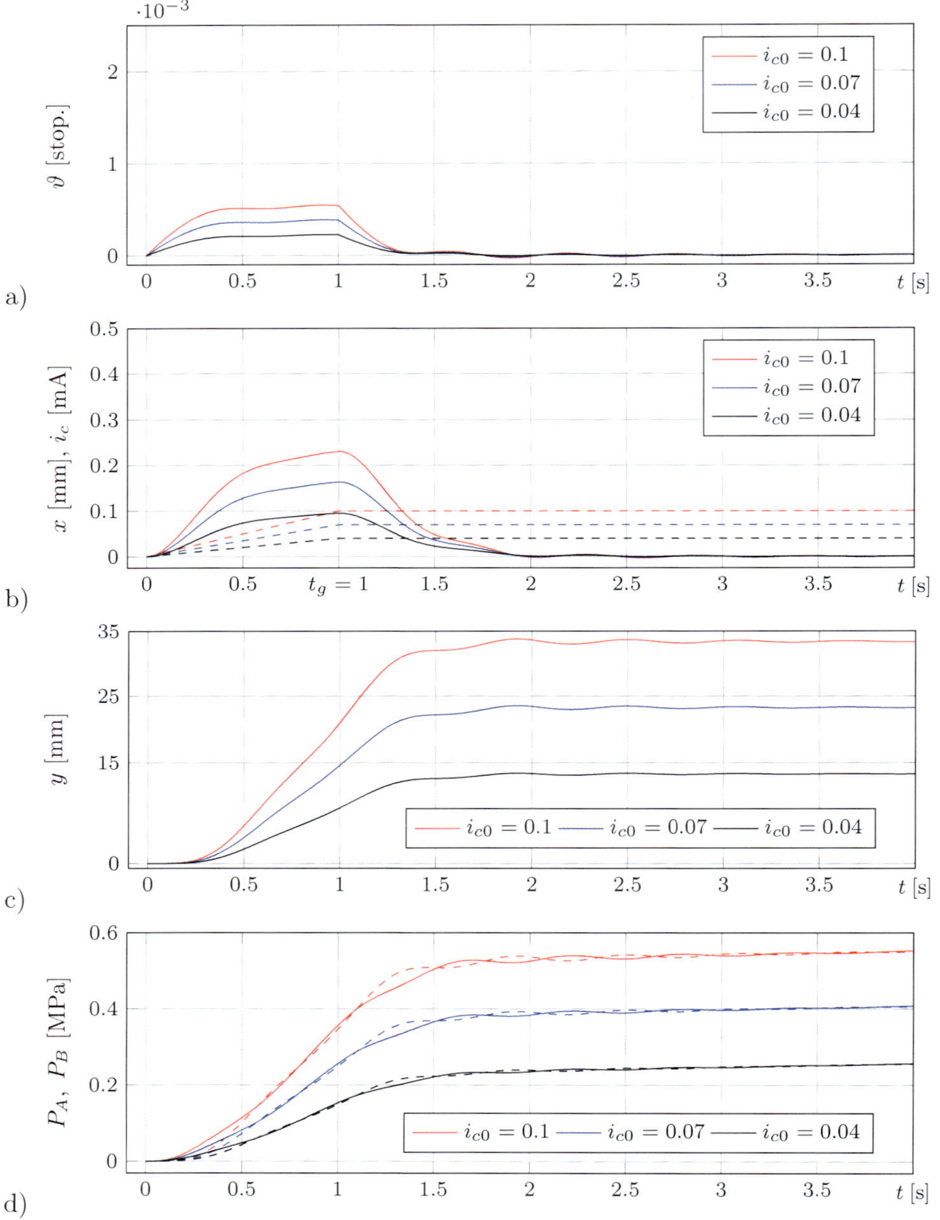

Fig. 6.18   Linear input responses of the servo due to a linearly rising reference value of control current at loading mass $m_o = 300\,\text{kg}$. The reference current characteristics, marked with a dashed line in figure b, is determined by a piecewise continuous function

$$i_c(t) = \begin{cases} i_{c0}t & \text{if} \quad t \in [0,1), \\ i_{c0} & \text{if} \quad t \in [1,4]. \end{cases} \tag{6.36}$$

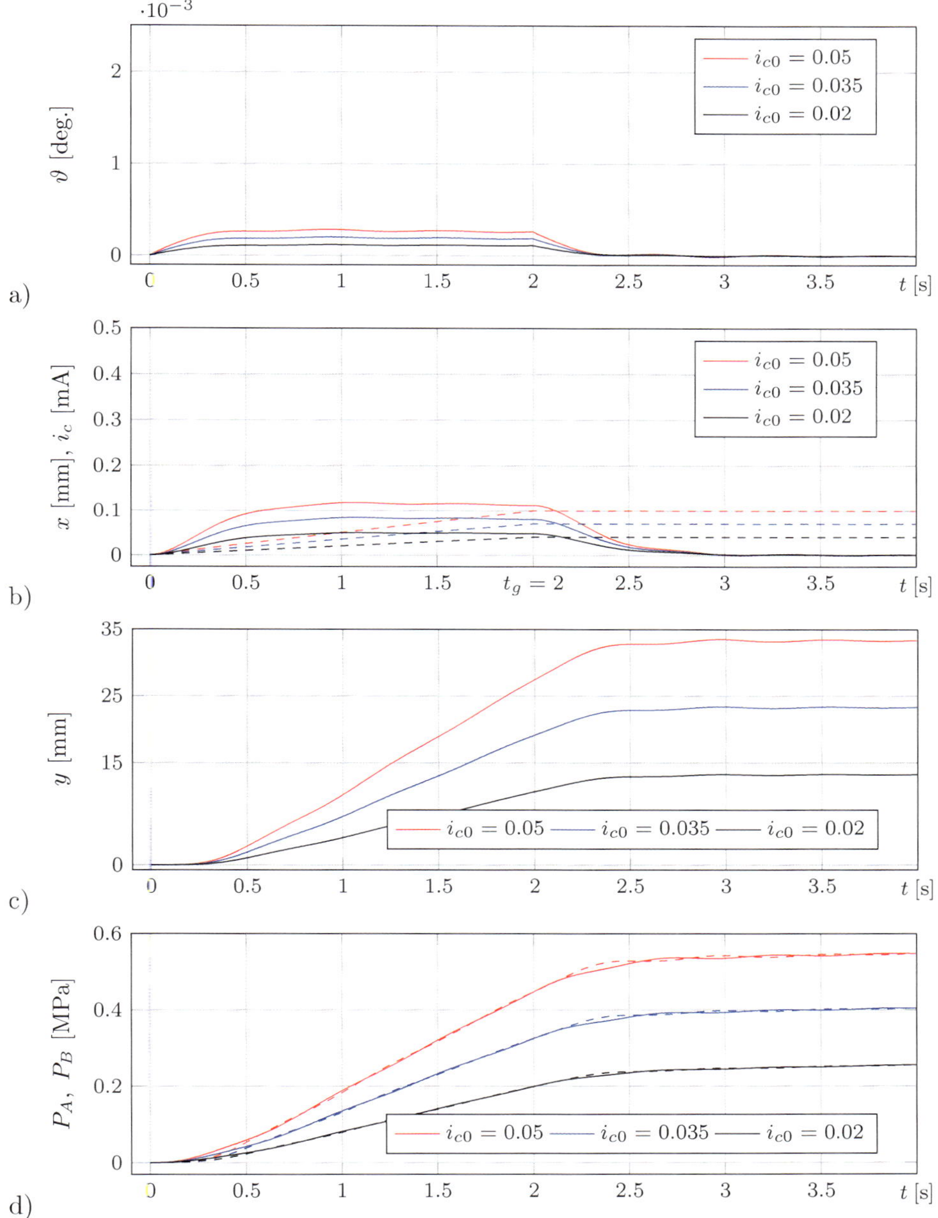

Fig. 6.19 Linear input responses of the servo due to a linearly rising reference value of the control current at loading mass $m_o = 300\,\text{kg}$. The reference current characteristics, marked with a dashed line in figure b, is also determined by a piecewise continuous function

$$i_c(t) = \begin{cases} i_{c0}t & \text{if} \quad t \in [0,2), \\ i_{c0} & \text{if} \quad t \in [2,4]. \end{cases} \tag{6.37}$$

Improvement of control quality is achieved after using a linearly increasing characteristic of the reference current $i_c(t)$ (see Eqs. (6.36) and (6.37)) on the system input, which is drawn using dashed lines in Figs. 6.18b and 6.19b, respectively. Up to time t_g, determined by a threshold value i_{c0} reached by the current i_c, the hydraulic cylinder response very stably follows the reference input. The higher the value of t_g (1 or 2 s during the numerical experiment), the more stable plant output is reported. Therefore, we meet a next optimization problem related to establishing a compromise between the duration of a transient response of the control system and stability of the resulting system response.

According to the above results, some wider scope of applications of the investigated system would require broader research. It could be oriented on more advanced optimization of the first-attempt control algorithm. One would expect to apply more effective algorithm of control for the purpose of shortening of the system transient response while maintaining its high robustness to uncertain disturbances.

To sum up, numerical simulation and control of some typical electrohydraulic servos has been carried out. Advantages and drawbacks of modeling of these kinds of mechatronic systems have been discussed as well as common types of servos have been virtualized to point out some areas of their applications. A few exemplary attempts of discretization of the selected physical models described by systems of ordinary differential equations have been provided. It has been also proved that the `LabVIEW` environment can successively serve as a basic tool to perform even highly complicated numerical simulations of discrete mechatronic systems.

A dependency between the reference current and a displacement time characteristics of the hydraulic cylinder has been particularly investigated. Depending on the value of the force loading cylinder's rod, a very possible step and linear input displacement responses of the investigated servos have been obtained. The closed loop of the considered control system has proved its usefulness, because loaded cylinder's rod has maintained the reference input values at good accuracy.

The designed control system works properly for a fast response positioning control of the hydraulic cylinder loaded by an attached solid body of a mass not greater than 150 kg. Increasing the loading mass above the estimated value results in some rapid decrease in the system stability reflected in undesirable rod oscillations of high amplitudes over the steady-state value. Therefore, the time of positioning of cylinder's rod is longer, causing its unstable convergence to the desired value. Positioning stabilization of loading of the mass equal to 300 kg and grater requires making the transient response of the control system longer by means of application of linearly growing characteristics of the reference input current. The introduced control procedure can be improved by a modification of structure of the control system or even the controller by replacing it with a more effective one based on the Proportional-Derivative (PD) action (see Chap. 12) or a Proportional-Integral-Derivative (PID) action that has been developed in Chap. 10.

Chapter 7

Atom Modeling

7.1 Newtonian Model

We assume a very simple electron model, in which the electron e of the mass m_e moves with velocity $\vec{v}$ around the nucleus along a circular orbit of the radius $\vec{r}$, as it is shown in Fig. 7.1.

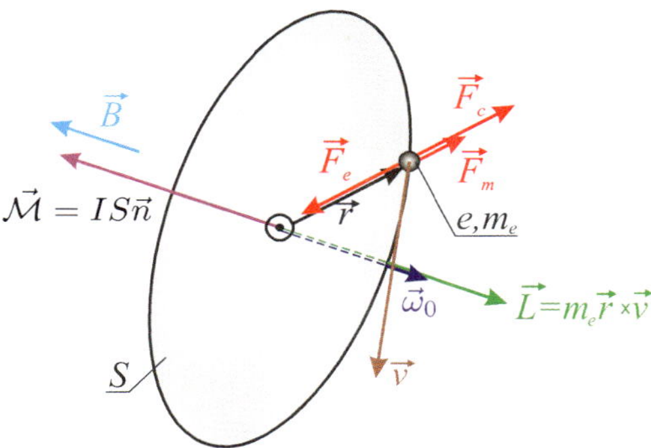

Fig. 7.1   Circular movement of an electron.

A magnetic moment generated by this electron movement is called the *orbital magnetic moment* $\vec{\mathcal{M}}$ of the electron. It can be presented as the following vector

$$\vec{\mathcal{M}} = I\vec{S} = IS\vec{n}, \tag{7.1}$$

where I stands for the molecular current intensity, S denotes the electron orbit surface (dashed in Fig. 7.1), $\vec{n}$ is the unit vector normal to the surface and

$$I = \frac{e}{T} = \frac{e\omega_0}{2\pi} = \frac{ev}{2\pi r}. \tag{7.2}$$

Above, T is the period of the electron circular movement ($[e] = \text{As}$). Since $S = \pi r^2$, we get

$$\mathcal{M} = \frac{erv}{2} = \frac{m_e erv}{2m_e}. \tag{7.3}$$

Equation (7.3) can be presented in its equivalent vector form

$$\vec{\mathcal{M}} = -\gamma \vec{L}, \tag{7.4}$$

where $\gamma = \frac{e}{2m_e}$, is the *gyromagnetic ratio*, and

$$\vec{L} = \vec{r} \times m_e \vec{v} = \vec{r} \times (m_e \vec{\omega}_0 \times \vec{r}) \tag{7.5}$$

denotes the electron angular momentum. The sign minus in Eq. (7.4) appears due to opposite direction of the current and electron movements.

When the vector $\vec{\mathcal{M}}$ (or the circular current loop, equivalently) is put into the uniform magnetic field $\vec{B}$, then $\vec{\mathcal{M}}$ experiences action of the following torque

$$\vec{T} = \vec{\mathcal{M}} \times \vec{B}, \tag{7.6}$$

being perpendicular to both $\vec{\mathcal{M}}$ and $\vec{B}$, what is schematically illustrated in Fig. 7.2.

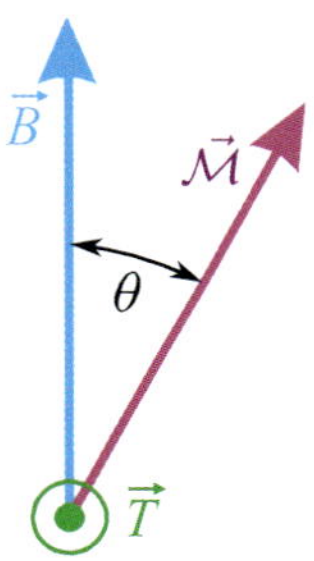

Fig. 7.2   Scheme of position vectors $\vec{\mathcal{M}}, \vec{B}, \vec{T}$.

The incremental potential energy required for the incremental rotation $d\theta$ follows

$$dU = -T d\theta, \tag{7.7}$$

and taking into account Eq. (7.6) we obtain

$$dU = -\mathcal{M}B \sin\theta d\theta. \tag{7.8}$$

Integrating Eq. (7.8) one gets

$$U = -\mathcal{M}B \cos\theta = -\vec{\mathcal{M}} \cdot \vec{B}. \tag{7.9}$$

The electric force F_e can be determined by Coulomb's law

$$F_e e = \frac{1}{4\pi\varepsilon} \frac{e^2}{r^2}. \tag{7.10}$$

The electric force F_e is balanced by the centrifugal force F_c, and hence

$$F_e = F_c = m\omega_0^2 r. \tag{7.11}$$

It means that the electron dynamical configuration is defined by $\vec{F}_e - \vec{F}_c = 0$.

Assume that we consider an electron embedded in the uniform magnetic field $\vec{B}$. The electron dynamics configuration is changed due to action of the Lorentz force

$$\vec{F}_m = -e\left(\vec{v} \times \vec{B}\right),\tag{7.12}$$

and the vectors $\vec{F}_e, \vec{F}_m, \vec{F}_c$ are shown in Fig. 7.1. Now, the electron movement is defined by the vector equation

$$\vec{F}_e = \vec{F}_c + \vec{F}_m.\tag{7.13}$$

It is known that $\vec{F}_m << \vec{F}_e$, and hence $\vec{B}$ does not influence the electron angular velocity significantly, which means that

$$\omega = \omega_0 + \Delta\omega.\tag{7.14}$$

We assume that, in spite of the action of $\vec{B}$, the electron centrifugal force is defined by Eq. (7.5) and, additionally, we assume that the vector $\vec{B}$ action does not change the electron orbit radius r (see (7.10)). Therefore, vector equation (7.13) has its following scalar counter-part

$$\omega^2 + \frac{eB}{m}\omega - \omega_0^2 = 0.\tag{7.15}$$

Substituting Eq. (7.14) into (7.15), we get

$$\omega_0^2 + 2\omega_0\Delta\omega + (\Delta\omega)^2 + \frac{eB}{m}\omega_0 + \frac{eB}{m}\Delta\omega - \omega_0^2 = 0,\tag{7.16}$$

which yields

$$\Delta\omega = -\frac{eB}{2m},\tag{7.17}$$

because $(\Delta\omega)^2$ and $\frac{eB\Delta\omega}{m}$ are higher order terms and they can be omitted. If we change the direction of $\vec{B}$, then, using Eq. (7.12), $\vec{F}_m$ changes its direction and instead of Eq. (7.17) we get $\Delta\omega = eB/(2m)$. The latter means that $\omega > \omega_0$, and hence the electron angular velocity is increased due to action of $\vec{B}$. We may estimate the influence of $\vec{B}$ on the magnetic moment (7.3). We have

$$\mathcal{M}_0 = \frac{er^2}{2}\omega_0, \qquad \mathcal{M} = \frac{er^2}{2}\omega,\tag{7.18}$$

and hence

$$\Delta\mathcal{M} = \mathcal{M} - \mathcal{M}_0 = \frac{er^2}{2}(\omega - \omega_0) = \frac{er^2\Delta\omega}{2}.\tag{7.19}$$

Substituting Eq. (7.17) into (7.19), we get

$$\Delta\vec{\mathcal{M}} = -\frac{e^2r^2}{4m}\vec{B}.\tag{7.20}$$

So far, we have considered the situation when $\vec{B} \parallel \vec{S}$. However, if the external magnetic field $\vec{B}$ is not normal to the electron orbital plane, then

$$\vec{\omega} = \vec{\omega}_0 + \Delta\vec{\omega},$$
$$\vec{M} = \vec{M}_0 + \Delta\vec{M}.\tag{7.21}$$

The following forces act on the electron $q = -e$. The central electric force

$$\vec{F}_e(r) = \frac{1}{4\pi\varepsilon}\frac{e^2}{r^3}\vec{r}.$$ (7.22)

This force, with the lack of B, is balanced by the centrifugal force

$$\vec{F}_c = m\vec{\omega}_0 \times (\vec{\omega}_0 \times \vec{r}).$$ (7.23)

If we apply the field $\vec{B}$, the electron is subjected to the action of the Lorentz force (7.12). The vector equation (7.13) governs the dynamical state of the electron, which has the following explicit form

$$\vec{\omega}_0 \times (\vec{\omega}_0 \times \vec{r}) = \vec{\omega} \times (\vec{\omega} \times \vec{r}) - \frac{e}{m}\left[(\vec{\omega} \times \vec{r}) \times \vec{B}\right].$$ (7.24)

The following formula sequence of steps holds:

$$1: \quad \vec{\omega}_0 \times (\vec{\omega}_0 \times \vec{r}) = (\vec{\omega}_0 + \Delta\vec{\omega}) \times [(\vec{\omega}_0 + \Delta\vec{\omega}) \times \vec{r}] - \frac{e}{m}\left\{[(\vec{\omega}_0 + \Delta\vec{\omega}) \times \vec{r}] \times \vec{B}\right\};$$

$$2: \quad \vec{\omega}_0 \times (\vec{\omega}_0 \times \vec{r}) = \vec{\omega}_0 \times [(\vec{\omega}_0 + \Delta\vec{\omega}) \times \vec{r}] + \Delta\vec{\omega} \times [(\vec{\omega}_0 + \Delta\vec{\omega}) \times \vec{r}] +$$
$$- \frac{e}{m}\left\{[(\vec{\omega}_0 \times \vec{r}) + (\Delta\vec{\omega} \times \vec{r})] \times \vec{B}\right\};$$

$$3: \quad \vec{\omega}_0 \times (\vec{\omega}_0 \times \vec{r}) = \vec{\omega}_0 \times (\vec{\omega}_0 \times \vec{r}) + \vec{\omega}_0 \times (\Delta\vec{\omega} \times \vec{r}) + \Delta\vec{\omega} \times (\vec{\omega}_0 \times \vec{r}) +$$
$$+ \underbrace{\Delta\vec{\omega} \times (\Delta\vec{\omega} \times \vec{r})}_{\approx 0} - \frac{e}{m}(\vec{\omega}_0 \times \vec{r}) \times \vec{B} - \underbrace{\frac{e}{m}(\Delta\vec{\omega} \times \vec{r}) \times \vec{B}}_{\approx 0};$$

$$\vec{\omega}_0 \times (\Delta\vec{\omega} \times \vec{r}) + \Delta\vec{\omega} \times (\vec{\omega}_0 \times \vec{r}) = \frac{e}{m}(\vec{\omega}_0 \times \vec{r}) \times \vec{B}.$$ (7.25)

We apply the following property $\vec{a} \times \left(\vec{b} \times \vec{c}\right) = \vec{b}\left(\vec{a} \cdot \vec{c}\right) - \vec{c}\left(\vec{a} \cdot \vec{b}\right)$, and from Eq. (7.25) we get

$$\Delta\vec{\omega}\underbrace{(\vec{\omega}_0 \cdot \vec{r})}_{=0} - \vec{r}(\vec{\omega}_0 \cdot \Delta\vec{\omega}) + \vec{\omega}_0\underbrace{(\Delta\vec{\omega} \cdot \vec{r})}_{=0} - \vec{r}(\Delta\vec{\omega} \cdot \vec{\omega}_0) = \frac{e}{m}(\vec{\omega}_0 \times \vec{r}) \times \vec{B};$$
$$- 2\vec{r}(\vec{\omega}_0 \cdot \Delta\vec{\omega}) = \frac{e}{m}(\vec{\omega}_0 \times \vec{r}) \times \vec{B}.$$ (7.26)

We again apply here the earlier used property of a double vector product:

$$-2\left[(\vec{\omega}_0 \times \vec{r}) \times \Delta\vec{\omega}\right] = 2\left[\vec{r}(\vec{\omega}_0 \cdot \Delta\vec{\omega}) - \Delta\vec{\omega}\underbrace{(\vec{\omega}_0 \cdot \vec{r})}_{=0}\right] \approx 2\vec{r}(\vec{\omega}_0 \cdot \Delta\vec{\omega}).$$ (7.27)

Taking into account Eqs. (7.26) and (7.27), we get

$$-2\left[(\vec{\omega}_0 \times \vec{r}) \times \Delta\vec{\omega}\right] = \frac{e}{m}(\vec{\omega}_0 \times \vec{r}) \times \vec{B}.$$ (7.28)

Finally, we obtain

$$(\vec{\omega}_0 \times \vec{r}) \times \Delta\vec{\omega} = (\vec{\omega}_0 \times \vec{r}) \times \left(-\frac{e}{2m} \cdot \vec{B}\right),$$ (7.29)

which means that

$$\Delta \vec{\omega} = -\frac{e}{2m} \cdot \vec{B}, \tag{7.30}$$

what coincides with formula (7.17), and hence we get electron magnetic moments

$$\vec{\mathcal{M}}_0 = -\frac{e}{2m}\vec{L}_0, \\ \vec{\mathcal{M}} = -\frac{e}{2m}\vec{L}, \tag{7.31}$$

where the corresponding angular moments are

$$\vec{L}_0 = \vec{r} \times m\left(\vec{\omega}_0 \times \vec{r}\right), \\ \vec{L} = \vec{r} \times m\left(\vec{\omega} \times \vec{r}\right), \tag{7.32}$$

and hence

$$\Delta \vec{L} = \vec{L} - \vec{L}_0 = m\left[\vec{r} \times (\vec{\omega} \times \vec{r}) - \vec{r}(\vec{\omega}_0 \times \vec{r})\right] = m\left[\vec{r} \times (\Delta \vec{\omega} \times \vec{r})\right]. \tag{7.33}$$

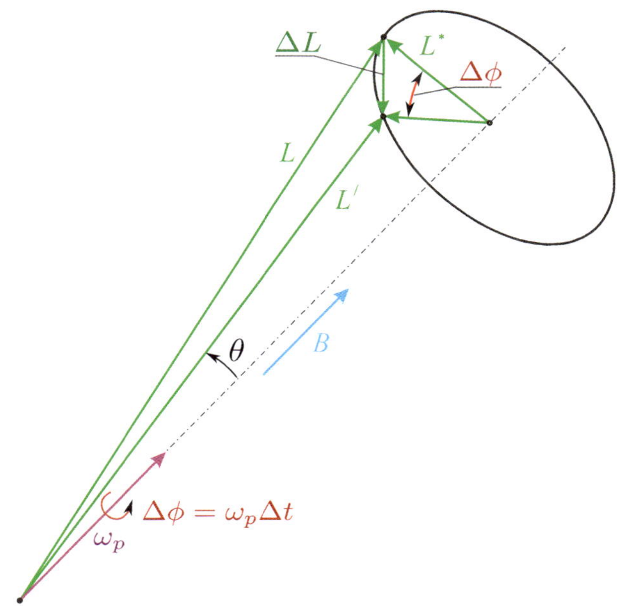

Fig. 7.3 Precession of the vector $\vec{L}$ about $\vec{B}$ with the angular velocity $\vec{\omega}_p$.

It follows from Fig. 7.3 that

$$\Delta L = 2 \cdot L \sin \theta \sin \frac{\Delta \phi}{2} \cong L\Delta\phi \sin\theta = L\left(\omega_p \Delta t\right)\sin\theta, \tag{7.34}$$

and hence

$$\Delta L = L^* \times \Delta\phi, \tag{7.35}$$

where $L^* = L \sin \theta$. Therefore, in the limits we get

$$\frac{dl}{dt} = \lim_{\Delta t \to 0} \frac{\Delta L}{\Delta t} = L^* \times \lim_{\Delta t \to 0} \frac{\Delta \phi}{\Delta t} = L^* \times \omega_p, \qquad (7.36)$$

where ω_p stands for the angular velocity of precession. If we look at Fig. 7.2 and formula (7.6), we can write

$$\left| \vec{T} \right| = \left| \frac{d\vec{\mathcal{J}}}{dt} \right| = \left| \vec{M} \times \vec{B} \right| = \mu B \sin \theta. \qquad (7.37)$$

Comparison of Eqs. (7.34) and (7.37) yields

$$\mathcal{J} \sin \theta \omega_p = MB \sin \theta, \qquad (7.38)$$

which allows to find (see Eq. (7.31)):

$$\omega_p = \frac{MB}{\mathcal{J}} = -\frac{e}{2m} B. \qquad (7.39)$$

Thus, owing to formula (7.17), we have shown that $\omega_p = \Delta \omega$.

The previously illustrated phenomenon is valid only for the orbital movement of electrons, which is associated with a central force field. It takes place for $M/\mathcal{J} = -e/(2m)$. This is implied by the so-called *Larmor's theorem* which states: *motion of a particle within the uniform magnetic field $\vec{B}$ is a superposition of its motion without $\vec{B}$ action and the additional rotational motion about vector $\vec{B}$ with the angular velocity $\vec{\omega}_L = e\vec{B}/(2m)$.*

7.2 Wave Model

In the *quantum mechanics*, the electron angular momentum L_1 (one electron) as well as the electron spin S_1 are multiplies of $\hbar = h/2\pi$, where h is Planck's constant. In general, an atom possesses a few electrons having different angular moments, and hence (see Eq. (7.4)) we get

$$\vec{\mathcal{M}}_L = -\vec{\gamma}\vec{L}, \qquad (7.40)$$

where $\gamma = \frac{e}{2m_e}$, $\vec{\mathcal{M}}_L$ is the *total orbital magnetic dipole moment (total magnetic moment)* and $Z\vec{L} = \sum_{z=1}^{Z} z\vec{L}$ is total orbital angular momentum of Z electrons (Ze would be the nucleus charge). If $\vec{L}$ stands for the quantum mechanical operator for the electron angular momentum, then the angular moment associated with $\vec{L}$ is

$$\vec{\mathcal{M}}_L = \gamma \hbar Z \vec{L}, \qquad (7.41)$$

where Z is the atomic number of electrons, and m is the electron mass, $\gamma = -|\gamma|$. Here, we have taken the motion reference point in which an observer is fixed to the electron. It means that the nucleus is moving around the electron with the orbital angular momentum $Z\hbar\vec{L}$. The produced $\vec{B}$ at the electron site by the nucleus is as follows

$$\vec{B} = \frac{\mu_0 \vec{\mathcal{M}}_L}{2\pi r^3}, \qquad (7.42)$$

or equivalently

$$\vec{B} = \frac{\hbar Z\,|\gamma|\,\mu_0}{2\pi r^3}\,\vec{L}, \tag{7.43}$$

where $[\mu_0] = \left[4\pi \cdot 10^{-7}\frac{H}{m}\right]$ is the permeability coefficient.

Even in the case of two electrons, assuming that one electron may have five possible orbits to be chosen to move around the nucleus (the so-called 3D electron), it can be shown [Vittoria (2011)] that there are 45 independent permutations of two electrons associated with 10 possible orbital and spin motions.

Taking into account one of the permutations, the equivalent representation of the multiple electron states with respect to the electron angular momentum $\vec{L}$ and its spin S relies on L, $\mathcal{M}_L$ and S, $\mathcal{M}_S$, where

$$\mathcal{M}_L = \sum m_L, \qquad \mathcal{M}_S = \sum m_S. \tag{7.44}$$

It occurs that a symmetry in the number of representations for $\pm\mathcal{M}_L$ and $\pm\mathcal{M}_S$ yields reduction of the representations number. For example, if we assign one electron with $m_S = \pm 1/2$ to the $m_L = 2$ orbit, and the second electron with $m_S = \pm 1/2$ to $m_L = 1$, then the *Pauli exclusion principle* yields the following new quantum numbers [Vittoria (2011)]: $\mathcal{M}_L = 3$, $\mathcal{M}_S = 1$.

The magnetic field generated by the nucleus acts on the electron spinal motion, and one may compute the potential energy (see Eq. (7.9))

$$U = -\vec{\mathcal{M}}_S \cdot \vec{B}, \tag{7.45}$$

where

$$\vec{\mathcal{M}}_S = 2\gamma\hbar\vec{S}, \tag{7.46}$$

and $\vec{S}$ is the spin angular momentum. The total spin $\vec{S}$ (total intrinsic angular momentum) was introduced in 1924 by W. Pauli.

From Eqs. (7.45) and (7.43) we get

$$U = \frac{\gamma^2\hbar^2\mu_0 Z}{\pi r^3}\vec{S}\cdot\vec{L} = \lambda\vec{S}\cdot\vec{L}, \tag{7.47}$$

where the spin-orbit interaction parameter λ, improved by Thomas by the factor $1/2$, follows

$$\lambda = \frac{\mu_0}{2\pi r^3}g_S\beta^2 Z\left(\frac{1}{2}\right), \tag{7.48}$$

and $g_S = 2$, $\beta = \gamma\hbar$, $\gamma = \frac{e}{2mc}$.

Here we have introduced *Landé g constant*, whereas *Landé g-formula* will be derived later. Dirac pointed out that the following ratio holds $g_S\mathcal{M}_S/L_S = \mathcal{M}/L$, where $\mathcal{M}_S$, LS $(\mathcal{M}, L)$ correspond to the electron spin (electron orbital movement). Since the resultant electron angular momentum includes components coming from the electrons orbital and spin motions, then $1 < g < 2$. Namely, for a purely orbital momentum we have $g = 1$, whereas for the purely spin motion in the latter case we have $g = 2$. In fact, if we take into account the relativistic improvement, then $g = 2.0023$.

It should be emphasized that protons and neutrons possess their own spin and may move along their orbits. Since the mass m_p of a proton is much larger than that of an electron m_e ($m_p/m_e = 1836.15$), then for an atom nucleus, we have

$$\vec{\mathcal{M}} = g \left(\frac{e}{2m_p} \right) \vec{L}, \tag{7.49}$$

where for the proton $g = 2 \cdot 2.79$, and for the neutron $g = 2 \cdot (-1.93)$. The neutron does not exhibit any charge but its magnetic momentum is negative.

The magnetic potential energy produced by $\vec{B}$ is given by the angular momentum $\vec{L}$ and the spin momentum $\vec{S}$ scalar product, and hence it is associated with the spin-orbit interaction.

There exist tables with ground state multiple for various electrons/ions, where the associated S and L are given [Vittoria (2011)].

In what follows, we take the coupling Hamiltonian $H = U$. It can be also written in the following two equivalent forms

$$H = -\vec{\mathcal{M}}_S \cdot \vec{B}_L = -\vec{\mathcal{M}}_L \cdot \vec{B}_S, \tag{7.50}$$

where

$$\vec{\mathcal{M}}_S = \gamma_S \hbar \vec{S}, \qquad \vec{B}_L = -\frac{\lambda \vec{L}}{\gamma_S \hbar}, \qquad \vec{\mathcal{M}}_L = \gamma_L \hbar \vec{L}, \qquad \vec{B}_S = -\frac{\lambda \vec{S}}{\gamma_S \hbar}. \tag{7.51}$$

The magnetic moment vector is governed by the following equation

$$\begin{aligned}
\frac{d\vec{\mathcal{M}}_S}{dt} &= \gamma_S \left(\vec{M}_S \times \vec{B}_L \right), \\
\frac{d\vec{\mathcal{M}}_L}{dt} &= \gamma_L \left(\vec{M}_L \times \vec{B}_S \right),
\end{aligned} \tag{7.52}$$

while taking into account Eqs. (7.51) and (7.52), we get

$$\frac{d\vec{S}}{dt} = -\frac{\lambda}{\hbar} \left(\vec{S} \times \vec{L} \right), \qquad \frac{d\vec{L}}{dt} = -\frac{\lambda}{\hbar} \left(\vec{L} \times \vec{S} \right). \tag{7.53}$$

Since the electron is subjected to action of the resultant angular momentum

$$\vec{J} = \vec{L} + \vec{S}, \tag{7.54}$$

then Eq. (7.53) yields

$$\begin{aligned}
\frac{d\vec{L}}{dt} &= -\frac{\lambda}{\hbar} \left[\vec{L} \times \left(\vec{J} - \vec{L} \right) \right] = -\frac{\lambda}{\hbar} \left(\vec{L} \times \vec{J} \right), \\
\frac{d\vec{S}}{dt} &= -\frac{\lambda}{\hbar} \left(\vec{S} \times \vec{J} \right).
\end{aligned} \tag{7.55}$$

The obtained vector equations imply that vectors $\vec{L}$ and $\vec{S}$ rotate about $\vec{\mathcal{J}}$ with angular precession speed $\lambda/\hbar$. Therefore, the quantization of $\vec{L}$, $\vec{S}$, $\vec{\mathcal{M}}_L$, $\vec{\mathcal{M}}_S$ can be introduced by the $\vec{\mathcal{J}}$ reference direction. The previously described situation is schematically depicted in Fig. 7.4.

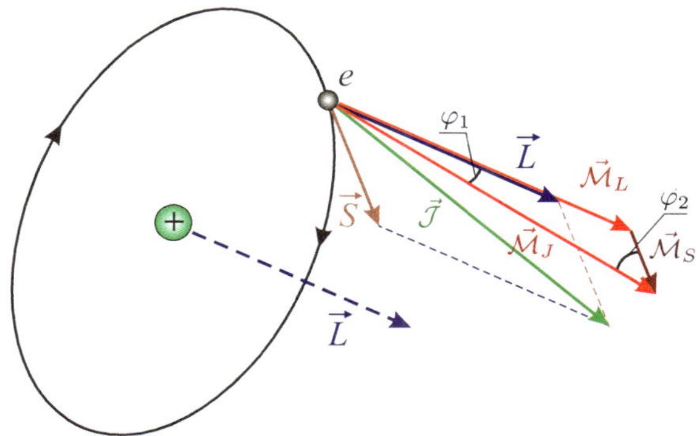

Fig. 7.4   Vectors of an electron angular orbital momentum, magnetic moment and their resultant vectors.

The relations between magnitudes of the vectors $\vec{\mathcal{M}}_L$, $\vec{\mathcal{M}}_S$, $\vec{\mathcal{M}}_J$, $L, S, \mathcal{J}$ follow:

$$\left|\vec{\mathcal{M}}_L\right| = g_L\lambda\hbar\sqrt{L(L+1)}, \quad \left|\vec{\mathcal{M}}_S\right| = g_S\lambda\hbar\sqrt{S(S+1)},$$
$$\left|\vec{\mathcal{M}}_J\right| = g_J\lambda\hbar\sqrt{\mathcal{J}(\mathcal{J}+1)}. \tag{7.56}$$

Projecting of $\vec{\mathcal{M}}_L$ and $\vec{\mathcal{M}}_S$ on their resultant $\vec{\mathcal{M}}_J$ direction yields

$$\left|\vec{\mathcal{M}}_J\right| = \left|\vec{\mathcal{M}}_L\right|\cos\varphi_1 + \left|\vec{\mathcal{M}}_S\right|\cos\varphi_2. \tag{7.57}$$

The *Carnot theorem* gives

$$\cos\varphi_1 = \frac{L(L+1)+\mathcal{J}(\mathcal{J}+1)-S(S+1)}{2\sqrt{L(L+1)J(J+1)}},$$
$$\cos\varphi_2 = \frac{S(S+1)+\mathcal{J}(\mathcal{J}+1)-L(L+1)}{2\sqrt{S(S+1)\mathcal{J}(\mathcal{J}+1)}}, \tag{7.58}$$

and hence

$$g_J\sqrt{\mathcal{J}(\mathcal{J}+1)} = \frac{g_L\left[L(L+1)+\mathcal{J}(\mathcal{J}+1)-S(S+1)\right]}{2\sqrt{\mathcal{J}(\mathcal{J}+1)}} +$$
$$+ \frac{g_S\left[S(S+1)+\mathcal{J}(\mathcal{J}+1)-L(L+1)\right]}{2\sqrt{\mathcal{J}(\mathcal{J}+1)}}. \tag{7.59}$$

For $g_L = 1$ and $g_S = 2$, the relationship (7.59) takes the following form

$$g_J = \frac{\left[L(L+1)+\mathcal{J}(\mathcal{J}+1)-S(S+1)\right]}{2\mathcal{J}(\mathcal{J}+1)} +$$
$$+ \frac{2\left[S(S+1)+\mathcal{J}(\mathcal{J}+1)-L(L+1)\right]}{2\mathcal{J}(\mathcal{J}+1)}, \tag{7.60}$$

which finally allows to get the Landé formula

$$g_J = \frac{3}{2} + \frac{S(S+1) - L(L+1)}{2J(J+1)}.\tag{7.61}$$

De Broglie pointed out that a particle motion is related to the wave-like motion due to the relation $\lambda = h/p$, where λ stands for the electromagnetic length of the wave associated with particle motion, p is the particle linear momentum and h is Planck's constant $[h] = [6.65 \cdot 10^{-32}\,\mathrm{J} \cdot \mathrm{s}]$. One electron moving around a wire loop generates maximum energy through the resonance phenomenon determined by the relationship $n\lambda = 2\pi r, n = 1, 2, 3,$, where λ is here the particle wavelength and r is the loop radius. Electron kinetic energy is

$$E = \frac{mr^2\omega^2}{2} = \frac{p^2}{2m}.\tag{7.62}$$

The particle position is described by the *Schrödinger equation*

$$i\hbar\frac{\partial\psi}{\partial t} = E\psi,\tag{7.63}$$

where ψ is the wave function defining the particle position in the probabilistic sense. A transition from Cartesian to cylindrical coordinates requires

$$p \to \hbar\frac{i}{r}\frac{\partial}{\partial\varphi}\tag{7.64}$$

to be satisfied, and from Eqs. (7.62) and (7.64) one gets:

$$E\psi = \frac{p^2}{2m}\psi = \frac{\left(\hbar\frac{i}{r}\frac{\partial}{\partial\varphi}\right)^2}{2m}\psi = -\frac{\hbar^2\frac{\partial^2\psi}{\partial\varphi^2}}{2mr^2}.\tag{7.65}$$

We are looking for a solution to Eq. (7.65) in the following form

$$\psi(\varphi) = C_1 e^{i\alpha\varphi} + C_2 e^{-i\alpha\varphi}.\tag{7.66}$$

Therefore

$$\begin{aligned}\frac{\partial\psi}{\partial\varphi} &= C_1 i\alpha e^{i\alpha\varphi} - C_2 i\alpha e^{-i\alpha\varphi}, \\ \frac{\partial^2\psi}{\partial\varphi^2} &= -C_1\alpha^2 e^{i\alpha\varphi} + C_2\alpha^2 e^{-i\alpha\varphi}\end{aligned}\tag{7.67}$$

and

$$\frac{\hbar^2}{2mr^2}\left(-C_1\alpha^2 e^{i\alpha\varphi} + C_2\alpha^2 e^{-i\alpha\varphi}\right) + E\left(C_1 e^{i\alpha\varphi} + C_2 e^{-i\alpha\varphi}\right) = 0.\tag{7.68}$$

Comparison of terms standing at $e^{i\alpha\varphi}$ and $e^{-i\alpha\varphi}$ yields

$$-\frac{C_1\hbar^2}{2mr^2}\alpha^2 + EC_1 = 0, \qquad \frac{C_2\hbar^2}{2mr^2}\alpha^2 + EC_2 = 0,\tag{7.69}$$

and hence

$$\alpha^2 = \frac{2mr^2 E}{\hbar^2} = \frac{r^2 p^2}{\hbar^2}.\tag{7.70}$$

The electrons repeat their movement with the period of 2π, and therefore

$$\psi(\varphi) = \psi(\varphi + 2\pi), \tag{7.71}$$

which means that

$$e^{i\alpha\varphi} = e^{i\alpha(\varphi+2\pi)}, \tag{7.72}$$

and consequently,

$$1 = e^{2\pi i\alpha}. \tag{7.73}$$

Therefore, we get

$$2\pi\alpha = 2\pi n, \qquad n = 0, 1, 2, \tag{7.74}$$

which means that α is quantized, and Eq. (7.70) with (7.74) imply

$$L \equiv pr = \hbar n. \tag{7.75}$$

It means that L is discrete and defined through the $\hbar$ units. The kinetic energy associated with the electron movement is

$$E = \frac{p^2}{2m} = \frac{\hbar^2 n^2}{2mr^2} = \frac{L^2}{2mr^2}. \tag{7.76}$$

Since L is quantized, then also E is discrete. This points out that within the classical mechanical modeling, we go beyond the classical description being valid for small scales. Furthermore, due to the linear relation between the magnetic dipole moment and the angular momentum, the latter one is also quantized (see Eq. (7.51)).

It should be emphasized that this nonclassical modeling allows to transit to the classical models when long wave exciatations tend to a limit. In what follows, we consider energy levels of atoms. For this purpose, we take the following Hamiltonian associated with the atom kinetic and potential energy in the Cartesian coordinates

$$H = -\frac{\hbar^2}{2m}\nabla^2 - \frac{Ze^2}{r}, \tag{7.77}$$

where r denotes the electron position (see Fig. 7.5), and Z denotes the number of electrons.

We extend here our previous approach by nonclassical modeling with the wave function ψ, using the following Schrödinger equation

$$H\psi = E\psi, \tag{7.78}$$

which in the spherical coordinates (r, θ, φ) takes the following form

$$\left[\frac{1}{r^2}\frac{\partial}{\partial r}\left(r^2 \frac{\partial}{\partial r} \right) + \frac{1}{r^2 \sin\theta}\frac{\partial}{\partial\theta}\left(\sin\theta \frac{\partial}{\partial r} \right) + \frac{1}{r^2 \sin^2\theta}\frac{\partial^2}{\partial\varphi^2} \right]\psi +$$
$$+ \frac{2m}{\hbar^2}\left[E + \frac{Ze^2}{r} \right]\psi = 0. \tag{7.79}$$

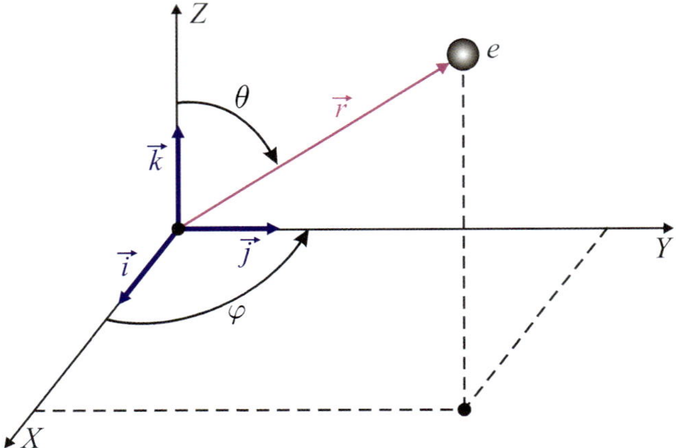

Fig. 7.5   Electron position in the Cartesian and spherical coordinates.

We look for a solution with the following property

$$\psi(r,\theta,\varphi) = A(r)B(\theta,\varphi). \tag{7.80}$$

From Eq. (7.79) we get

$$\lambda \equiv \frac{1}{A}\frac{\partial}{\partial r}\left(r^2\frac{\partial A}{\partial r}\right) + \frac{2mr^2}{\hbar^2}\left(E + \frac{Ze^2}{r}\right) =$$
$$-\frac{1}{B\sin\theta}\frac{\partial}{\partial\theta}\left(\sin\theta\frac{\partial B}{\partial\theta}\right) - \frac{1}{B\sin^2\theta}\frac{\partial^2 B}{\partial\varphi^2}, \tag{7.81}$$

where λ is a certain constant. Although there are many solutions to Eq. (7.81), one of them is

$$\lambda = k(k+1), \qquad k = 0,1,2,3,\dots, \tag{7.82}$$

i.e., we go from a continuous system governed by PDE equation (7.81) to its discrete solution form (7.82). From Eqs. (7.81) and (7.82), we find that

$$\frac{1}{r^2}\frac{d}{dr}\left(r^2\frac{dA}{dr}\right) + \frac{2m}{\hbar^2}\left(E + \frac{Ze^2}{r}\right)A = \frac{k(k+1)}{r^2\theta}A,$$
$$\frac{1}{\sin\theta}\frac{\partial}{\partial\theta}\left(\sin\theta\frac{\partial B}{\partial\theta}\right) + \frac{1}{\sin^2\theta}\frac{\partial^2 B}{\partial\varphi^2} = -k(k+1)B, \tag{7.83}$$

which means that we have separated functions $A = A(r)$ and $B = B(\theta,\varphi)$.

The obtained second order differential equation allows to find the eigenvalues of the problem, and hence to define the quantized energy

$$E_n = -\frac{Z^2e^2}{2n^2a_0}, \qquad n = 1,2,3,\dots, \tag{7.84}$$

where now n stands for the principal quantum number, and a_0 is the so-called *Bohr radius*. For $Z = 1$, we have

$$a_0 = \frac{\hbar^2}{me^2}, \tag{7.85}$$

which can be validated by the following simple consideration given by Bohr. In the case of dynamical electrons equilibrium state, we have

$$\frac{e^2}{r^2} = \frac{mv^2}{r},\tag{7.86}$$

where e^2/r^2 stands for the Coulomb force in the central nuclear field (centrifugal force).

Owing to the quantized energy form (7.84), we have

$$E_n = -\frac{1}{2}m\frac{c^2\beta^2}{n^2} = \frac{1}{2}mv^2 - \frac{e^2}{r},\tag{7.87}$$

where

$$\beta^2 = \left(\frac{e^2}{\hbar c}\right)^2 = \frac{1}{137},\tag{7.88}$$

$$a_0 = \frac{e^2}{mc^2\beta^2} = \frac{\hbar^2}{me^2}.\tag{7.89}$$

Comparison of Eqs. (7.84) and (7.87) yields (7.85). Formula (7.87) implies that the total electron energy E_n is negative.

The so far introduced quantum numbers k and n are responsible for a variety of periodic electron planar motions. The total number of electrons can be calculated with the use of the relationship

$$N = \sum_{l=0}^{n-1}(2l+1),\tag{7.90}$$

where each of the electron occupies its associated periodic orbits, i.e., the total number of electrons equals the number of independent circular electron motions (in the above equation $l = 0, 1, 2, \ldots, n-1$). For instance, for $n = 3$ we have $N = 9$. In other words, there exist 9 different periodic orbits with the same total energy E_3. On the other hand, the values of $\mathcal{M}_l$ are defined by l, and they are equal to $l, l-1, l-2, \ldots, -l$ (see also [Vittoria (2011)]).

The chosen set $\{n, l\}$ and $\mathcal{M}_l$ correspond to each electron 3D space dynamical state governed by r, θ, φ. On the other hand, each periodic orbit corresponds to its quantized associates of a discrete angular momentum and a discrete magnetic momentum.

This gives rights to introduce the discrete angular momentum and magnetic (dipole) momentum associated with each periodic orbit by the relationship (7.55).

7.3 Magnetic Field vs. Free Atom

The total energy of an atom includes kinetic energy $p_0^2/2m$ and potential energy $V(\vec{r})$, where $\vec{r}$ is the atom center position vector, and $\vec{p}_0 = m\vec{v}$.

Action of $\vec{B}$ yields larger Hamiltonian energy $H > H_0$, the old (without $\vec{B}$) and the new Hamiltonians have the following form

$$H_0 = \frac{p_0^2}{2m} + V(\vec{r}),$$
$$H = \frac{p^2}{2m} + V(\vec{r}), \tag{7.91}$$

and $p \neq p_0$, the $V(r)$ does not change.

We have

$$\vec{\nabla} \times \vec{E} = -\frac{\partial \vec{B}}{\partial t} = -\frac{\partial}{\partial t} \vec{\nabla} \times \vec{P} = -\vec{\nabla} \times \frac{\partial \vec{P}}{\partial t}, \tag{7.92}$$

and hence

$$\vec{E} = -\frac{\partial \vec{P}}{\partial t} + C, \tag{7.93}$$

where $\vec{P}$ is the vector magnetic potential and C is a constant. We take $C = -\vec{\nabla} U$, where U is externally applied voltage. In our case, however, $U = 0$, and hence we have $C = 0$. Therefore, from Eq. (7.93) we obtain that

$$\vec{E} = -\frac{\partial \vec{P}}{\partial t}. \tag{7.94}$$

On the other hand, we have

$$\frac{\partial \vec{p}}{\partial t} = q\vec{E} = -\frac{\partial \vec{P}}{\partial t} q, \tag{7.95}$$

which means that

$$\vec{p} = -q\vec{P} + \vec{p}_0. \tag{7.96}$$

Since

$$\vec{p}(0) = \vec{p}_0, \tag{7.97}$$

and for $q = -e$, we have

$$\vec{p} = e\vec{P} + \vec{p}_0, \tag{7.98}$$

and taking into account Eq. (7.98) in (7.91), we obtain

$$H = \frac{\left(\vec{p}_0 + e\vec{P}\right)^2}{2m} + V(\vec{r}). \tag{7.99}$$

Formula (7.99) implies that

$$H = \frac{p_0^2}{2m} + V(\vec{r}) + \frac{e}{m}\vec{P} \cdot \vec{p}_0 + \frac{e^2}{2m}P^2 = H_0 + \underline{\frac{e}{m}\vec{P} \cdot \vec{p}_0 + \frac{e^2}{2m}P^2}, \tag{7.100}$$

and the underlined term corresponds to the kinetic energy change due to action of the magnetic field $\vec{B}$. In the case of a steady magnetic field $\left(\dot{\vec{B}} = 0\right)$, we have

$$\vec{P} = \frac{1}{2}\vec{B} \times \vec{r} = -\frac{1}{2}\vec{r} \times \vec{B}, \qquad \left|\vec{P}\right| = -\frac{1}{2}\sqrt{(r^2 \cdot B^2) - \left(\vec{r} \cdot \vec{B}\right)^2}. \tag{7.101}$$

On the other hand, we get

$$\vec{P} \cdot \vec{p_0} = -\frac{1}{2}\left(\vec{r} \times \vec{B}\right) \cdot \vec{p_0} = -\frac{1}{2}\vec{B} \cdot (\vec{r} \times \vec{p_0}) = -\frac{1}{2}\vec{B} \cdot \vec{J_0}. \tag{7.102}$$

Consider now the case when $\vec{J_0} = \vec{L}$ or $\vec{J_0} = \vec{S}$. The angular momentum vector dynamics is governed by the following equation

$$\dot{\vec{J}} = \vec{M} \times \vec{B} = \gamma \vec{J} \times \vec{B}. \tag{7.103}$$

We apply the following Cartesian coordinates

$$\begin{aligned} \vec{J} &= \mathcal{J}_1 \vec{i} + \mathcal{J}_2 \vec{j} + \mathcal{J}_3 \vec{k}, \\ \vec{B} &= B\vec{k}. \end{aligned} \tag{7.104}$$

Putting Eq. (7.104) into (7.103) gives

$$\dot{\mathcal{J}}_1 = \gamma B \mathcal{J}_2, \qquad \dot{\mathcal{J}}_2 = -\gamma B \mathcal{J}_1, \qquad \dot{\mathcal{J}}_3 = 0. \tag{7.105}$$

The first two equations (7.105) yield

$$\ddot{\mathcal{J}}_1 + \gamma^2 B^2 \mathcal{J}_1 = 0. \tag{7.106}$$

Since the formula (7.106) governs the dynamics of a linear conservative oscillator, its solution follows

$$\mathcal{J}_1 = a \sin\left(\gamma B t + \psi_0\right), \tag{7.107}$$

where a, ψ_0 are constants to be defined by the initial conditions. Then, we have

$$\mathcal{J}_2 = \frac{1}{\gamma B}\dot{\mathcal{J}}_1 = a \cos\left(\gamma B t + \psi_0\right). \tag{7.108}$$

It means that the end of the vector $\vec{J}$ moves in a plane Π parallel to the OXY plane and there is a distance $C = \mathcal{J}_3$ between two planes. The end of the vector $\vec{J}$ moves uniformly along the circle with a radius "a" and is characterized by a period $2\pi/\gamma B$ (see Fig. 7.6).

The situation illustrated in Fig. 7.6 shows that the angle $(\vec{k}, \vec{J})$ has continuous values. However, in quantum mechanics, $\vec{J}$ takes discrete values and, for instance, if $n = 2$ (n is the principal quantum number being equal $1, 2, 3, \dots$) and $L = 1$, we have

$$\left|\vec{L}\right| = \hbar\sqrt{l\left(l+1\right)} = \sqrt{2}\hbar. \tag{7.109}$$

It is known that $\vec{L} \cdot \vec{k}$ may take three different values $\hbar, 0, -\hbar$ associated with $m_L = 1, 0, -1$, respectively. One may easily find the corresponding discrete θ angle values from the formula $\vec{L}\cos\theta = \vec{L} \cdot \vec{k}$, which are as follows: $\theta = 45°, 90°, 135°$. Similar situation takes place in the case of the spin dynamics. We have $\left|\vec{S}\right| = \hbar/\sqrt{2}$ and $\vec{S} \cdot \vec{k} = \pm\hbar/2$. The electron may occupy three orbitals associated with $\vec{L}$ and two orbitals associated with $\vec{S}$. If all of them are occupied, then the net vector $\vec{J} = \vec{S} + \vec{L}$ is zero. Two mentioned orbital moments imply two magnetic moments $\vec{M}_L = \gamma_L \vec{L}$, $\vec{M}_S = \gamma_S \vec{S}$.

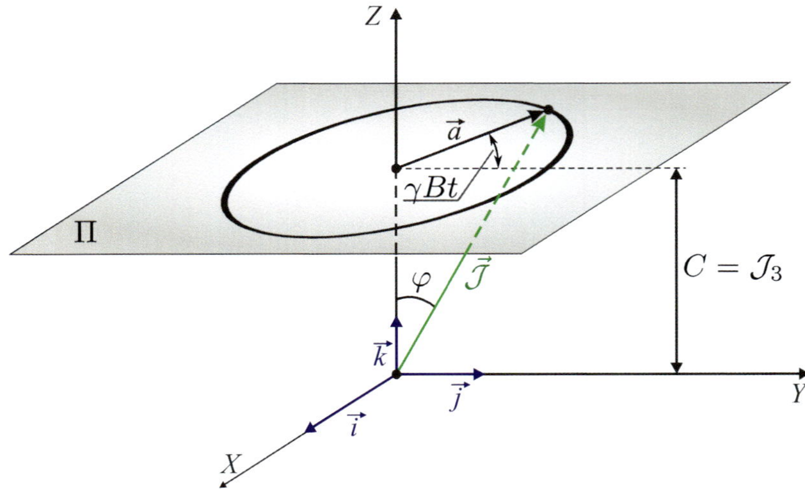

Fig. 7.6   Movement of the vector $\vec{\mathcal{J}}$ end along a circle.

The increase in the Hamiltonian, caused by the magnetic field B, follows from Eq. (7.100), so

$$H^{/} = \left(r_1^2 + r_2^2\right) B^2,\tag{7.110}$$

because

$$\vec{r} \cdot \vec{B} = \left(r_1\vec{i} + r_2\vec{j} + r_3\vec{k}\right) \cdot B\vec{k} = r_3 B,$$

and

$$\vec{r} = r_1\vec{i} + r_2\vec{j} + r_3\vec{k}.$$

Introducing the following averaging quantities

$$\left\langle r_1^2\right\rangle = \frac{1}{3}\left\langle r^2\right\rangle, \qquad \left\langle r_2^2\right\rangle = \frac{1}{3}\left\langle r^2\right\rangle,\tag{7.111}$$

it can be written that

$$\left\langle r_1^2 + r_2^2\right\rangle = \frac{2}{3}\left\langle r^2\right\rangle,\tag{7.112}$$

and the formula (7.110) can be cast to the following form

$$H^{/} = -\left(\vec{M}_S + \vec{M}_L\right) \cdot \vec{B} + \frac{e^2}{12m}\left\langle r^2\right\rangle.\tag{7.113}$$

The net magnetic moment in $\vec{k}$ direction is

$$M_3 = \vec{M}_S \cdot \vec{k} + \vec{M}_L \cdot \vec{k} + \vec{M}_D,\tag{7.114}$$

where

$$M_D = -\frac{e^2 B}{6m}\left\langle r^2\right\rangle = I_D\pi\left\langle r^2\right\rangle.\tag{7.115}$$

Here, M_D is referred to as the *diamagnetic* contribution to the total magnetic moment. The current I_D induced by the field $\vec{B}$ can be estimated from Eq. (7.115)

$$I_D = -\frac{e^2 B}{6\pi m}. \tag{7.116}$$

Introducing the diamagnetic energy

$$E_D = \frac{e^2 \phi_B^2}{12m} \langle r^2 \rangle = \frac{e^2 (BS)^2}{12m} \langle r^2 \rangle, \tag{7.117}$$

the induced current I_D can be found from

$$I_D = -\frac{dE_D}{d\phi_B} \langle r^2 \rangle = -\frac{e^2 \phi_B}{6m} \langle r^2 \rangle = -\frac{e^2 BS}{6m} \langle r^2 \rangle. \tag{7.118}$$

Comparison of formulas (7.116) and (7.118) allows to get

$$S = \pi \langle r^2 \rangle. \tag{7.119}$$

If all the electric orbitals are occupied by a free atom, then its net angular momentum and the magnetic moment is zero. There exist $(2l + 1)$ ways of the electron movements around the free atom nucleus, but the atom total energy does not change. The electron spends equal time on each of its orbits, which are defined by $\mathcal{M}_l = l, l - 1, \ldots, -l$. In other words, the electron motion is degenerated for assigned l. However, action of an external magnetic field removes the degeneracy, since the electron motion achieves the lowest energy level corresponding to the orbit associated with $\mathcal{M}_l$. Presence of the magnetic field may polarize free atoms.

Finally, it should be noted that simple considerations shown in this section include neither electron-electron interactions nor internal field generated by spin-orbit interaction and electrostatic interactions.

7.4 Electron Orbital Perturbation by a Proton Moving in a Magnetic Field

We consider an electron movement in a homogeneous and uniform magnetic field $\vec{B}\left(\vec{B} = \vec{k}B\right)$ being caused by the moving proton. In the initial moment, both particles (electron and proton) lie in a horizontal plane OXY, and the distance between them is equal to R – Larmor's radius of the electron moving with velocity $\vec{v}$. We assume that for $t = 0$, $\vec{v}_e(0) \perp \vec{R}$, $\vec{v}_p(0) \perp \vec{R}$, $v_e(0) = v_p(0)$, where $R = |\vec{r}_p - \vec{r}_e|$ and $\vec{v}_p(\vec{v}_e)$ denotes the proton (electron) position in the given Cartesian coordinates.

The equation governing the coupled dynamics of the electron and proton has the following form

$$m_e \ddot{\vec{r}}_e = e^2 \frac{\vec{R}}{R^3} - e\left(\vec{v}_e \times \vec{B}\right),$$

$$m_p \ddot{\vec{r}}_p = -e^2 \frac{\vec{R}}{R^3} + e\left(\vec{v}_p \times \vec{B}\right). \tag{7.120}$$

Let us introduce the following characteristic quantities

$$m_e \frac{v_0^2}{r_0} = \frac{e^2}{r_0^2}, \qquad \omega_0 r_0 = v_0, \tag{7.121}$$

and hence the following nondimensional variables:

$$\tau = \omega_0 t, \quad r_e' = r_e/r_0, \quad v_e' = v_e/v_0, \quad r_p' = r_p/r_0, \quad v_p' = v_p/r_0. \tag{7.122}$$

The introduced initial conditions force both particles to move in the plane OXY, and in order to get scalar equations, we multiply the vector equation (7.120) by the unit vectors $\vec{i}, \vec{j}$, so we get

$$m_e \ddot{\vec{r}}_e \cdot \vec{i} = e^2 \frac{\vec{R}}{R^3} \cdot \vec{i} - e\left(\vec{v}_e \times \vec{B}\right) \cdot \vec{i},$$

$$m_e \ddot{\vec{r}}_e \cdot \vec{j} = e^2 \frac{\vec{R}}{R^3} \cdot \vec{j} - e\left(\vec{v}_e \times \vec{B}\right) \cdot \vec{j},$$

$$m_p \ddot{\vec{r}}_p \cdot \vec{i} = -e^2 \frac{\vec{R}}{R^3} \cdot \vec{i} + e\left(\vec{v}_p \times \vec{B}\right) \cdot \vec{i}, \tag{7.123}$$

$$m_p \ddot{\vec{r}}_p \cdot \vec{j} = -e^2 \frac{\vec{R}}{R^3} \cdot \vec{j} + e\left(\vec{v}_p \times \vec{B}\right) \cdot \vec{j}.$$

We have

$$\left|\vec{v} \times \vec{B}\right| = \begin{vmatrix} \vec{i} & \vec{j} & \vec{k} \\ v_x & v_y & 0 \\ 0 & 0 & B \end{vmatrix} = \vec{i} v_y B - \vec{j} v_x B, \tag{7.124}$$

and hence Eq. (7.123) can be cast to the following set of nondimensional scalar equations:

$$\dot{v}_{ex} = \frac{e^2}{m_e} \frac{(x_p - x_e)}{R^3} - \frac{e}{m_e} v_{ey} B,$$

$$\dot{v}_{ey} = \frac{e^2}{m_e} \frac{(y_p - y_e)}{R^3} + \frac{e}{m_e} v_{ex} B,$$

$$\dot{v}_{px} = -\frac{e^2}{m_p} \frac{(x_p - x_e)}{R^3} + \frac{e}{m_p} v_{py} B, \tag{7.125}$$

$$\dot{v}_{py} = -\frac{e^2}{m_p} \frac{(y_p - y_e)}{R^3} - \frac{e}{m_p} v_{px} B,$$

$$v_{ex} = \dot{x}_e, \quad v_{ey} = \dot{y}_e, \quad v_{px} = \dot{x}_p, \quad v_{py} = \dot{y}_p.$$

$$\frac{dx_e}{d\tau} = v_{ex}, \qquad \frac{dy_e}{d\tau} = v_{ey},$$

$$\frac{dv_{ex}}{d\tau} = -v_{ey} + \frac{x_p - x_e}{R^3}, \qquad \frac{dv_{ey}}{d\tau} = v_{ex} + \frac{y_p - y_e}{R^3},$$

$$\frac{dx_p}{d\tau} = v_{px}, \qquad \frac{\partial y_e}{\partial \tau} = v_{py}, \tag{7.126}$$

$$\frac{dv_{px}}{d\tau} = \frac{1}{1836}\left(v_{py} - \frac{x_p - x_e}{R^3}\right), \qquad \frac{dv_{py}}{d\tau} = -\frac{1}{1836}\left(v_{px} + \frac{y_p - y_e}{R^3}\right),$$

where primes have been omitted.

7.5 Planar Dynamics of a Particle in a Magnetic Field

We consider a particle of the mass m and charge q governed by the following equation

$$m\frac{d\vec{v}}{dt} = q\left(\vec{v} \times \vec{B}\right),\tag{7.127}$$

where $\vec{B}$ *(const)* is the magnetic field induction.

We take two components of the vector $\vec{v} = \vec{v}_{||} + \vec{v}_{\perp}$, as well as the Lorentz vector $e\left(\vec{v} \times \vec{B}\right) = e\left(\vec{v} \times \vec{B}\right)_{||} + e\left(\vec{v} \times \vec{B}\right)_{\perp}$, where $\vec{v}_{||}\,||\vec{B}$. We obtain

$$m\left(\frac{d\vec{v}_{||}}{dt} + \frac{d\vec{v}_{\perp}}{dt}\right) = q\left[(\vec{v}_{||} + \vec{v}_{\perp}) \times \vec{B}\right],\tag{7.128}$$

and hence

$$m\frac{d\vec{v}_{||}}{dt} = q\left(\vec{v}_{||} \times \vec{B}\right) = 0, \qquad m\frac{d\vec{v}_{\perp}}{dt} = q\left(\vec{v}_{\perp} \times \vec{B}\right) = 0.\tag{7.129}$$

The first equation in Eq. (7.129) implies

$$\vec{v}_{||} = const,\tag{7.130}$$

which means that the point charge is moving in the $\vec{B}$ direction with the constant vector $\vec{v}_{||}$, i.e., having constant magnitude and direction. The second equation of (7.129) yields

$$m\frac{dv_{\perp}}{dt} = \frac{mv_{\perp}^2}{R} qv_{\perp}B,\tag{7.131}$$

and therefore, *Larmor's radius* is defined

$$R = \frac{mv_{\perp}}{qB}.\tag{7.132}$$

The associated *Larmor's period* follows

$$T = \frac{2\pi}{\omega} = \frac{2\pi R}{v_{\perp}} = \frac{2\pi m}{qB}.\tag{7.133}$$

It means that the point charge moves with a constant velocity $v_{||}$, simultaneously rotating with the angular velocity $\omega = \frac{qB}{m}$. The resultant charge motion is a screw motion. The situation is schematically illustrated in Fig. 7.7.

So far, we have considered the uniform and homogeneous field. In what follows, we assume that $B = B(z)$, i.e., we have a nonhomogeneous magnetic field. In this case, the charge q will rotate also in plane OXZ, but this time its instantaneous center of rotation will move with a certain drift velocity v_d.

Let us consider now a magnetic field generated by an infinite length straight wire with the current i (see Fig. 7.8).

Initial charge velocity $\vec{v}_0 \parallel \vec{j}$, and the charge dynamics equation follow

$$m\frac{d\vec{v}}{dt} = q\left(\vec{v} \times \vec{B}\right) \equiv \vec{F}_L,\tag{7.134}$$

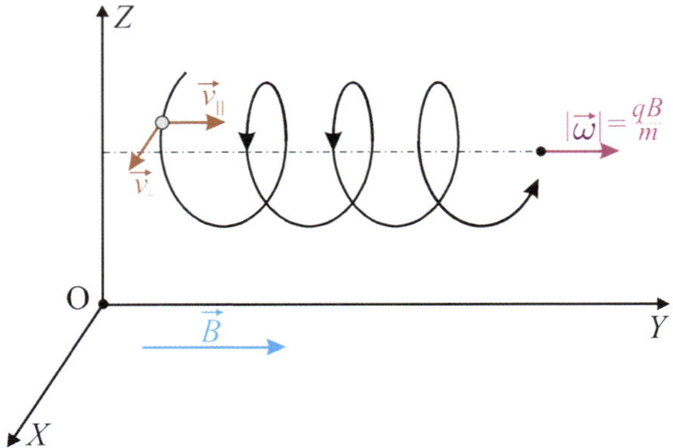

Fig. 7.7   Screw motion of the point charge.

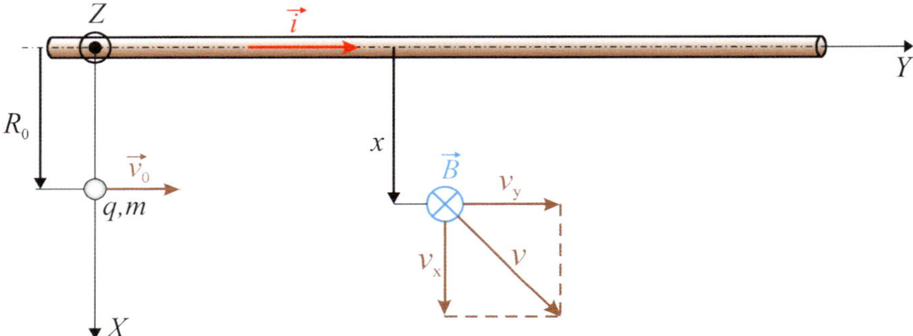

Fig. 7.8   The point charge q in the magnetic field produced by the current i.

or equivalently ($\vec{F}_L$ stands for the Lorentz force),

$$m\frac{d\left(v_x\vec{i}+v_y\vec{j}+v_z\vec{k}\right)}{dt}=q\left(v_x\vec{i}+v_y\vec{j}+v_z\vec{k}\right)\times B(x)\vec{k}. \qquad (7.135)$$

The vector equation (7.135) implies the following scalar equations:

$$\begin{aligned}
m\frac{dv_x}{dt}&=qv_yB(x), & v_x&=\frac{dx}{dt}, \\
m\frac{dv_y}{dt}&=-qv_xB(x), & v_y&=\frac{dy}{dt}, & B(x)&=\frac{\mu_0}{2\pi x}.
\end{aligned} \qquad (7.136)$$

7.6   3D Dynamics of a Charge

We consider a charge q of the mass m in a homogeneous magnetic field of the density H, where $\vec{H}\parallel\vec{k}$. We assume that the particle Lagrangian is

$$L=\frac{m}{2}\left(\dot{x}^2+\dot{y}^2+\dot{z}^2\right)+\frac{m}{2}\left(\omega_1^2x^2+\omega_2^2y^2+\omega_3^2z^2\right)+\frac{m}{2}\omega_H\left(y\dot{x}-x\dot{y}\right), \qquad (7.137)$$

where $\omega_H = \frac{eH}{mc}$, c is the light speed.

The Lagrange's equations have the form

$$\frac{d}{dt}\left(\frac{\partial L}{\partial \dot{q}_n}\right) - \frac{\partial L}{\partial q_n} = 0, \qquad (7.138)$$

and hence

$$\frac{\partial L}{\partial \dot{x}} = m\dot{x} + \frac{m}{2}\omega_H y, \qquad \frac{\partial L}{\partial x} = m\omega_1^2 x - \frac{m}{2}\omega_H \dot{y},$$

$$\frac{\partial L}{\partial \dot{y}} = m\dot{y} - \frac{m}{2}\omega_H x, \qquad \frac{\partial L}{\partial y} = m\omega_2^2 y + \frac{m}{2}\omega_H \dot{x}, \qquad (7.139)$$

$$\frac{\partial L}{\partial \dot{z}} = m\dot{z}, \qquad \frac{\partial L}{\partial z} = m\omega_3^2 z.$$

Finally, the following second order ODEs are obtained:

$$\ddot{x} = -\omega_1^2 x + \omega_H \dot{y}, \qquad \ddot{y} = -\omega_2^2 y - \omega_H \dot{x}, \qquad \ddot{z} = -\omega_3^2 z. \qquad (7.140)$$

Chapter 8

Maxwell's Equations

Complete analysis of the electromagnetic phenomena, including static and induced electric fields as well as magnetic effects, and others, can be carried out using a compact set of equations called Maxwell's equations. Maxwell's equations contribute to other important sets of equations regarding Newton's laws of classical mechanics and three laws of thermodynamics (see [Edminister and Hahvi (2011); Fleisch (2008); Schwartz (1972)]).

Maxwell's equations not only allow to explain and validate numerous experimental observations found in Maxwell's time (Maxwell's theory was published in 1873), but also allow to predict novel physical phenomena including the existence of electromagnetic waves and electromagnetic theory of light which was discovered by H. Hertz in 1888.

It should be emphasized that Maxwell's equations are universal, since they do not require any improvements due to the theory of relativity, as Newton's equation do.

In what follows, we present *Maxwell's equations in the integral form.*

(i) Gauss' law for the electric field

$$\oint \vec{E} \cdot d\vec{S} = \frac{Q}{\varepsilon_0}. \tag{8.1}$$

It governs the total electric field flux passing through a closed surface and represents the correspondence between the electric charge distribution and the resulting static electric field.

(ii) Gauss' law for the magnetic field

$$\oint \vec{B} \cdot d\vec{S} = 0. \tag{8.2}$$

It governs the total magnetic field flux passing through a closed surface, and the resulting zero net flux represents the experimentally deduced absence of magnetic monopoles.

217

(iii) *Faraday's induction law*

$$\oint \vec{E} \cdot \vec{dl} = -\frac{d\phi_B}{dt}.$$ (8.3)

It governs the circulation along a closed loop of the electric field induced by the changes introduced by the magnetic field flux $\dot{\phi}_B$. Since the circulation is nonzero ($\dot{\phi}_B \neq 0$), the field is *vortex*. The induced electric field $\vec{E}$ is different than that produced by static charges Q (see item (i)).

(iv) *Extended Ampére's law*

$$\oint \vec{B} \cdot \vec{dl} = \mu_0 i + \mu_0 \varepsilon_0 \underline{\frac{d\phi_E}{dt}}$$ (8.4)

or in the case of a magnetic sample,

$$\oint \vec{B} \cdot \vec{dl} = \mu_0 (i + i_D + \underline{i_M}).$$ (8.5)

If the term underlined in Eq. (8.4) equals zero, we obtain the already discussed *Ampére's law*. Since $\mu_0 \neq 0$, then Ampére's law exhibits the vorticity of the magnetic field generated by the current. The underlined term has been introduced by Maxwell, and it implies induction of the magnetic field by changing the electric field flux. In other words, changes in the magnetic field induce the electric field normal to the magnetic field. Extended Ampére's law allows us to avoid the previous right-hand side symmetry of the four equations (8.1)–(8.4).

In what follows, we briefly discuss transition from Eq. (8.4) to (8.5). One may introduce a purely conventional name of the "displacement current" $i_D = \varepsilon_0 \frac{d\phi_E}{dt}$, since $\left[\varepsilon_0 \frac{d\phi_E}{dt}\right] = [A]$. However, although the dimension of i_D coincides with a current, i_D does not produce any movement of charges. Therefore, omitting the underlined term in Eq. (8.5), one may deduce that the magnetic field may be either induced by the conduction current i or by the displacement current i_D. The underlined term presents the third way of magnetic field production. The so-called *magnetization current* i_M implies the additional conduction current producing the magnetic effect, being equivalent to that generated by the magnetic material.

The so far reported integral set of Maxwell's equations is complete. Both its left and right hand sides are symmetrical in parts. Recall that the magnetic flux equals zero in Eq. (8.2) (no magnetic monopoles exist in our matter), whereas in Eq. (8.1), the flux passing through a closed surface equals the electric charge. The term underlined in Eq. (8.4) is analogous to the term $\dot{\phi}_B$ occurring in Eq. (8.3).

The following auxiliary formulas must be added to complete the fundamental set of Maxwell's equations:

(v) The electric field $\vec{E}$ induces the current $\vec{J}$ in a conductor with conductivity σ:

$$\vec{J} = \sigma \vec{E}.$$ (8.6)

(vi) The electric induction $\vec{D}$ is associated with the electric field $\vec{E}$ by the formula

$$\vec{D} = \varepsilon\varepsilon_0\vec{E}. \tag{8.7}$$

(vii) The above formula implies the relationship between a magnetic induction $\vec{B}$ in a magnetic material and $\vec{H}$:

$$\vec{B} = \mu\vec{H}. \tag{8.8}$$

($viii$) The following relativity formula presents the relationship between the light speed in vacuum and the electric and magnetic constants

$$c^2 = \mu_0\varepsilon_0. \tag{8.9}$$

In addition, we present Maxwell's equations in the operator form (constitutive equations) and in the SI units. They follow:

$$\vec{\nabla} \times \vec{E} = -\frac{\partial\vec{B}}{\partial t}, \tag{8.10}$$

$$\vec{\nabla} \times \vec{H} = \vec{J} + \frac{\partial\vec{D}}{\partial t}, \tag{8.11}$$

$$\vec{\nabla} \cdot \vec{D} = \rho, \tag{8.12}$$

$$\vec{\nabla} \cdot \vec{B} = 0, \tag{8.13}$$

where $\vec{J} = \sigma\vec{E}$ is the current density.

Two auxiliary constitutive equations are as follows

$$\vec{B} = \mu_0\left(\vec{H} + \vec{M}\right) = \mu\vec{H}, \tag{8.14}$$

$$\vec{D} = \varepsilon_0\vec{E} + \vec{P} = \varepsilon\vec{E}, \tag{8.15}$$

where ε is the permittivity constant, and μ is the permeability constant. The quantities occurred in Eqs. (8.10)–(8.15) and their SI units are as follows: current density $[J] = [\text{A/m}^2]$, electric field density $[E] = [\text{V/m}]$, magnetic flux density $[B] = [\text{Wb/m}^2]$, electric displacement $[\text{D}] = [\text{C/m}^2]$, magnetization $[M] = [\text{A/m}]$, electric polarization $[P] = [\text{C/m}^2]$, charge density $[\rho] = [\text{C/m}^3]$, conductivity $[\sigma] = [\text{S/m}]$, free space permeability $[\mu_0] = [4\pi \cdot 10^{-7}\text{H/m}]$, free space dielectric constant $[\varepsilon_0] = [(1/36\pi) \cdot 10^{-9}\text{F/m}]$.

In addition, the following relations hold for $\varepsilon(\mu)$ and the corresponding magnetic and electric susceptibilities $\chi_m(\chi_e)$:

$$\begin{aligned} \mu &= \mu_0(1 + \chi_m), \\ \varepsilon &= \varepsilon_0(1 + \chi_e), \end{aligned} \tag{8.16}$$

where: $[\mu] = [\text{H/m}]$, $[\varepsilon] = [\text{F/m}]$.

For the free space set (for $\rho = 0$ and no conduction currents $\vec{J}_c = 0$), Maxwell's equations take the following point (or differential) form:

$$\nabla \times \vec{H} = \frac{\partial \vec{D}}{\partial t},$$

$$\nabla \times \vec{E} = -\frac{\partial \vec{B}}{\partial t}, \tag{8.17}$$

$$\nabla \cdot \vec{D} = 0,$$

$$\nabla \cdot \vec{B} = 0,$$

and the following integral form:

$$\oint \vec{H} \cdot d\vec{l} = \int_S \left(\frac{\partial \vec{D}}{\partial t} \right) \cdot d\vec{S},$$

$$\oint \vec{E} \cdot d\vec{l} = \int_S \left(-\frac{\partial \vec{B}}{\partial t} \right) \cdot d\vec{S}, \tag{8.18}$$

$$\oint_S \vec{D} \cdot d\vec{S} = 0,$$

$$\oint_S \vec{B} \cdot d\vec{S} = 0.$$

One may deduce that time-varying $\vec{E}$ and $\vec{H}$ cannot exist independently. If $\vec{E} = \vec{E}(t)$, then $\vec{D}(t) = \varepsilon_0 \vec{E}(t)$, and $\frac{\partial \vec{D}}{\partial t} \neq 0$ (see first two point form equations). It means that $\nabla \times \vec{H} \neq 0$, which means that $\vec{H}(t) \neq 0$ is generated beginning with $\vec{H} = \vec{H}(t)$ and showing that $\vec{E}$ should be present.

Maxwell's equations exhibit a set of Faraday's law, Ampére's law, and Gauss' law which govern electrostatic and magnetostatic as well the dynamic time-varying electromagnetic fields.

In the case of static electric and magnetic fields, the equations corresponding to a field vector are not coupled. They follow:

(i) electric field:

$$\nabla \times \vec{E} = 0 \quad \text{(Faraday' law)}, \qquad \nabla \cdot \vec{D} = \rho \quad \text{(Gauss' law)}; \tag{8.19}$$

(ii) magnetic law:

$$\nabla \times \vec{H} = \vec{J} \quad \text{(Ampére's law)}, \qquad \nabla \cdot \vec{B} = 0 \quad \text{(Gauss' law)}. \tag{8.20}$$

Integral form of Faraday's law is as follows

$$\oint_C \vec{E} \cdot d\vec{l} = -\frac{d}{dt} \int_S \vec{B} \cdot d\vec{S}, \tag{8.21}$$

where the right-hand rule holds to properly define the positive sense around C and the normal direction $d\vec{S}$. When $\vec{B}(t)$ increases, the right-hand side of Eq. (8.21) becomes negative. It means that the direction of $\vec{E}$ must be opposite to that of the contour to get the negative value of the left-hand side of Eq. (8.21).

This observation has allowed Lenz to formulate his law:

The induced voltage and the change in magnetic flux have opposite signs. It is a qualitative law that refers to the direction of the induced current in relation to the effect which produces it, without quantitatively relating their magnitudes.

A *motional* electric field intensity in time-independent fields $\vec{E}_m$ can be defined as the force per unit charge

$$\vec{E}_m = \frac{\vec{F}}{q} = \vec{V} \times \vec{B}. \tag{8.22}$$

Now, when a lot of free charges move in a conductor through a field $\vec{B}$, $\vec{E}_m$ induces a voltage difference between two ends A and B of the conductor. Voltage of A with respect to B follows:

$$V_{\mathrm{AB}} = \int_{\mathrm{B}}^{\mathrm{A}} \vec{E}_m \cdot d\vec{l} = \int_{\mathrm{A}}^{\mathrm{B}} \left(\vec{V} \times \vec{B} \right) \cdot d\vec{l}. \tag{8.23}$$

Assuming that the conductor is normal to $\vec{V}$ and $\vec{B}$, and $\vec{V} \perp \vec{B}$, we get

$$V = BLv.$$

In the case of a closed loop conductor equation (8.23),

$$V = \oint \left(\vec{v} \times \vec{B} \right) d\vec{l}.$$

The situation is more complicated in the case of time-dependent fields. Since Faraday's law takes the following form

$$V = -\frac{d}{dt} \int_S \vec{B} \cdot d\vec{S} = -\int_S \frac{\partial \vec{B}}{\partial t} \cdot d\vec{S} + \oint \left(\vec{v} \times \vec{B} \right) d\vec{l}. \tag{8.24}$$

The right-hand side of Eq. (8.24) has two terms. The first term contributes to the voltage amount regarding the change of $\vec{B}$, whereas the second term presents voltage generated by the motion of the loop for fixed $\vec{B}$.

Chapter 9

Optimization

9.1 Introduction

Optimization aims at obtaining the best result (solution) at given conditions and a criterion of assessment. The obtained result is called the *optimal solution.*

Optimization is applied to the inanimate matter, flora and fauna. A man consciously conducting optimization is counting on both the positive personal and social effects.

The main objective of considerations of engineers and manufacturers is to seek the optimal solutions. The search for optimal solutions by means of some empirical methods is labor-intensive and consumes large amount of resources, and thus it is also very costly.

It has become a very important ability to develop mathematical models of the designed or already existing objects. The analytical description of such objects facilitates finding the optimal solution and significantly reduces costs and time of optimization. The modern level of information technology makes the process of optimization less abstract – more transparent – observable, and less time consuming. Special computer programs help in leading the optimization process for engineering purposes.

According to many fields of application, the optimization can be static, dynamic, single- or multi-objective.

Static optimization is a process of searching for the optimum values of decision variables (control signals) of the investigated stationary object that meet the given *criterion* – the objective function or quality index.

Dynamic optimization is a process of searching for the optimum time histories of decision variables that would guarantee obtaining the assumed objective function.

Single-objective (scalar) optimization is based on looking for an optimal (minimum or maximum) value of only one scalar-valued objective function.

Multi-objective (vector) optimization focuses on searching for the best solution by analyzing a few vector-valued objective functions. Choosing the best solution from a set of optimal solutions requires to make a decision being a compromise resulting from the assumed system of values. To a certain multi-objective optimiza-

223

tion problem one can attach a scalar one, the optimal solution of which leads to solutions to the original problem [Lorenz and Wanka (2012)].

The selection has to be done by the designer, who can can perform the following actions:

- choose the *best solution* (in their opinion);
- choose a subset from the set of *optimal values*;
- create a *ranking list* for re-analysis of the obtained solutions.

9.2　Methods of Optimization

Optimization can be conducted with the use of experimental or mathematical methods [Borzi and Schulz (2012); Butenko and Pardalos (2014); Delfour (2012); Kanno (2011); Kelley (1999)].

9.2.1　*Experimental Methods*

Experimental methods belong to the oldest, and are used in current industrial applications along with the modern ones. In the vast majority, forest roads and lanes delineated years before, hiking and water trails or routes to mountain peaks do not require any correction. Achievements of old technology are still admired nowadays.

The experimental methods can be used for:

(i) direct searching for the optimal solutions,

(ii) carrying out experiments focused on finding parameters necessary to an empirical derivation of the mathematical model of the optimized physical object.

Usually, the empirical model is based on functions introduced in Chap. 2 (p. 7), given by formulas (2.1) and (2.2).

In particular, `Scilab` provides a function `polyfit` which computes values of an approximating function with the use of the least squares method. A `Scilab` numerical procedure given by Algorithm 9.1 computes values of square function approximating a series f_1 of test data.

Algorithm 9.1. A procedure created in `Scilab`, computing the approximate function $f_2(x)$ of the experimentally obtained function $f_1(x)$.

$x = $ **linspace**$(0, 1, 11)$'
$f_1 = [$-0.447, 1.978, 3.28, 6.16, 7.8, 7.34, 7.66, 9.56, 9.02, 8.3, 7.2$]$
//polyfit() and polyval() require packages 'Linear Algebra' and 'Stixbox'
$p = $ **polyfit**$(x, f_1, 2)$
$f_2 = $ **polyval**(p, x)
plot$(x, f_1$', "g-", x, f_2, "b-"$)$

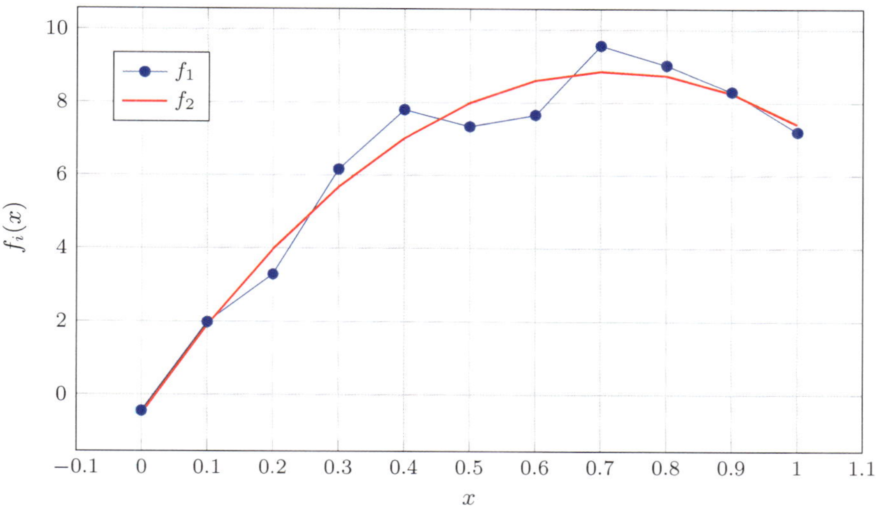

Fig. 9.1 Graphical representation of the function $f_1(x)$ (blue line) acquired from experimental measurements and its polynomial approximation $f_2(x)$ (red line) of the second degree.

9.2.2 *Mathematical Methods*

In general, some analytical, numerical and geometric programming methods of optimization are distinguished. For each of them, a mathematical model is necessary to elaborate an optimization procedure.

Conducting many optimization processes using the mathematical methods [Leondes (1981)], the following steps are usually made:

1. Choosing the criterion by determining the objective function

$$f = f(x_1, x_2, \ldots, x_j), \tag{9.1}$$

   where x_j are the decision variables. A particular objective function can be written in the form $f = c_1 x_1 + c_2 x_2 + \cdots + c_j x_j$.
2. Indication of the type of the sought optimum (minimum or maximum) of the objective function.
3. Formulation of problem constraints in the form of inequality or equality functions:

$$\begin{vmatrix} f_1(x_1, x_2, \ldots, x_j) \\ f_2(x_1, x_2, \ldots, x_j) \\ \cdots\cdots\cdots\cdots \\ f_2(x_1, x_2, \ldots, x_j) \end{vmatrix} \begin{matrix} \leq \\ = \\ \geq \end{matrix} \begin{vmatrix} b_1 \\ b_2 \\ \cdots \\ b_n \end{vmatrix}, \tag{9.2}$$

   where $b_{1,\ldots,n}$ are the constraints.

The following short formulation is also used:

$$\sum_{j}^{n} a_{nj}x_j \overset{\leq}{\underset{\geq}{=}} b_n \,, \tag{9.3}$$

where: n – row number, j – column number, a_{nj} – components.
4. Writing the constraints for the decision variables $x_j \geq 0$.

Table 9.1   Analytical description of elements of constraint functions.

COMPONENTS	PRODUCTS				CONNECTIONS	RESTRICTIONS
S	x_1	x_2	$\ldots$	x_j	$\leq = \geq$	b
S_1	$a_{11}x_1$	$a_{12}x_2$	$\ldots$	$a_{1j}x_j$	$\leq$	b_1
S_2	$a_{21}x_1$	$a_{22}x_2$	$\ldots$	$a_{2j}x_j$	$=$	b_2
$\ldots$	$\ldots$	$\ldots$	$\ldots$	$\ldots$	$\geq$	$\ldots$
S_n	$a_{n1}x_1$	$a_{n2}x_2$	$\ldots$	$a_{nj}x_j$		b_n
f – objective function	c_1x_1	c_2x_2	$\ldots$	c_jx_j	**min** or **max**	f_{opt}

If symbol $\geq$ appears in Table 9.1, then coefficients a and b take opposite values. If relations (9.1) and (9.2) are linear, then the optimization reduces itself to a linear problem. This kind of optimization is called *linear programming*. Otherwise, if there appear products, quotients or powers of one or many decision variables in relations (9.1) and (9.2), then the optimization states a nonlinear problem, which in this case is called *nonlinear programming*. Table 9.1 contains analytical description of elements of the constraint functions.

Among the numerical methods of optimization, the following can be enumerated [Ostanin (2009)]:

a) searching for optimum values of functions,
b) linear programming,
c) square programming,
d) nonlinear programming with or without constraints,
e) least square optimization,
f) multi-objective optimization (also known as vector optimization, multi-criteria optimization or Pareto optimization),
g) evolutionary optimization.

9.3   Examples

9.3.1   *Linear and Nonlinear Programming*

Problem 9.1. *A company produces two kinds of paving, i.e., in red and grey color [Neogy et al. (2009)]. Calculate a proportion between the number of tons of red* *and grey paving that will guarantee the lowest production costs at the following constraints:*

1. Production of 1 ton of red paving requires 2 working hours of machines, 3 working hours of humans and 2 liters of pigment.
2. Production of 1 ton of grey paving requires 1 working hour of machines and 3 working hours of humans. Both kinds of paving require the same amount of cement and gravel.
3. The company is at disposal of 10 working hours of machines, 24 working hours of humans and 8 liters of pigment.
4. Production of 1 ton of red paving makes 300 PLN of profit, but production of 1 ton of grey paving makes 200 PLN of profit.

Derivation of a mathematical description to the Problem 9.1 will be easier after creating the Table 9.2.

Table 9.2   Description of the optimization Problem 9.1.

COMPONENTS	PRODUCTS (paving)		CONSTRAINTS $b_{1,\dots,3}$
Means S consumption of ti1me and resources			
x_1 [tons]	red		
x_2 [tons]		grey	
Machines [h/ton]	2	1	10 [h]
Employees [h/ton]	3	3	24 [h]
Pigment [l/ton]	2	0	8 [h]
Profit Z [PLN/ton]	300	200	none [PLN]

On the basis of the objective function of profit defined in the last row of Table 9.2, the following equation can be written

$$Z = 300x_1 + 200x_2 \,. \tag{9.4}$$

Using the data given in Table 9.2, the equations defining some permissible area are found:

$$2x_1 + x_2 \leq 10,$$
$$3x_1 + 3x_2 \leq 24, \tag{9.5}$$
$$2x_1 \leq 8,$$

with constraints for the decision variables:

$$x_1 \geq 0, \quad x_2 \geq 0. \tag{9.6}$$

Replacing inequality symbols in (9.5) and (9.6) by equality symbols, the linear paired equations are obtained. Graphs of linear functions present points of intersection of functions, which define the unknown solutions as well as give information about parallelism or mutual coverage of these functions. This, however, is not suitable to determine the permissible area, which is necessary to find an optimum of the objective function (9.4). Looking for the optimum relies on choosing the point having the greatest value from all values determined by intersections. Additionally, the point must lie in the permissible area. In Table 9.3, the sought optimal value is marked in red.

Table 9.3   Looking for the optimal value of the objective function of profit Z.

INTERSECTION POINTS	PERMISSIBLE AREA	OBJECTIVE FUNCTION
Equations: a) $2x_1 + x_2 = 10$ [M] b) $3x_1 + 3x_2 = 24$ [L] c) $2x_1 = 8$ [F] d) $x_1 = 0$ [axis] e) $x_2 = 0$ [axis]	1) $2x_1 + x_2 \leq 10$ [M] 2) $3x_1 + 3x_2 \leq 24$ [L] 3) $2x_1 \leq 8$ [F] Are the conditions of permissible area met? Y – yes, N – no	Eq. (9.4)

No.	Pair of Eqs.	x_1	x_2	(1)	(2)	(3)	Z – profit
		Point		Inequality			
1	a b	2	6	**Y**	**Y**	**Y**	1800
2	a c	4	2	Y	Y	Y	1600
3	a d	0	1 0	Y	N	Y	2000
4	a e	5	0	Y	Y	N	1500
5	b c	4	4	N	Y	Y	2000
6	b d	0	8	Y	Y	Y	1600
7	b e	8	0	N	Y	N	2400
8	c d	‖	–	–	–	–	–
9	c e	4	0	Y	Y	Y	1200
10	d e	0	0	Y	Y	Y	0

In the case of two independent variables, the permissible area and the optimal value of the objective function are found geometrically by drawing all functions expressed by particular constraints. The optimal value is to be estimated by comparing values of the constraint functions in all characteristic points. The geometrical method can be used for linear and nonlinear problems.

Now, a solution to the Problem 9.1 can be proposed.

The company will generate the lowest costs by producing 2 tons of red paving per 6 tons of grey paving. The ratio of tons does not take the account of economic factors, e.g., a demand for products of this company.

A numerical and graphical solution to the linear programming problem, described by relations (9.4)-(9.6) and realized in the simulation model depicted in Fig. 9.2, is presented in Fig. 9.3.

Let us extend our task. Focusing ourselves only on the algebraic equations defin-

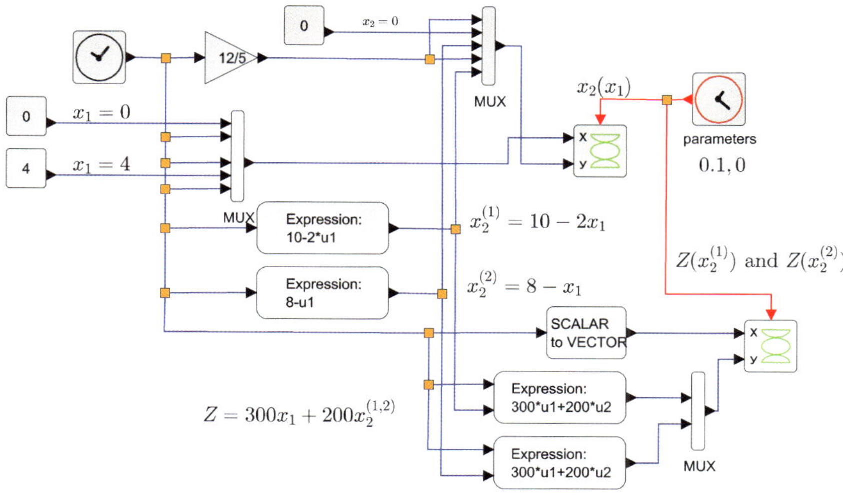

Fig. 9.2   Simulation model of a linear optimization problem.

ing the objective function and constraints for 2 decision variables of the nonlinear problem (compare with relations (9.4)-(9.6)), i.e.,

$$Z = 2x_1 + 1.5x_2, \quad 2x_1 + x_2 = 10, \quad 0.5x_1^2 + x_2 = 8, \quad x_1 = x_2 = 0, \qquad (9.7)$$

a simulation model shown in Fig. 9.4 has been created for obtaining a solution to the nonlinear programming problem in the geometrical form shown in Fig. 9.5.

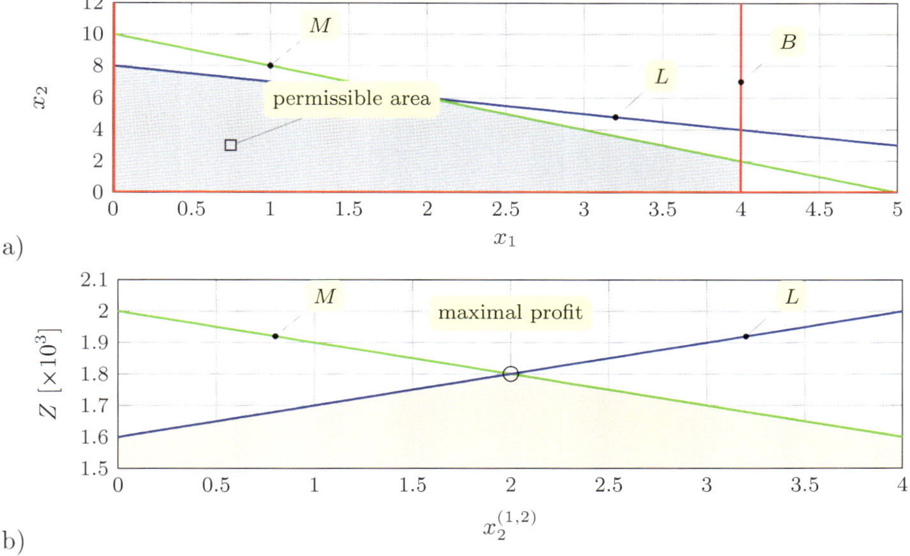

Fig. 9.3   A graphical solution to the linear programming problem: (a) the permissible area; (b) the objective function $Z(x_2^{(1,2)})$, created in `Scilab`.

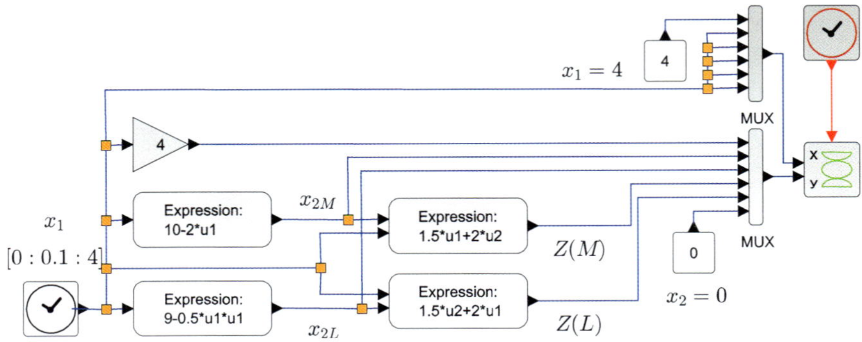

Fig. 9.4   Simulation model of a nonlinear optimization.

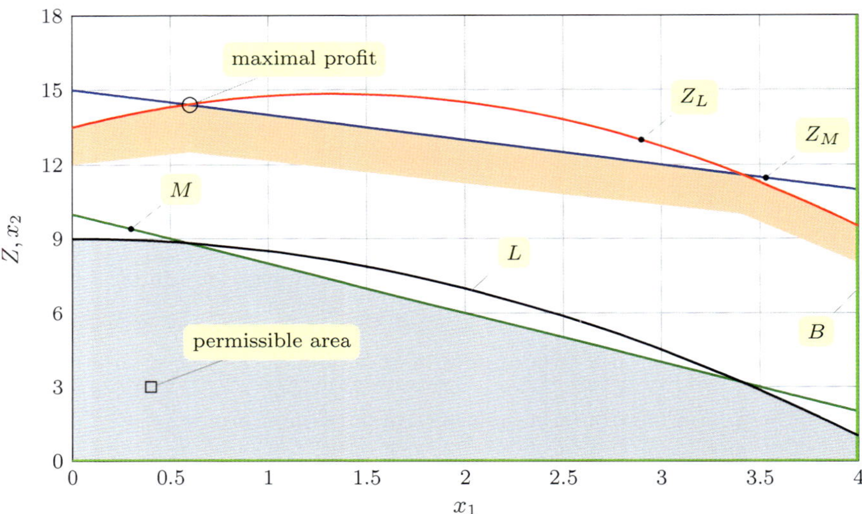

Fig. 9.5   Solutions $Z(x_1)$ and $x_2(x_1)$ to the nonlinear optimization problem solved in `Scilab` by means of a geometrical method.

Among others, geometrical methods are used to solve linear programming problems with the number $n > 2$ of independent variables. A geometrical evaluation of the constraints determined by 4 planes is shown in Fig. 9.6.

Simplex, gradient and programming methods are useful in searching for optimal solutions to the linear problems given by the general system of algebraic equations:

$$
\begin{aligned}
S_1 \quad & a_{11}x_1 + a_{12}x_2 + a_{13}x_3 + \cdots + a_{1j}x_j = b_1, \\
S_2 \quad & a_{21}x_1 + a_{22}x_2 + a_{23}x_3 + \cdots + a_{2j}x_j = b_2, \\
& \vdots \qquad\qquad \vdots \qquad\qquad \vdots \\
S_n \quad & a_{n1}x_1 + a_{n2}x_2 + a_{n3}x_3 + \cdots + a_{nj}x_j = b_n,
\end{aligned}
\tag{9.8}
$$

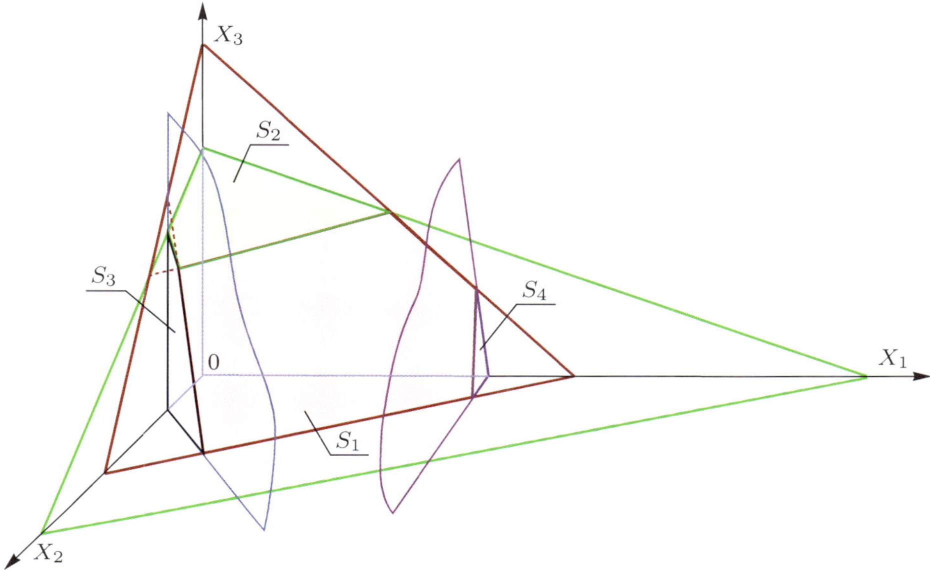

Fig. 9.6   Example of the geometrical method of linear programming with the constraints determined by planes $S_1 \dots S_4$.

In literature, the simplex method is a common technique for solving many linear problems [Pardalos *et al.* (2003); Shi (2001); Ostanin (2009)]. The proposed algorithm is complex, but for multi-dimensional purposes one can use some programs that could be helpful in searching for its solution. The representative examples follow: LPSolve, LIPSOL, Matlab Optimization Toolbox, Java ILP, Linear Program Solver, GLPK.

Problem 9.2. *A gardening company is at disposal of 3 types of fertilizers x_1, x_2, x_3, which can be used to fertilize the soil. The fertilizers contain up to 5 components $S_1 \dots S_5$. It has been calculated that the minimum number of units of individual components assigned to fertilize the field under cultivation is: $S_1 = 90$, $S_2 = 40$, $S_3 = 80$, $S_4 = 150$, $S_5 = 100$. Unit prices of the fertilizers are given: $x_1 = 20$, $x_2 = 16$, $x_3 = 10$.*

The method of solution should be based on such an assignment of particular fertilizers that while complying with the prescribed requirement, the fertilization costs would be minimal.

Let us initiate a mathematical description of that problem by putting all knows into the Table 9.4. In addition to the data in the problem task, some values of components $S_1, \dots, S_5$ written on fertilizers' packaging have been assumed.

One searches for a minimum of the objective function f which, according to the data in Table 9.4, has the form

$$f(\bar{x}) = 20x_1 + 16x_2 + 10x_3. \tag{9.9}$$

Table 9.4   Description of the optimization problem 9.2.

COMPONENTS S_i – means of prod.	PRODUCTS			CONSTRAINTS b_n
	x_1	x_2	x_3	
S_1	4	5	1	90
S_2	2	1	6	40
S_3	6	4	0	80
S_4	7	3	4	150
S_5	0	5	3	100
Unit cost	20	16	10	min. of obj. fun.

Constraints determined from the amount of components S_i included in the fertilizers are expressed by the relations:

$$4x_1 + 5x_2 + x_3 \geq 90,$$
$$2x_1 + x_2 + 6x_3 \geq 40,$$
$$6x_1 + 4x_2 + 0 \geq 80, \tag{9.10}$$
$$7x_1 + 3x_2 + 4x_3 \geq 150,$$
$$0 + 5x_2 + 3x_3 \geq 100,$$

which should be supplemented with the constraints that the decision variables have to satisfy:

$$x_1 \geq 0, \quad x_2 \geq 0, \quad x_3 \geq 0. \tag{9.11}$$

Solution to the described problem will be found using a simplex method in the Devex strategy implemented in program **LPSolve** (see Algorithm 9.2).

Algorithm 9.2. Input data for the procedure optimizing the objective function given by Eq. 9.9.

// Objective function f
min: $20x_1 + 16x_2 + 10x_3$
// Constraints for the decision variables
C1: $4x_1 + 5x_2 + 1x_3 >= 90$
C2: $2x_1 + 1x_2 + 6x_3 >= 40$
C3: $6x_1 + 4x_2 + 0x_3 >= 80$
C4: $7x_1 + 3x_2 + 4x_3 >= 150$
C5: $0x_1 + 5x_2 + 3x_3 >= 100$
$x_1 >= 0$
$x_2 >= 0$
$x_3 >= 0$

After execution of the numerical procedure taking $0.01\,$s, we find the solution $\bar{x}^* = [7.2, 9.2, 18]$ at which the minimum of the objective function $\min\{f(\bar{x})\} = f(\bar{x}^*) = 543.2$.

Problem 9.3. *Compare the numerical and geometrical method of solution of a 3-dimensional optimization problem. Find a minimum of the objective function*

$$f(\bar{x}) = 2x_1 + 1.5x_2 + 3x_3, \tag{9.12}$$

subject to the constraints:

$$5x_1 + 10x_2 + 25x_3 \geq 250,$$
$$2x_1 + 2x_2 + 3x_3 \geq 60, \tag{9.13}$$
$$5x_1 + 4x_2 + 2x_3 \geq 100.$$

General equations of planes are obtained by replacing in Eq. (9.13) the inequality with equality symbols. The current numerical procedure is provided in Algorithm 9.3 by applying the same method of numerical solution used for the previous problem.

Algorithm 9.3. Input data for the `LPSolve` procedure optimizing the objective function expressed by Eq. (9.12).

```
// Objective function f
min: 2x1 + 1.5x2 + 3x3
// Constraints for the decision variables
C1: 5x1 + 10x2 + 25x3 >= 250
C2: 2x1 + 2x2 + 3x3 >= 60
C3: 5x1 + 4x2 + 2x3 >= 100
x1 >= 0
x2 >= 0
x3 >= 0
```

After execution of the Algorithm 9.3 taking $0.007\,$s, we find the solution $\bar{x}^* = [0, 30, 0]$ at which the minimum of the objective function $\min\{f(\bar{x})\} = f(\bar{x}^*) = 45$.

In the first step of the graphical method, the system of equations (9.13) is transformed to 3 basic equations of planes $S_{1,\dots,3}$ as follows:

$$S_1: \quad \frac{1}{50}x_1 + \frac{1}{25}x_2 + \frac{1}{10}x_3 = 1,$$
$$S_2: \quad \frac{1}{30}x_1 + \frac{1}{30}x_2 + \frac{1}{20}x_3 = 1, \tag{9.14}$$
$$S_3: \quad \frac{1}{20}x_1 + \frac{1}{25}x_2 + \frac{1}{5}x_3 = 1.$$

In the second step, planes $S_{1,\dots,3}$ have to be drawn in the Cartesian system of coordinates. Vertices of the planes have to be placed on the axes, at values being the reciprocals of coefficients at subsequent variables $x_{1,\dots,3}$. For example, plane S_1 is defined by lines beginning at values 50, 25 and 10 on the axis x_1, x_2 and x_3, respectively. As a result, we get a graph of planes shown in Fig. 9.7.

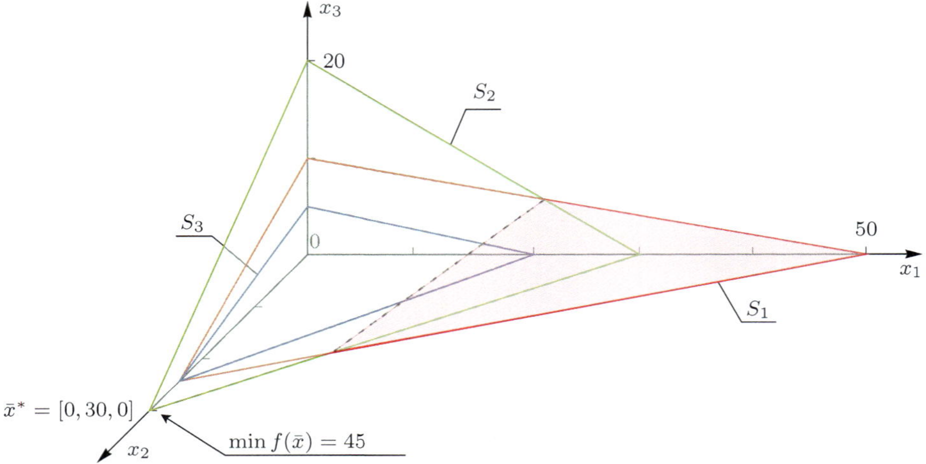

Fig. 9.7   A graphical method of solution to an optimization problem minimizing the objective function defined in Problem 9.3.

9.3.2   *Dynamic Programming*

A walled cantilever beam carries an uniformly distributed load q as presented in Fig. 9.8. The load q acts transversely on the beam, and a next load P is concentrated in a point and moves in x-axis along the beam.

Problem 9.4. *Find the solution (a corrective force P_k) in which, independently of the placement of the concentrated load P, any deformation of the beam in the direction of y-axis does not exist. To meet the requirement, we have to assume that the objective function $y = 0$.*

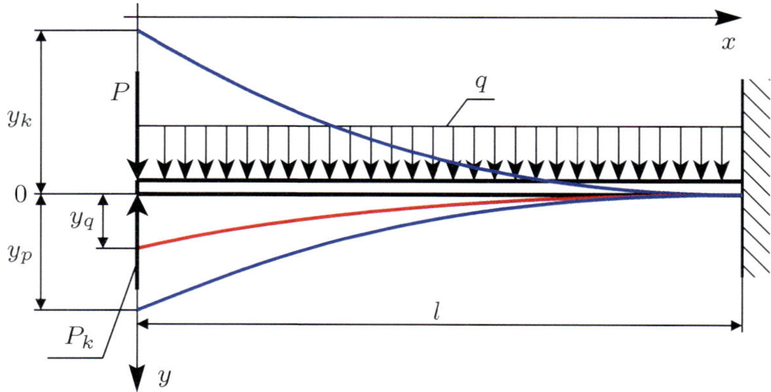

Fig. 9.8   Distribution of loading of the analyzed cantilever beam.

A proposition for solving the stated problem follows:

> Compute a function of corrective force $P_k(x)$ reacting in opposite direction to the action of load P and moving in x-axis, respectively to the load displacement.

It results from the proposition that

$$y = y_q + y_p + y_k = 0, \qquad (9.15)$$

where:

$$y_q = \frac{q}{24EJ}\left(x^4 - 4l^3x + 3l^4\right),$$

$$y_p = \frac{P}{6EJ}\left(x^3 - 3lx^2 + 2l^3\right), \qquad (9.16)$$

$$y_k = -\frac{P_k}{6EJ}\left(x^3 - 3lx^2 + 2l^3\right).$$

Combining Eqs. (9.15) and (9.16), a relation for the corrective force is determined as follows

$$P_k(x) = P + q\frac{x^4 - 4l^3x + 3l^4}{4\left(x^3 - 3lx^2 + 2l^3\right)}. \qquad (9.17)$$

A mechanical engineer should now ask a question:

> What kind of actuating system would be needed to produce such a mechanical forcing that will react perpendicularly to the analyzed beam's bottom surface, in a response to the concentrated load P moving in x-axis of the beam?

Partial solution to the task, relying on generation of the perpendicular corrective force given by Eq. 9.17, can be realized in a hydraulic system schematically presented in Fig. 9.9. During the corrective forcing, a tip end of hydraulic cylinder's rod should be in sliding contact with the bottom surface of the beam and move in the x-axis along its length accordingly to the displacement of load P.

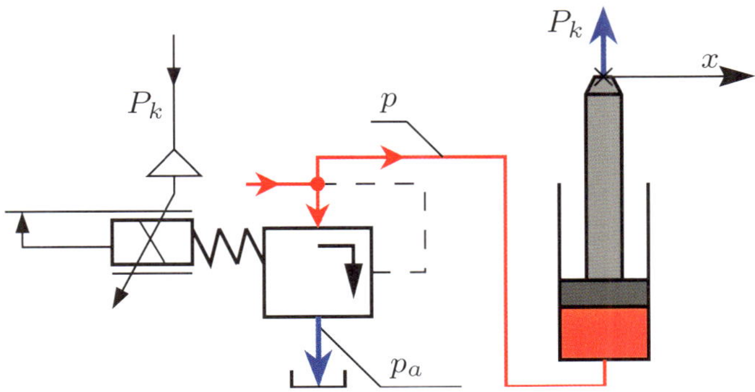

Fig. 9.9   A hydraulic system generating the corrective force P_k.

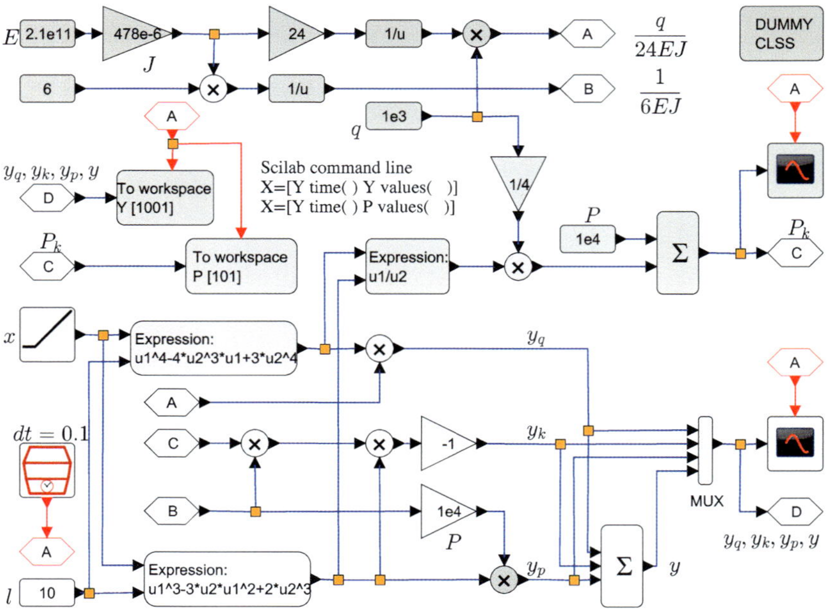

Fig. 9.10   A simulation model solving the optimization Problem 9.4.

In the problem of interest, formulas (9.15)-(9.16) state the mathematical model. A simulation model created on this basis, as well as some results of its execution, are presented in Fig. 9.11 on x-dependent plots.

To conclude, we computed the corrective force $P_k(x)$ reacting in the direction opposite to the action of load P and moving in x-axis respectively to the load displacement. The simulation has confirmed highly optimized response of the actuating system, because function $y(x)$ of displacement is equal to zero on the whole length of the beam.

9.3.3   *Geometric Programming Methods*

Geometric programming is devoted to a type of a mathematical optimization problem characterized by objective and constraint functions that have a special form [Boyd *et al.* (2007)]. The importance of geometric programming comes from two recent developments.

1. New solution methods can solve even large-scale geometric programming problems extremely efficiently and reliably.
2. A number of practical problems in electric circuit design have recently been found to be well approximated by geometric programs.

An example devoted to generalized geometric programming touches a problem of floor planning [Rosenberg (1989)]. In the problem, there exist some rectangles

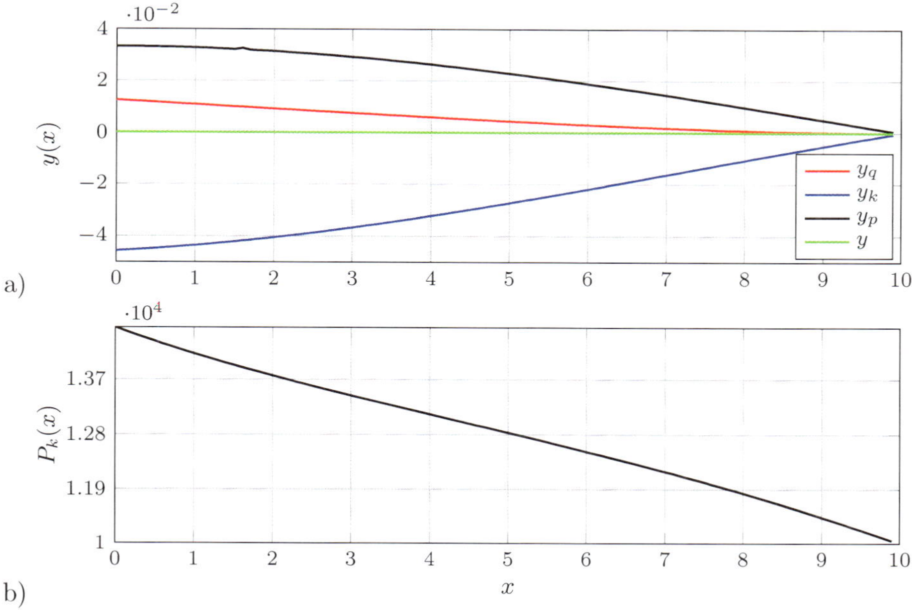

a)

b)

Fig. 9.11   Results of exploration of the simulation model in Fig. 9.10. Plots of the deformation $y(x)$ (a) of the cantilever beam shown in Fig. 9.8 and the corrective force $P_k(x)$ (b) of the beam deformation versus x displacement. Model parameters: $E = 2.1 \cdot 10^{11}$, $J = 478 \cdot 10^{-6}$, $P = 10^4$, $q = 10^3$, $l = 10$.

to be configured and placed in such a way that they do not mutually overlap. The objective is usually to minimize the area of the bounding box, which is the smallest rectangle that contains the rectangles to be configured and placed. Each rectangle can be reconfigured within some limits. An extension to the research on floor planning can be found in [Sherwani (1999)].

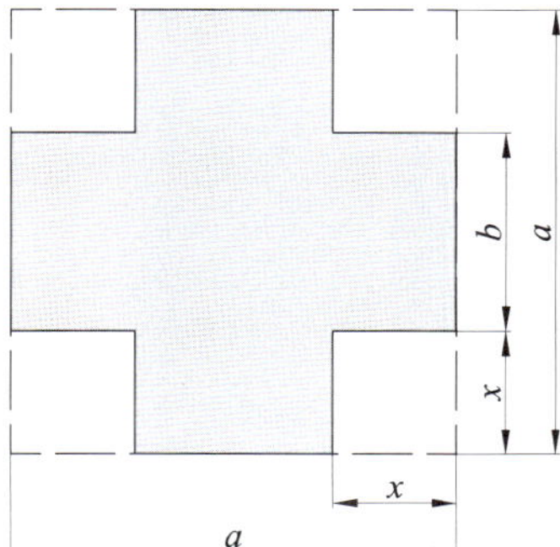

Fig. 9.12   Dimensions of the unfolded surface of the box.

Looking for the dimensions of a box having a maximal volume at given bounding surface of the box is a next good example of usage of the geometric programming method of optimization [Neogy *et al.* (2009)].

Problem 9.5. *Calculate dimensions a, b and h of a box having the maximal volume at given bounding (total) surface of the box. The unfolded surface of the box is shown in Fig. 9.12.*

In the case of the provided unfolded surface, the objective function is expressed by the formula for volume of the box and some relations between unknown dimensions a, b and h in function of the decision variable x:

$$V(x) = (a - 2x)^2 h, \quad b(x) = a - 2x, \quad h(x) = x,$$

subject to a constraint for the decision variable, i.e., $0 < x < a/2$.

The maximal volume of the box is determined by a maximum of the objective function of variable x, i.e., $\max\{(a - 2x)^2 x\}$. This extreme value falls in one of the points of the zero of first derivative if $V'(x) = 12x^2 - 8x + 1 = 0$, where we get two solutions $x_1 = 1/6$ and $x_2 = 1/2$. The function $V(x)$ has the maximum $V_{\max} = V(x_1) = 2/27$ in the point x_1. If one calculates $b(x_1)$ at $a = 1$ (the total surface is given) and $h(x_1)$, then one gets some optimal dimensions of the box: $b_{\mathrm{opt}} = 2/3$, $h_{\mathrm{opt}} = 1/6$.

It is possible to read from Fig. 9.13a the maximum $V_{\max}$ of the objective function in point x_{opt} at the assumption that $a = 1$. On the other hand, in Fig. 9.13b points of intersection of graphs and the dashed green line match values of requested dimensions b_{opt} and h_{opt} of the box having the largest volume.

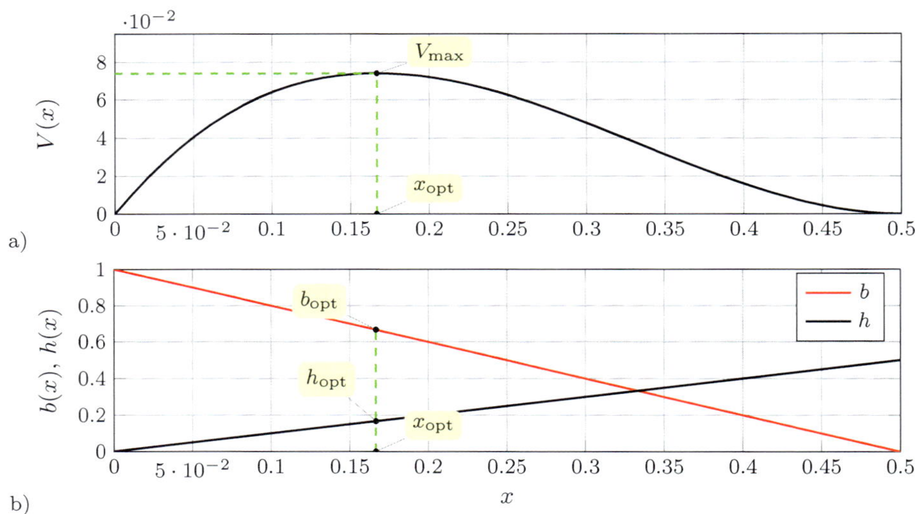

Fig. 9.13   Results of solution to the optimization Problem 9.5: (a) function of volume $V(x)$; (b) linear functions of dimensions $h(x)$ and $b(x)$.

Problem 9.6. *Calculate lateral dimensions of an unilaterally mounted cantilever beam of a rectangular cross section (see Fig. 9.14), loaded at the end [Neogy et al. (2009)]. The material used to make the beam has a shape of a solid circular shaft of the length l and diameter D. The lateral dimensions should guarantee:*

    a) the smallest bending stress σ_g,
    b) the largest stiffness j,
    c) and the smallest compression stresses σ_s.

The following definitions initiate the optimization:

$$\sigma_g = \frac{Fl}{W}, \quad j = \frac{F}{l} = \frac{3E}{l^3}I, \quad \sigma_s = \frac{F}{S}, \tag{9.18}$$

where: F is a fixed bending or compression force, l – known length of the beam, W – section modulus, E – Young's modulus, S – cross-sectional area of the beam.

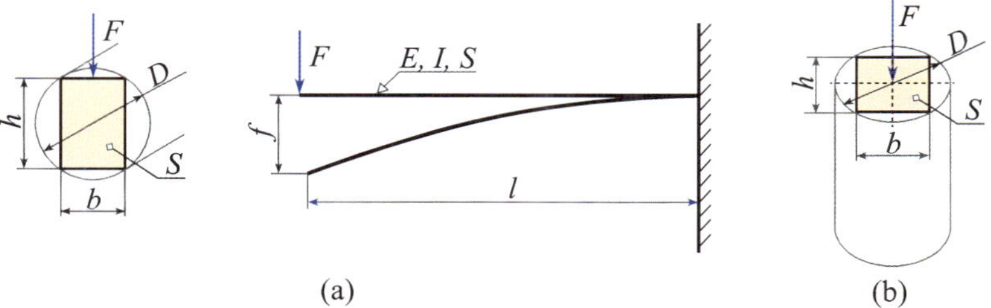

(a)   (b)

Fig. 9.14   Physical model and dimensions of the cantilever beam subjected to pure bending (a) and compression (b).

Fulfillment of the problem requirements (a-c) leads to the following:

   – section modulus should reach a maximum value:

$$W = \max\left\{bh^2/6\right\},$$

   – moment of inertia about the bending axis should reach a maximum value:

$$I = \max\left\{bh^3/12\right\},$$

   – area of cross section has to be the largest:

$$S = \max\left\{bh\right\}.$$

At the assumed diameter $D = 1$ of the shaft of which the beam is to be made, one takes into account that the first lateral dimension $b = x$ will be the decision variable subject to constraints $0 < x < 1$. Then, using the geometrical relation between the diameter of the shaft and dimensions b and h of the rectangular cross section inscribed in a circle of the diameter D, one finds:

$$D^2 = h^2 + x^2 \quad \Rightarrow \quad h = \sqrt{D^2 - x^2}.$$

Continuing with the introduced requirements (a-c), the optimization problem of searching for some lateral dimensions of the beam will be solved if the following functions have maximum values defined as below:

* section modulus:

$$W(x) = x\left(1 - x^2\right)/6\,, \tag{9.19}$$

* moment of inertia:

$$I(x) = x\left(1 - x^2\right)^{3/2}/12\,, \tag{9.20}$$

* area of cross section:

$$S(x) = x\sqrt{1 - x^2}\,. \tag{9.21}$$

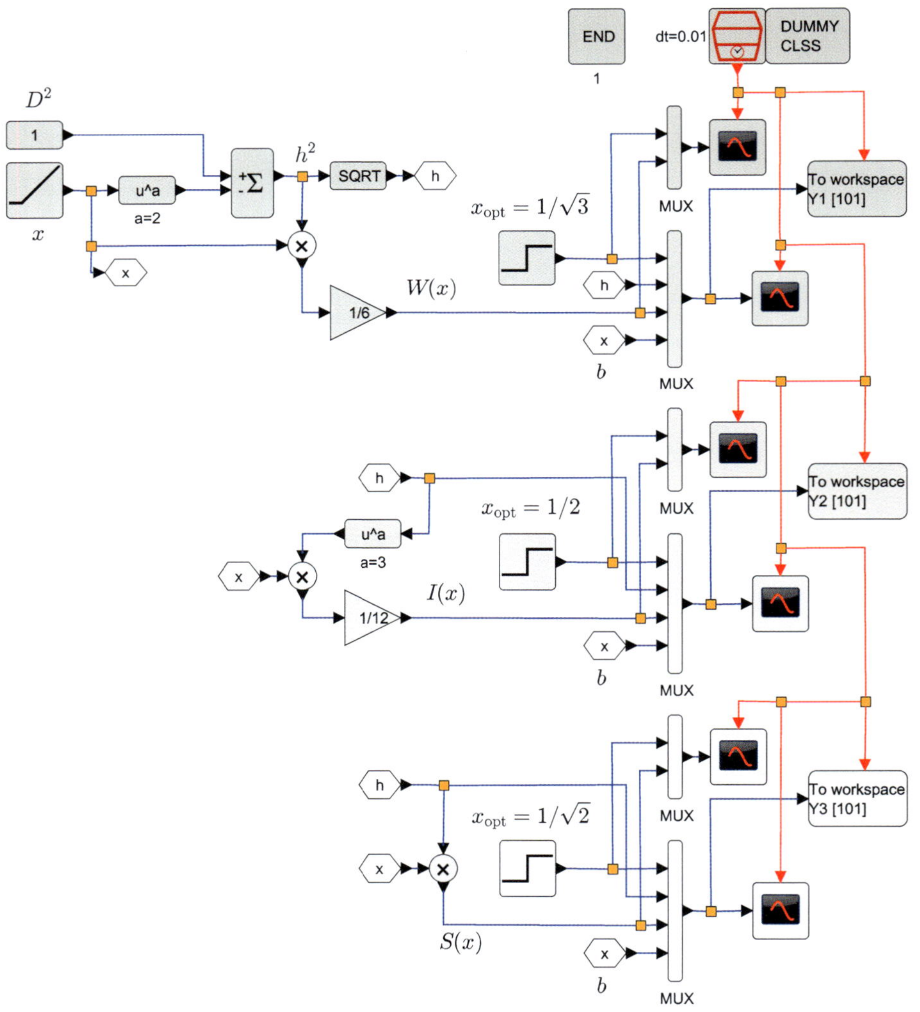

Fig. 9.15   A simulation model solving the optimization Problem 9.6.

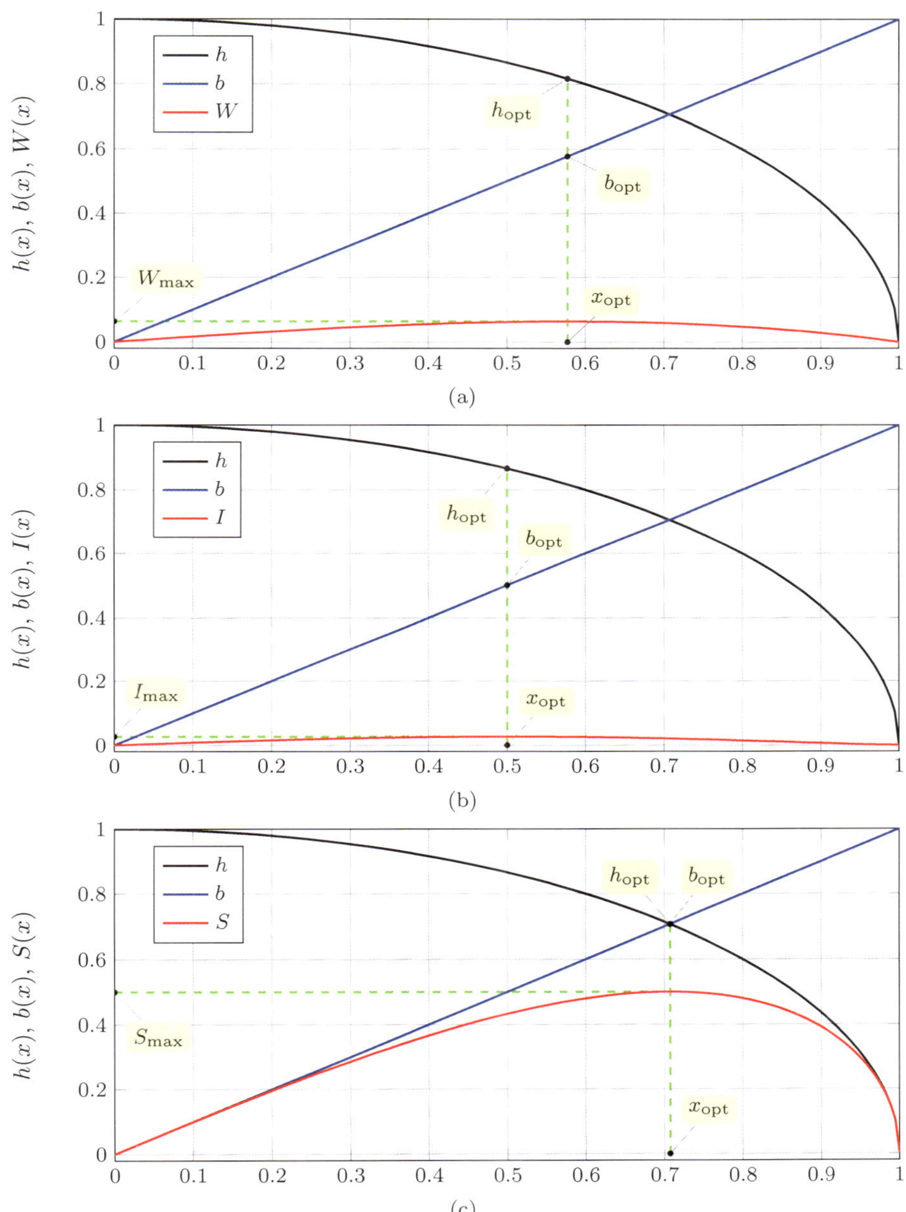

Fig. 9.16   Results of solution to the optimization Problem 9.6.

Functions (9.19)-(9.21) will be maximized if their first derivatives with respect to the decision variable x are equal to zero, i.e.,

$$W'(x) = -\left(3x^2 - 1\right)/6 = 0\,,$$

$$I'(x) = \left(4x^2 - 1\right) \sqrt{1 - x^2}/12 = 0\,,$$

$$S'(x) = -\frac{2x^2 - 1}{\sqrt{1 - x^2}} = 0\,. \qquad (9.22)$$

Solving equations (9.22), a set of optimal decision variables is found:

$$x_W^{\text{opt}} = 1/\sqrt{3}, \quad x_I^{\text{opt}} = 1/2, \quad x_S^{\text{opt}} = 1/\sqrt{2}\,. \qquad (9.23)$$

With the requirements (a-c), the obtained optima (9.23) of the decision variable x guarantee:

- the smallest bending stress σ_g if

$$b = x_W^{\text{opt}} \quad \text{and} \quad h = \sqrt{2/3}\,,$$

- the largest stiffness j if

$$b = x_I^{\text{opt}} \quad \text{and} \quad h = \sqrt{3}/2\,,$$

- and the smallest compression stresses σ_s if

$$b = x_S^{\text{opt}} \quad \text{and} \quad h = 1/\sqrt{2}\,.$$

Correctness of the above calculations is confirmed by the results presented in Fig. 9.16, that were obtained after numerical solution to the optimization problem 9.6 with the use of the simulation model in Fig. 9.15.

9.3.4 *Stiffness Optimization of a Spindle System*

A spindle shaft is the weakest point in machine tools structure. Increasing its stiffness will further increase the machine tools accuracy and the product quality as well. Moreover, high productivity requires some machine tools with high speed machining capability, which leads into unavoidable dynamic effects that occur in the machine tool spindle during the production process, such as regenerative chatter [Prakosa *et al.* (2013)].

The effects of spindle overhang and bearing span on the frequency responses of a spindle system are studied in [Gao and Meng (2011)]. Not less important is the stiffness of the system based on a hydrostatic bearing investigated in this section, having the direct connection with mentioned effects.

Here, we take into investigation an optimization of stiffness of grinder's spindle. Spindle shaft's stiffness has to be optimized in the place of mounting of the grinding wheel by choosing proper spanning of headstock bearings. Balance of forces acting on the spindle shaft and the bearings is illustrated in Fig. 9.17, and its physical model in Fig. 9.18.

From the definition of linear stiffness, we have

$$j = \frac{P}{y}\,, \qquad (9.24)$$

where: j – stiffness of the spindle, y – transverse displacement of spindle's head part (free end), P – force exerted on spindle's head part.

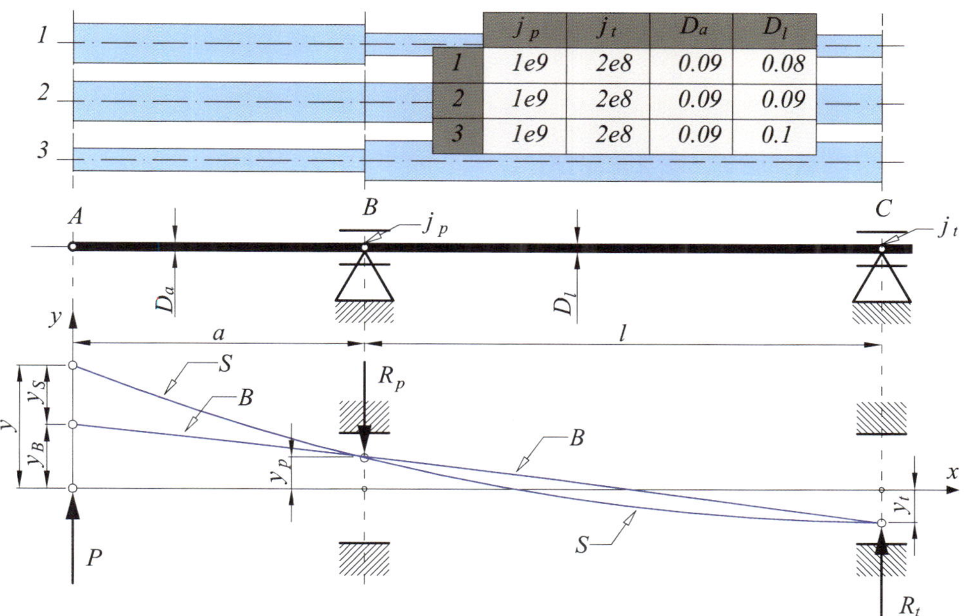

Fig. 9.17 Balance of forces acting in the spindle system.

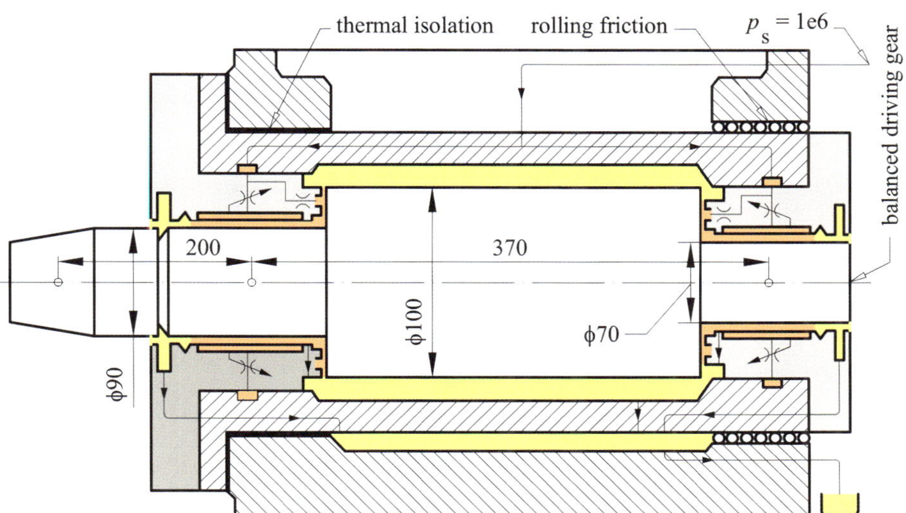

Fig. 9.18 Physical model of a spindle system.

Relations for the strength, determined from the block diagram 9.21 by means of the superposition method at the assumed stiffness of the front and rear bearing, yield

$$y = y_a + y_l + y_{\mathrm{jp}} + y_{\mathrm{jt}} = \left[Aa^3 \left(\frac{1}{D_a^4} + \frac{1}{D_l^4 \alpha} \right) + \frac{1 + 2\alpha + \alpha^2}{j_p} + \frac{\alpha^2}{j_t} \right] P, \quad (9.25)$$

where: $y_a = \frac{A*a^3}{D_a^4} P$, $y_l = \frac{A*a^3}{D_l^4 \alpha} P$, $y_{\mathrm{jp}} = \frac{1+2\alpha+\alpha^2}{j_p} P$, $y_{\mathrm{jt}} = \frac{al^2}{j_t} P$, constant $A = \frac{64}{3\pi E}$, $a = 0.2$ – length of the spindle shaft head (spindle overhang), l – spacing of bearings (bearing span), D_a – diameter of the spindle shaft head, D_l – diameter of the spindle between bearings, $\alpha = \frac{a}{l} \in (0,1)$ – a coefficient, j_p, j_t – front and rear bearing stiffness, respectively.

Stiffness $j(\alpha)$ will be the maximized objective function calculated from Eq. (9.24), and for y being expressed by Eq. (9.25), i.e.,

$$j(\alpha) = \frac{1}{Aa^3 \left(\frac{1}{D_a^4} + \frac{1}{D_l^4 \alpha} \right) + \frac{1+2\alpha+\alpha^2}{j_p} + \frac{\alpha^2}{j_t}}. \quad (9.26)$$

Let the factor α of the spindle overhang with a constructional constraint given by the inequality $\alpha_{(\mathrm{lb.})} \leq \alpha \leq 1$ be a decision variable. The displacement of the spindle head will be investigated during searching for a minimum value of the relation (9.26) for the same decision variable, identical constraints resulting from a technological conditions as well as at the transverse displacement constraint $y \leq y^{(\mathrm{ub.})}$ of the spindle head at the place of mounting of the grinding wheel.

Summing up, some constraints resulting from constructional capabilities of the bearing arrangement in the case of the system visible in Fig. 9.18 were subject to an optimization. Figure 9.19c illustrates some configuration plots of displacements y_p and y_t of the front and rear bearing journals, respectively, versus the factor α of the spindle overhang on a background of the permissible area bounded by the allowable displacement $y^{(\mathrm{ub.})}$ of the spindle shaft mounted in a hydrostatic bearing. On the basis of Fig. 9.17, functions of displacements $y_p(\alpha)$ and $y_t(\alpha)$ are calculated as below:

$$y_p(\alpha) = \frac{(1+\alpha)P}{j_p}, \quad y_t(\alpha) = \frac{\alpha P}{j_t}. \quad (9.27)$$

In Algorithm 9.4 and in Fig. 9.19, a computational scheme regarded to the physical model shown in Fig. 9.17 as well as some results of its numerical exploration are presented. From the three propositions of the diameter D_l of the spindle shaft between bearings, the most optimal with respect to the objective function $j(\alpha)$ has been selected.

Algorithm 9.4. A `Scilab` numerical procedure for computation of solutions $y(\alpha)$ and $j(\alpha)$.

$A = 3.23e - 11$, $a = 0.2$, $D_a = 0.09$
$D_l = [0.08, 0.09, 0.1]$
$j_p = 1e9$, $j_t = 2e8$
$P = 1000$, $n = 50$

$y_a = \mathbf{ones}(1,\, n)*(Aa^3/D_a^4)$
$y_l = \mathbf{ones}(3,\, n),\ y_p = \mathbf{ones}(3,\, n),\ j = \mathbf{ones}(3,\, n)$
$\alpha = \mathbf{linspace}(1/n,\, 1,\, n)$
$y_{\mathrm{jp}} = (1 + 2\alpha + \alpha^2)/j_p,\ y_{\mathrm{jt}} = \alpha^2/j_t$
for $i = 1:3$
   $y_l(i,\, :) = A*(a^3)/(D_l^4(i))*\alpha^{-1}$
   $y_p(i,\, :) = y_a + y_l(i,\, :) + y_{\mathrm{jp}} + y_{\mathrm{jt}}$
   $j(i,\, :) = y_p(i,\, :)^{-1}$
end
//Displacements of front and rear bearing journals
$y_p = (1 + \alpha)*P/j_p$
$y_t = \alpha*P/j_t$
figure(1)
plot$(\alpha(1:n)',\ [y_p(:,\, 1:n)*P]'\,,"-")$
figure(2)
plot$(\alpha(1:n)',\ [j(:,\, 1:n)]',\ "-")$

9.3.5 *Minimization of Total Power Loss in a Hydrostatic Bearing*

Figure 9.21 presents minimization of total power loss N_s due to friction N_t and power supply N_p in the hydrostatic bearing shown in Fig. 9.13 as a function of the decision variables, like the oil film width h_0 and the oil dynamic viscosity η. There is no need to create any simulation or computational models, because the solutions are possible to obtain just by plotting functions of the particular power losses N_t and N_p with respect to η and h_0:

$$N_t = \frac{\pi^3 \eta \beta D^3 L n^2}{h_0}, \quad N_p = \frac{\pi \alpha D p^2 h_0^3}{6\eta l}, \quad N_s = N_t + N_p.$$

Model parameters:

β – ratio of sizes of the friction and bearing surfaces in the analyzed hydrostatic bearing,
D – diameter of bearing [m],
L – length of bearing [m],
l – parameter of the bearing surface (see races in the hydrostatic pad shown in Fig. 9.20) [m],
h_0 – width of the clearance for oil outflow from the bearing [m],
n – rotational velocity of the spindle shaft [rev/min],
η – dynamic viscosity of oil.

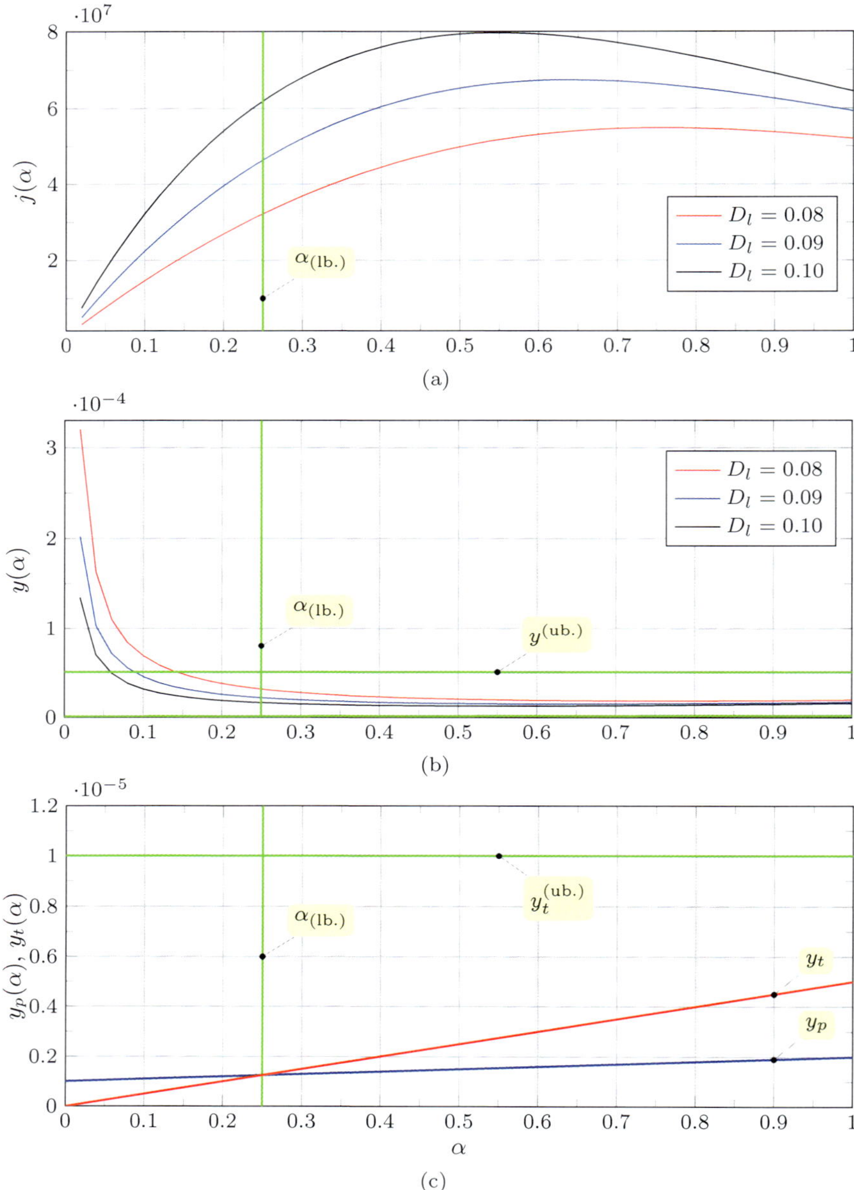

Fig. 9.19   Results of numerical exploration of the computational scheme presented in Table 9.4: (a) stiffness of the spindle system $j(\alpha)$; (b) transverse displacement $y(\alpha)$ of the free end of the spindle; (c) displacements $y_p(\alpha)$ and $y_t(\alpha)$ of front and rear bearing journals. The free end of the spindle overhang has been loaded by the concentrated force $P = 1000$ N. Constraints: $\alpha_{(\text{lb.})} = 0.25$ – left boundary of factor α, $y^{(\text{ub.})} = 50\,\mu$m – upper boundary of the transverse displacement y of the end of the spindle shaft, $y_t^{(\text{ub.})} = 10\,\mu$m – upper boundary of the displacement y_t in the rear bearing.

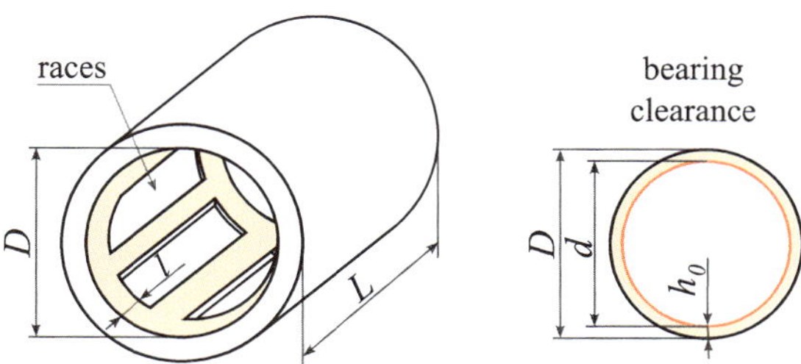

Fig. 9.20   Dimensions of the hydrostatic bearing.

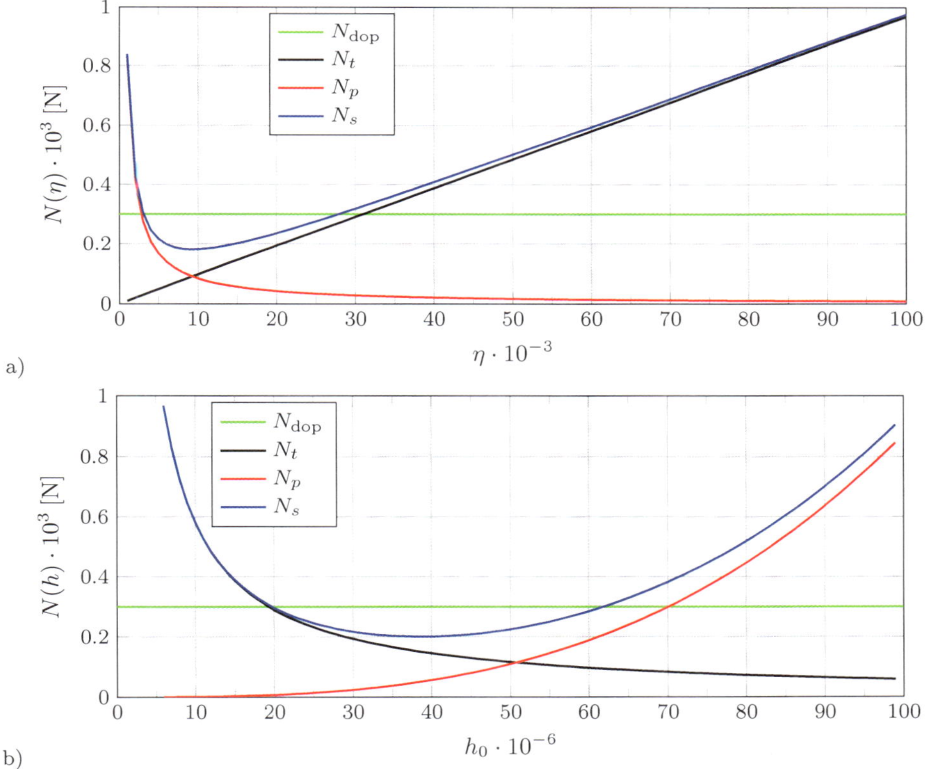

a)

b)

Fig. 9.21   Minimization of total power loss N_s due to friction N_t and power supply N_p in the hydrostatic bearing: (a) as a function of h_0 at $\eta = 15\,\mathrm{e}{-3}$; (b) as a function of oil viscosity η (the decision variable) at $h_0 = 40\,\mathrm{e}{-6}$. Parameters: $\beta = 0.2$, $D = 0.1$, $L = 0.1$, $l = 4\,\mathrm{e}{-3}$, $q = 1\,\mathrm{e}3$, $n = 25$.

Chapter 10

Fuzzy Logic in Numerical Algorithms

Fuzzy logic control algorithms are regarded as a relatively new concept in modern control theory. This chapter delivers a comparative analysis of two qualitatively different approaches used for angular velocity control of a DC (direct current) motor subject to chaotic disturbances coming from a gear with a transmission belt carrying a vibrating load. The purpose is to achieve an accurate control of the speed of the DC motor (a plant), especially when the motor parameters and some external loading conditions are partially unknown. Firstly, the classical approach based on the PID control is considered and then, a fuzzy logic based alternative is proposed. Two different controllers are presented for the purpose of completion of the classical PID controller and a Takagi-Sugeno type fuzzy logic PI controller. Both control algorithms were implemented on an 8-bit AVR ATmega644PA microcontroller. On the basis of step responses of the plant, an analysis as well as an interesting comparison of controllers performance has been presented.

Fuzzy logic is characterized by a fresh and innovative approach to the problem of control. In the assumption, it has to approximate the human process of reasoning and perception of physical phenomena. Adopting a linguistic variables makes it possible to verbally write laws of automatic regulation without the need of derivation of any complex mathematical formulas. Therefore, it can settle the construction of the control system on an incomplete information about the controlled object (the plant). The use of fuzzy logic can be found in some situations in which the described phenomena are ambiguous and difficult in modeling by means of the classical two-state logic. Fuzzy controllers are applied to unknown mathematical models of miscellaneous systems or even when their restoration is very time-consuming and complicated.

Fuzzy systems can be found in the areas of technology such as image and shape recognition, speech recognition, searching of databases, optimization of banking systems, automotive braking systems, medical systems, and many other. Some literature overview devoted to the topic studied in this chapter is done below.

Fuzzy logic is a mathematical concept striving to imitate human perception. Instead of numerical values, some linguistic descriptions are used to characterize input and output variables. The control strategy is derived from expert knowledge and stored in a base of fuzzy rules. This enables the design engineer to describe the behavior of the object under control with the use of words (linguistic variables) rather than with the use of complex mathematical expressions. Another advantage of the intelligent approach in control of uncertain dynamical systems is the ability of fuzzy logic controllers to handle control by means of an incomplete portion of information. Fuzzy logic has found many applications in a wide variety of applications like process control, electrical engineering, information technology, image recognition, telecommunications, banking [Zadeh (1989); Bai and Wang (2006); Jager (1995)].

In paper [Neto *et al.* (2010)], a PI fuzzy force controller was applied to control movement of industrial robot's end-effector. After generating the robot program, a "foreign" object was introduced to the environment. Without any obstruction during its operation, the robot moves along its programmed path. However, when a contact with the foreign object appears along the way of motion, the force control system controls the robot by adjusting the end-effector position. The force control ensures that the contact forces and moments converge to a desired value. Comparing to the classical PI algorithm, a smaller overshoot and average constant force were achieved.

In the work [Velagić and Galijasević (2009)], a robust fuzzy controller application has been described for a permanent magnet DC motor. The system parameters concerning the load and DC motor's constants were unknown. The algorithm was implemented on dSPACE rapid prototyping controller board and connected to `Simulink` programming environment. The fuzzy logic controller (FLC) has two inputs, a voltage error and its derivative. The control action generated by the controller is the actual voltage supplying the DC motor. It was concluded from the experimental data that the elaborated FLC has achieved shorter response times on pulse input signals and smaller oscillations about the setpoint than the adequate PID controller.

Paper [Chan and Chu (2009)] describes an application of a fuzzy controller in Internet traffic management. Web servers suffer from extremely varying load parameters. They are sometimes very lightly loaded, but occasionally suffer from enormous connection requests. Designing web servers for peak load is not profitable, because even the most efficient web servers may still be overloaded by the ever growing population of Internet users. During an overload period, not all users can receive services in a timely manner without latency, but on the other hand, it is possible to provide a faster connection for premium users. Due to the nonlinear properties of web servers and difficulties in constructing their accurate mathematical model, a fuzzy PI controller was proposed. Incoming connections were divided into two classes: premium and basic. The task of the fuzzy controller is to maintain a delay ratio between these two classes by assigning suitable number of processes to handle incoming requests. Experimental data shows that significantly lower oscillation and shorter settling time of the delay ratio was achieved comparing to the classical PI controller. The fuzzy controller has improved the control quality by approximately 35%.

A hybrid solution of a fuzzy logic and a PI controller was presented in work [Teeter *et al.* (1996)]. The object under control consists of a DC motor with a metal disc mounted on its shaft. The load is generated by applying a magnetic field to the disc. The fuzzy logic component is responsible for calculating a gain coefficient for the PI controller. The fuzzy part has three inputs: the reference speed, measured speed and the control action of the whole controller from the previous time period. The base of fuzzy rules consists only of one rule, and its aim is to reduce output of the PI controller in low setpoint speeds. This is due to the nonlinear behavior caused by frictional effects in the mentioned region. This friction compensation method yields faster response of the system and smaller settling time.

An application of a FLC in electrical engineering was presented in the article [Bašić *et al.* (2013)]. The controlled object is a self-excited induction generator. The control system was given a task to maintain the steady level of output voltage. Mamdani and Takagi-Sugeno type fuzzy logic PI controllers were developed in the discussed work. Both controllers have two inputs (a voltage error and its derivative) and one output. Performance of the proposed solutions was tested against the classical PI controller. Both controllers were programmed in `Simulink` and used to control a model of the self-excited induction generator. Due to high processing power required by the Mamdani controller, the Takagi-Sugeno type FLC was solely implemented on a dSPACE real-time system. It has been concluded from the simulation and experimental data that the fuzzy logic controllers offer significantly better performance compared to the optimally tuned PI type controller in terms of the response time, settling time and robustness. The downside of the described application is reflected in an increase in the computational performance requirements, especially for the Mamdani type FLC.

In work [Jang (1991)], a self-learning Takagi-Sugeno controller was used for identification purposes. The mentioned controller has two input variables with three bell shaped membership functions each. Output of every rule is a linear function of input variables. To implement a gradient-descent learning algorithm, the controller was designed in a form of a generalized neural network. The task of the algorithm was to modify the weighting coefficients of the neural network to ensure convergence of the output from the fuzzy-neural network to three sets of training data. One of them was obtained from a real object and the other from different mathematical functions. After 200 iterations of the algorithm execution for each set of data, an average percentage error obtained for the first set of training data was reduced to 1.57%. For the remaining sets, the error was reduced to 0.47% and 0.014%, respectively. Although the number of fuzzy rules and input membership functions was preliminarily specified, the algorithm achieved satisfactory results regarding the error and tuning time.

Another example of a self-tuning FLC was presented in [Jee and Korem (2004)]. A Takagi-Sugeno fuzzy logic PD controller was applied to a 3-axis milling machine for contour milling. Basing on the position error, the change in the error, the velocity feedback and the fuzzy control action from the previous time period, output is calculated and sent to the amplifier which drives the motor. The performance of the controller affects the adaptive algorithm that expands or contracts the input fuzzy sets or even shifts the position of the output sets in the numeric domain. The described FLC was then compared to a well-tuned PID controller. The comparison showed that for cutting straight lines, the adaptive FLC achieves twice lower root mean square contour error and three times lower maximum contour error than the PID controller.

An interesting fuzzy logic controller implemented in an industrial controller was presented in [Arrofiq and Saad (2008)]. The object of control (a plant in the theory of control) consists of a DC motor subject to varying load parameters. The Takagi-Sugeno fuzzy logic PI controller was implemented on a PLC. A self-tuning fuzzy algorithm, which calculates a gain coefficient for the main controller, was also implemented. The investigated system performance was tested for three types of variance conditions: load parameters, setpoint velocity and setpoint velocity with changing load parameters. In all of these cases the FLC has provided satisfactory good results.

In this chapter, a fuzzy logic PI controller and its classical PID equivalent is analyzed. We take into investigation efficiency of both solutions in controlling a multi degree-of-freedom discontinuous dynamical system with friction subject to an irregular external excitation. Experimental stand's components are discussed in the second part of this chapter. In relation to the experiment carried out in [Kunikowski *et al.* (2013)] and being continued here, a comparative study of test results of efficiency of the presented fuzzy logic approach is experimentally verified in Sec. 10.3.4.

10.1 Basic Concepts

One of the main concepts of fuzzy logic is a *linguistic variable*. It is such an input/output or state variable, the values of which are words or even sentences in a natural or artificial language. In examples one finds: height, speed, temperature. The use of linguistic variables in many applications reduces the overall computation complexity of the application.

The *linguistic value* is an assessment in the form of verbal description applied to a linguistic variable. The values of the linguistic variables from the previous example can be respectively: very small, small, not very large, large (in terms of growth), very low, low, not very high, high (in terms of speed), cold, warm, very warm (with respect to temperature).

Another important concept of fuzzy logic is a *fuzzy set* defined in the selected area. The considered elements may be in the region in whole, in part, or may not belong to the region at all. A membership function of the output values from the interval [0, 1] defines the degree of membership of these elements to the relevant fuzzy set as well as defines the shape of this set.

A fuzzy set can be described by the relation:

$$A = \{(x,\, \mu_A(x)): \quad x \in X,\, \mu_A(x) \in [0,\, 1]\}, \tag{10.1}$$

where: x – variable, X – numerical space of the variable x, μ_A – membership function of the variable x to the fuzzy set A.

The use of fuzzy sets allows for a formal writing of uncertainty. For instance, in a two-state logic (low/high), the definition of high speed equal to or greater than 10 km/h raises the question of whether the speed 9.9 km/h is certainly low or already high. In fuzzy logic, we can express such value as 99 percent high and in 1 percent low. This allows for a smoother transition between the intervals of values compared to the classic two-state logic.

10.1.1 *Membership Functions of Fuzzy Sets*

A fuzzy membership function characterizes a fuzzy set. It transforms a specific numerical value of a physical quantity to its degree (level) of membership to the considered fuzzy set. It can be an arbitrary function having some output values in the interval $[0,\, 1]$ equal to 1 – if fully belonging to the given fuzzy set, and 0 – if not belonging.

In Fig. 10.1, common shapes of membership functions [Rutkowski *et al.* (2015)] are depicted.

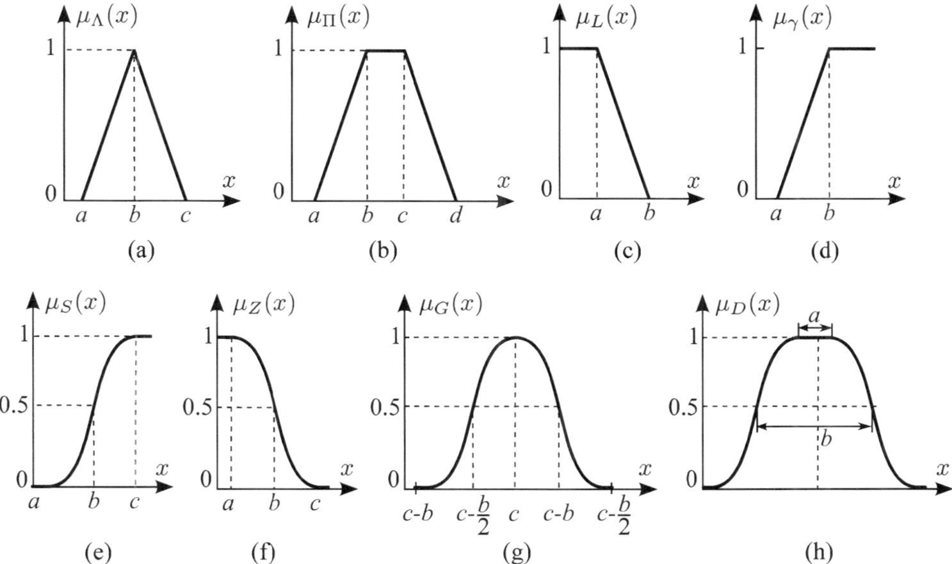

Fig. 10.1   Shapes of membership functions defining a fuzzy set.

The membership functions (MF) shown in Fig. 10.1a-h are expressed by the following equations:

a) triangular

$$\mu_\Lambda(x,a,b,c) = \begin{cases} 0, & x \le a, \\ \dfrac{x-a}{b-a}, & a < x \le b, \\ \dfrac{c-x}{c-b}, & b < x \le c, \\ 0, & x > c; \end{cases} \tag{10.2}$$

b) trapezoidal

$$\mu_\Pi(x,a,b,c,d) = \begin{cases} 0, & x \le a, \\ \dfrac{x-a}{b-a}, & a < x \le b, \\ 1, & b \le x \le c, \\ \dfrac{d-x}{d-c}, & c < x \le d, \\ 0, & x > d; \end{cases} \tag{10.3}$$

c) L-shaped

$$\mu_L(x,a,b) = \begin{cases} 1, & x \le a, \\ \dfrac{b-x}{b-a}, & a < x \le b, \\ 0, & x > b; \end{cases} \tag{10.4}$$

d) γ-shaped

$$\mu_\gamma(x,a,b) = \begin{cases} 0, & x \le a, \\ \dfrac{x-a}{b-a}, & a < x \le b, \\ 1, & x > b; \end{cases} \tag{10.5}$$

e) S-shaped

$$\mu_S(x,a,b,c) = \begin{cases} 0, & x \le a, \\ 2\left(\dfrac{x-a}{c-a}\right)^2, & a < x \le b, \\ 1-2\left(\dfrac{x-c}{c-a}\right)^2, & b < x \le c, \\ 1, & x > c; \end{cases} \tag{10.6}$$

f) Z-shaped

$$\mu_Z(x,a,b,c) = \begin{cases} 1, & x \le a, \\ 1-2\left(\dfrac{x-a}{c-a}\right)^2, & a < x \le b, \\ 2\left(\dfrac{x-c}{c-a}\right)^2, & b < x \le c, \\ 0, & x > c; \end{cases} \tag{10.7}$$

g) Gaussian

$$\mu_G(x, b, c) = \exp\left[-\left(\frac{x-c}{b}\right)^2\right];$$ (10.8)

h) generalized bell-shaped (denoted by D below Fig. 10.1h)

$$\mu_D(x, a, b, c) = \frac{1}{1 + \left|\dfrac{x-c}{a}\right|^{2b}}.$$ (10.9)

10.1.2 *Operations on Fuzzy Sets*

Similarly to the two-state logic, the basic operations on fuzzy sets are as follows:

- sum (OR),
- product (AND),
- negation (NOT).

Graphical representation of the product of fuzzy sets is a common part of the joint fuzzy sets. A sum of fuzzy sets is the sum of the areas described by membership functions of these sets.

One could enumerate a lot of mathematical dependencies allowing to perform basic operations on fuzzy sets. They are called s-norm operator for the sum of fuzzy sets and t-norm operator for the product of fuzzy sets. The most common and simplest s-norm operator is a function $\max(\mu_A(x), \mu_B(x))$, and for t-norm, a function $\min(\mu_A(x), \mu_B(x))$, where μ_A and μ_B are some membership functions of fuzzy sets A and B. In Tables 10.1 and 10.2, the examples of other kinds of fuzzy logic operators are presented [Ruano (1999)].

Table 10.1   The operators of s-norm $\mu_{A \cup B}(x)$ in the fuzzy logic.

Operator name	Mathematical notation
Maximum	$\max(\mu_A(x), \mu_B(x))$
Algebraic sum	$\mu_A(x) + \mu_B(x) - \mu_A(x) \cdot \mu_B(x)$
Hamacher sum	$\dfrac{\mu_A(x) + \mu_B(x) - 2\mu_A(x) \cdot \mu_B(x)}{1 - \mu_A(x) \cdot \mu_B(x)}$
Einstein sum	$\dfrac{\mu_A(x) + \mu_B(x)}{1 + \mu_A(x) \cdot \mu_B(x)}$
Drastic sum	$\begin{cases} \max(\mu_A(x), \mu_B(x)), & \min(\mu_A(x), \mu_B(x)) = 0 \\ 1, & \text{otherwise} \end{cases}$
Bounded sum	$\min(1, \mu_A(x) + \mu_B(x))$

Table 10.2   The operators of t-norm $\mu_{A \cap B}(x)$ in the fuzzy logic.

Operator name	Mathematical notation
Minimum	$\min(\mu_A(x), \mu_B(x))$
Product	$\mu_A(x) \cdot \mu_B(x))$
Hamacher product	$\dfrac{\mu_A(x) \cdot \mu_B(x)}{\mu_A(x) + \mu_B(x) - \mu_A(x) \cdot \mu_B(x)}$
Einstein product	$\dfrac{\mu_A(x) \cdot \mu_B(x)}{2 - (\mu_A(x) + \mu_B(x) - \mu_A(x) \cdot \mu_B(x))}$
Drastic product	$\begin{cases} \min(\mu_A(x), \mu_B(x)), & \max(\mu_A(x), \mu_B(x)) = 1 \\ 0, & \text{otherwise} \end{cases}$
Bounded product	$\max(0, \mu_A(x) + \mu_B(x) - 1)$

Operation of negation of a membership function is expressed by the mathematical relation

$$\bar{A} = 1 - \mu_A(x). \tag{10.10}$$

10.1.3 *Construction of a Fuzzy Controller*

In technical systems, measurement devices provide some discrete measurements (like $10\,\text{km/h}$). These crisp values must be transformed into fuzzy sets (linguistic terms). The transformation process is called *fuzzification*.

The following components included in fuzzy controllers are found:

(i) fuzzification,
(ii) decision-making (inference),
(iii) Rule Base,
(iv) defuzzification.

In Fig. 10.2, a schematic model of a fuzzy controller is presented.

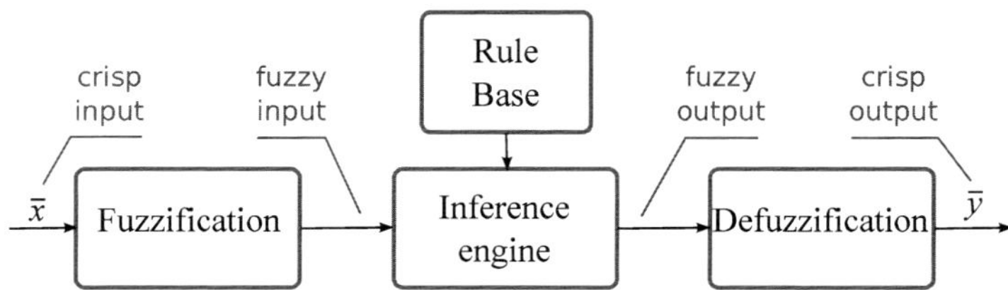

Fig. 10.2   Basic scheme of a fuzzification-defuzzification system.

The fuzzification block is responsible for translation of a *crisp* input value, represented by a numeric value, into the format acceptable by the fuzzy regulator.

Fuzzification involves assigning a value to one or many fuzzy sets (depending on their definition) and calculating the value of membership function. The most often used membership functions are the triangular and Gaussian functions.

Rule Base contains cause and effect dependencies existing between input and output variables of the fuzzy controller. It states a representation of knowledge about the considered control system. An exemplary rule yields

$$R_n : \text{IF } (x_1 = A_1) \text{ AND } (x_2 = B_1) \text{ THEN } (y_1 = C_1),$$

where x_1 and x_2 are input variables, y_1 – output variable, A_1, B_1, C_1 – fuzzy sets.
The above rule is read as follows:

> If the first input value belongs to the set A_1 and the second input value belongs to the set B_1, then an output y_1 of n-th rule belongs to the set C_1.

The components in parentheses after the IF operator are called *premise* (antecedence). The components after the word THEN are called *conclusion* (consequence).

Reasoning is applied by an inference engine to compute fuzzy output. It aims at checking each rule the premises of which are satisfied. Based on the conclusions, a resulting fuzzy set is created.

If several rules have the same variable in conclusion, they have to be fired in parallel and all their conclusion fuzzy sets have to be *aggregated* in one fuzzy set to be used as entry for others rules. Conclusion fuzzy sets are aggregated with the min operator to take into account uncertainty in the final fuzzy set.

Defuzzification is a process aiming at transformation of the conclusion fuzzy set to a numeric value, treated as the control signal.

10.1.4 *Mamdani Model*

The most commonly used and the most natural from the point of view of fuzzy logic is an inference with a Mamdani minimum operation rule. Rule Base (RB) of this reasoning is based on expert knowledge devoted to the investigated system. Fuzzy models of the Mamdani type [Mamdani (1974)] have most often a few inputs and one output. A scheme of a fuzzy logic controller created according to the described reasoning is shown in Fig. 10.3.

In Fig. 10.3, an inference scheme of deducing output y^* of a controller having 2 inputs x_1 and x_2 has been presented. In the first step, a *fuzzifier* acts on crisp input values and assigns them to some individual fuzzy sets. Then, a membership function τ_Δ is computed. In the last step, the scheme makes reasoning by checking some rules described by the formulas

$$R_i : \text{IF } (x_1 = B_{i1}) \text{ AND } (x_2 = B_{i2}) \text{ THEN } (y = D_i), \quad i = 1, 2.$$

Conclusions determine the shapes of fuzzy sets generated on outputs of individual rules. Operation AND (*t*-norm) is responsible for the membership functions of output sets. From a geometric point of view, conclusion of a rule determines placement of a fuzzy set on the *y*-axis, and operation AND (product) cuts the set to a proper height. A value of membership function of the fuzzy set is called *firing strength* of the rule. If this value is equal to 0, then one says that the rule has not been fired.

The next important step is focused on *aggregation of conclusions* of all rules. This operation does the *s*-norm (sum) of all fuzzy sets resulting from the fired rules leading to creation of one fuzzy set.

To obtain a specific numerical value on the output, one requires to use some of many known methods of defuzzification. The most common are listed below.

1. *Discrete center-of-gravity method (COG).* The output value of the regulator is related to the placement of the center of gravity of a figure described by the resultant fuzzy set. It is the most often used method of defuzzification. It is the most "democratic" in generation of output value. Its drawback is related to a relatively high computational complexity.

2. *Mean of maximums method (MeOM).* The output value of the regulator

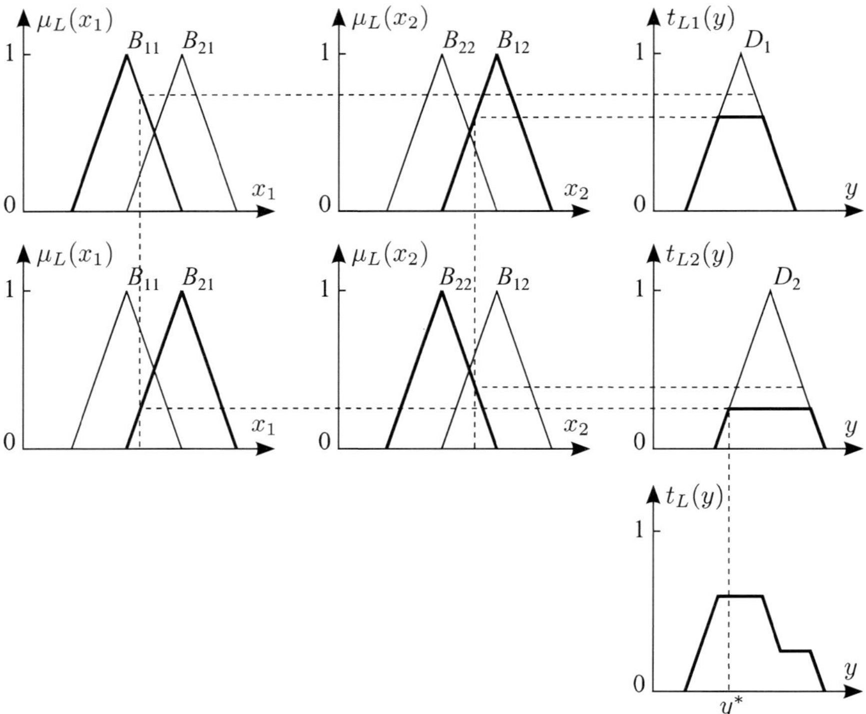

Fig. 10.3   Inference scheme with the Mamdani minimum operation rule.

is defined by the center of a maximum of a membership function of the resultant fuzzy set. This method guarantees a relatively low computational complexity at the cost of omission of smaller parts of the output fuzzy set.

3. *First of maximum method (FOM)*. The output value of the regulator depends on the placement of the first maximum of a membership function. This method is characterized by low computational complexity.

4. *Last of maximum method (LOM)*. The output value of the regulator is defined by the placement of the last maximum of a membership function. This method is also characterized by low computational complexity.

In Fig. 10.4, a graphical form of mentioned defuzzification methods is depicted.

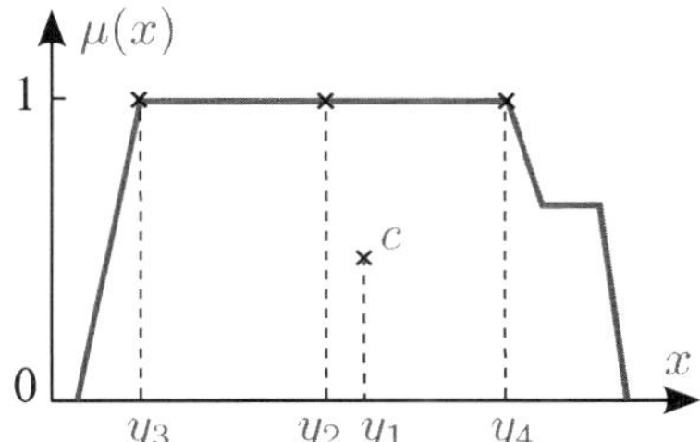

Fig. 10.4   A graphical representation of results of the defuzzification methods 1-4: y_1 – COG, y_2 – MeOM, y_3 – FOM, y_4 – LOM, c – center of gravity of the figure describing the output fuzzy set.

As can be noticed, there exists a big difference between the output crisp values y_i of the analyzed *defuzzifiers*.

Selection of the defuzzifier depends on current knowledge about the control system and on the experience of the engineer designing the controller. It is often observed that selection of the defuzzifier is guided by the trial-and-error method.

10.1.5   *Takagi-Sugeno Model*

The Takagi-Sugeno model [Sugeno and Kang (1988)] differs from the Mamdani model in the method of creating the conclusions from premises in the definitions of rules. In contrast to the Mamdani model [Mamdani (1974)], in which a fuzzy set is obtained from rule's conclusion, in the Takagi-Sugeno model the conclusion of a rule is determined by a linear function of crisp input values or a constant value. Constant values in conclusions are called *singletons* – one-element fuzzy sets. Below, an exemplary rule is shown:

$$R_1 : \text{ IF } (x_1 = B_{11}) \text{ AND } (x_2 = B_{12}) \text{ THEN } (y_1 = b_{10} + b_{11}x_1 + b_{12}x_2),$$

where x_1 and x_2 are input variables, y_1 – output variable, B_{ij} – fuzzy sets, b_{10}, b_{11}, b_{12} – coefficients of output function.

A crisp value y^* of the controller is computed from the weighted average of output values y_i of each rule, but the weight is determined by a firing level of the specific rule τ_i, i.e.,

$$y^* = \frac{\sum_{i=1}^n \tau_i \cdot y_i}{\sum_{i=1}^n \tau_i}. \tag{10.11}$$

Fuzzy systems based on the Takagi-Sugeno models are used as a simplification of more complex Mamdani models, as they have a lower computational effort. They can be also used for linearization of nonlinear models. In the classical approach, after the local approximation of a trajectory by a linear function, a nonsmooth switching between simplifications appears. Application of Takagi-Sugeno models allows for smoothing of approximation result in the points of switching.

10.2   Experimental Stand

An experimental stand [Olejnik (2002); Awrejcewicz and Olejnik (2005b, 2002); Kunikowski (2014)] depicted in Fig. 10.5 served in tests of two types of a discrete controller: a) standard PID, b) fuzzy PI.

Fig. 10.5   Experimental stand: 1 – a DC motor with a 15:1 worm gear, 2 – RN12 DC motor driver 50 V, 3 – the physical object under control, 4 – IVO GI333 incremental encoder, 5 – Atmel Testing Board 1.03 with ATmega644PA microcontroller.

Regulation of the angular velocity of the PZTK 62-42J DC motor (1) is made by a control system composed of the Atnel Testing Board (ATB) with ATmega644PA microcontroller (5) and the RN12 driver (2). The real system under control (3) influences the speed of the transmission belt measured by the incremental encoder GI333 (4). The L293D IC is supplied by 15 V power source. The DC motor is powered by the MATRIX 60V – a regulated direct current source for the RN12 power circuit and 30V for the RN12 logical circuit.

Figure 10.6 presents a wiring diagram of the described control system. For the purpose of noise cancellation in the measuring circuit, a Schmitt inverter

M74HCT14N was used. It damps voltage peaks that could cause incorrect counting of impulses sent to the microprocessor – the physical control unit.

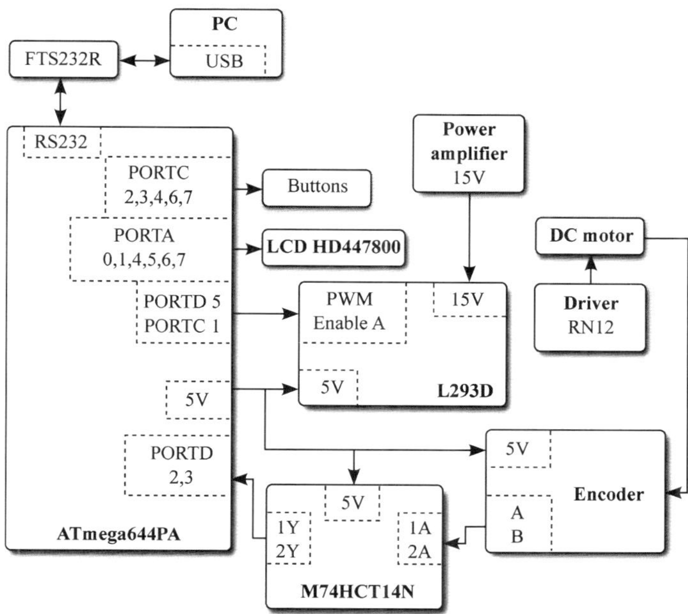

Fig. 10.6   Wiring diagram of the closed-loop control system.

To enable compatibility between the microcontroller and the RN12 motor driver, a L293D integrated circuit (IC) was implemented. The RN12 driver accepts a PWM signal with amplitude ranging from 12 V to 15 V, but the microprocessor is equipped only with 5 V TTL outputs. The L293D served as an amplifier for the PWM signal. The known parameters of the DC motor are as follows: electrical constant $k_e = 0.104\,\mathrm{V/rad/s}$, mechanical constant $k_m = 0.39\,\mathrm{Nm/A}$, armature resistance $R_w = 1.1\,\Omega$, armature inductance $L_w = 0.001\,\mathrm{H}$.

A model of the dynamical system being a source of instability of the belt driven by the controlled DC motor with a gear (compare with element 1 in Fig. 10.5) has been shown in Fig. 10.7. The friction-induced vibration of the mass m on the moving belt is responsible for the dynamically changing loading of DC motor's shaft that drives the transmission belt.

The controlled object consists of a conveyor belt with a block of the mass m oscillating on it in x-direction, a bracket of the mass M and of the mass moment of inertia J, rotating around point S about φ angle. The bracket is attached to the block by means of two linear springs (see Fig. 10.7). Depending on the linear displacement of the block m and the angular displacement φ of the bracket J (a pendulum), friction force on a contact surface between the block and the belt

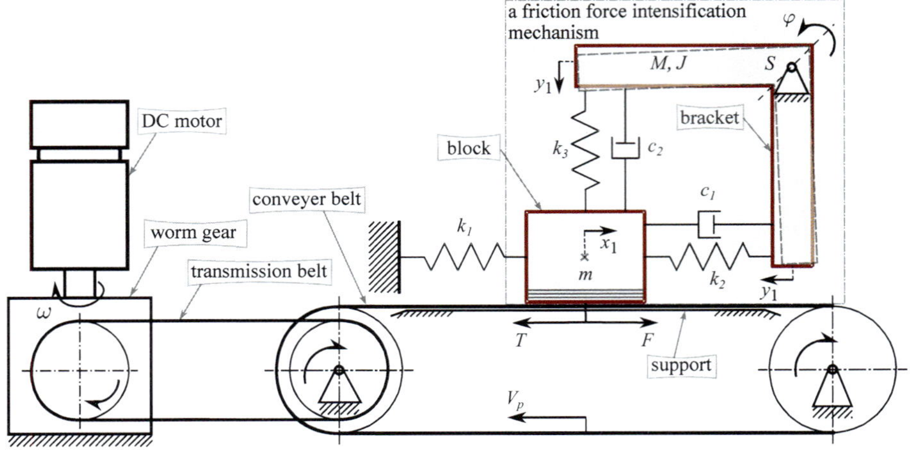

Fig. 10.7   A schematic view of the mechanical system (ω – process variable).

changes rapidly.

Moreover, the friction force characteristics switches itself between its kinetic and static form. The experimentally observed irregular changes in frictional force in the block-on-belt model are responsible for the significantly varying load transferred on the DC motor's shaft driving the mechanical system shown in Fig. 10.7. Constant changing of the amplitude of the load affects the speed of gear's conveyor belt. A more detailed description of the model can be found in previous works of the authors [Awrejcewicz and Olejnik (2005c, 2007); Olejnik (2013, 2002)].

Many physical phenomena, such as wear of the worm gear, viscous friction in all of the bearings of the mechanical system, stiffness of the transmission belt, radial

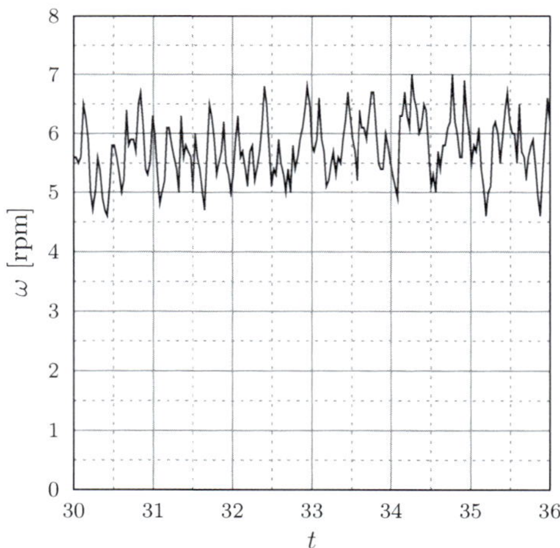

Fig. 10.8   Speed variations of an uncontrolled system.

run-out of belt pulleys shafts and an unevenly distributed dry friction coefficient on the surface of transmission belt, have the direct impact on stabilization of the belt linear velocity of movement.

To inspect the behavior of the conveyor belt shown in Fig. 10.7, the belt pulley's angular velocity was measured by an incremental encoder.

Uncontrolled velocity of the belt pulley is very irregular, as depicted in Fig. 10.8. Figure 10.9 shows exemplary Fourier transform of the system's step response. In that trial, the system was controlled by a P controller with the proportional gain $K_P = 2.25$. The presented spectrum of many amplitudes visible in Fig. 10.9 ends at the Nyquist frequency: $f_c = 1/(2h) \approx 16.666\,\text{Hz}$.

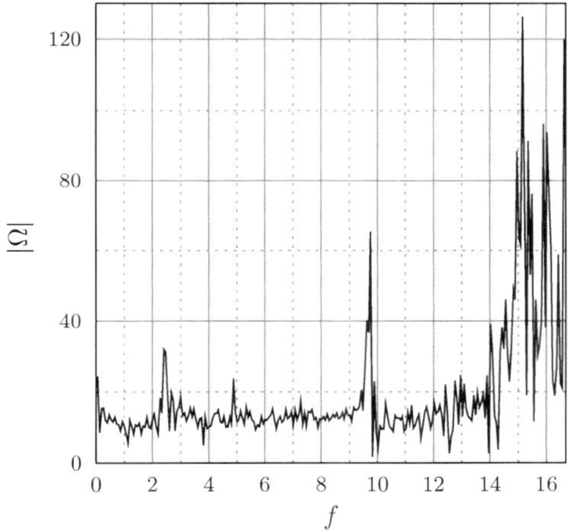

Fig. 10.9   Fourier transform of the system step response.

To measure the speed of the transmission belt on which the oscillating mass vibrates, an incremental encoder of $5000\,\text{imp/rev}$ was used. The 4x encoding was applied for the angular velocity measurement. This means that in a constant time period, both rising and falling edges of two shifted measurement lines A and B of the encoder are counted. This method allows to virtually increase the base resolution of the sensor [Petrella *et al.* (2007)]. Using the 4x encoding and setting the acquisition time to $30\,\text{ms}$ yields a measurement error of $0.1\,\text{rpm}$. This means that each subsequent pulse counted with the mentioned sampling time increases output of the velocity recognition algorithm by $0.1\,\text{rpm}$. The encoder is not an ideal sensor here, so the length of the high state tends to vary due to the manufacturing tolerances. The used method of encoding could magnify that phenomenon, hence the speed measurement error may be bigger than the assumed $0.1\,\text{rpm}$. The contribution of that error to the shape depicted in Fig. 10.8 is unknown. No filtering technique was applied to the measurement to capture the dynamics of the analyzed system as accurately as possible.

10.3   Control Algorithms

Due to complexity of the control task, a black-box approach to the control problem was used while any mathematical model of the controlled object was omitted. Tuning of the controller was done manually by observing the real time plot of the encoded angular velocity measurement. The goal here was to achieve the lowest possible oscillations of velocity of the belt pulley after reaching the desired setpoint. Both presented algorithms were implemented in C programming language on the ATmega644PA microcontroller [Kunikowski *et al.* (2015)].

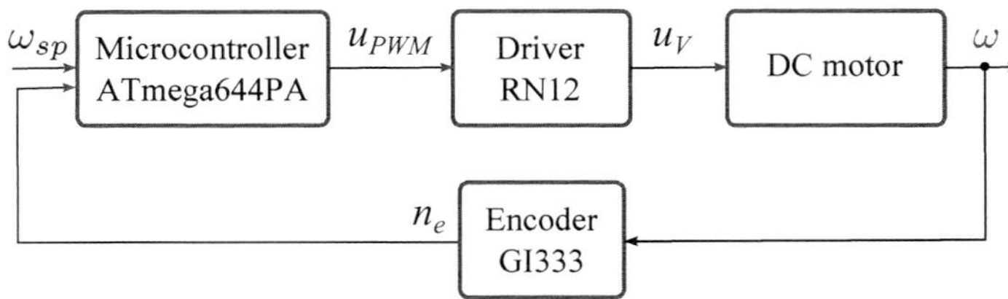

Fig. 10.10   The closed-loop control system (ω_{sp} – reference input, ω – process variable).

In Fig. 10.10, the assumed closed-loop control system has been presented. Basing on the setpoint speed (reference angular velocity ω_{sp}) and counted number of impulses y from the encoder, the proposed control algorithm calculates the adequate duty cycle u_{PWM} of the PWM signal. This information is sent to the RN12 driver which generates appropriate voltage u_V for the DC motor. A 10-bit resolution PWM signal with 9.7 kHz frequency was used. Therefore, the variable describing its duty cycle can be changed from 0 to 1023. The control algorithm corrects the PWM duty cycle every 30 ms (the feedback loop time is restricted by resolution).

10.3.1   *PID Controller*

Classical discrete PID controller's formula in n-th iteration follows

$$u_{\mathrm{PWM}}(n) = K_p e(n) + K_i \sum_{k=0}^{n} e(k) - K_d(y_n - y_{n-1}), \qquad (10.12)$$

where: u_{PWM} – output of the regulator, e – error of regulation, gains: K_p – proportional, $K_i = K_p T_s / T_i$ – integral, $K_d = K_p T_d / T_s$ – derivative; y – measured angular velocity, T_s – sampling time, T_i – integral time coefficient, T_d – derivative time coefficient, n – iteration.

To eliminate rapid responses of the derivative part when the setpoint speed changes with regard to the so-called "derivative kick", regulator's variable was changed from the error of regulation to the difference between two successive measurements of speed. PID controller's output was limited to the range [0, 1023] and treated as a new value of the PWM duty cycle.

10.3.2 *Fuzzy Logic PI Controller*

The second considered regulator is the Takagi-Sugeno-type fuzzy logic PI controller. It takes two inputs, i.e., the error of regulation e and the difference δe between the current value of error and the value of error from the last cycle. The output value is $\Delta u_{\mathrm{PWM}}(n)$ – the increase in the PWM duty cycle. A classical alternative that would match the earlier described regulator is given

$$\Delta u_{\mathrm{PWM}}(n) = K_i e(n) + K_p(e(n) - e(n-1)), \qquad (10.13)$$

where: e – error, K_p – proportional gain, K_i – integral gain.

Numerical range of each input variable was divided into 5 triangular T and a 2 piece-wise linear L and g fuzzy sets. The output variable was divided into 7 singletons. Graphical interpretation of this classification is presented in Figs. 10.11–10.12. Formulas for calculation of the value of membership function of each type of fuzzy set are given in Eqs. (10.14)–(10.16). The Takagi-Sugeno controller was chosen due to its defuzzification simplicity and lower requirements regarding processing power than the Mamdani type controller.

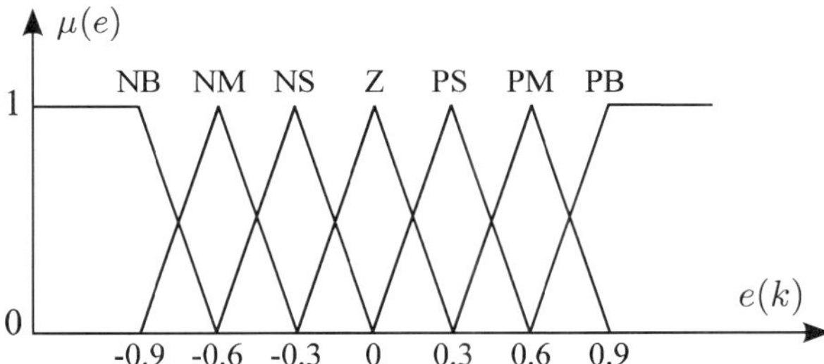

Fig. 10.11   Fuzzy sets of the first input variable e – error of regulation.

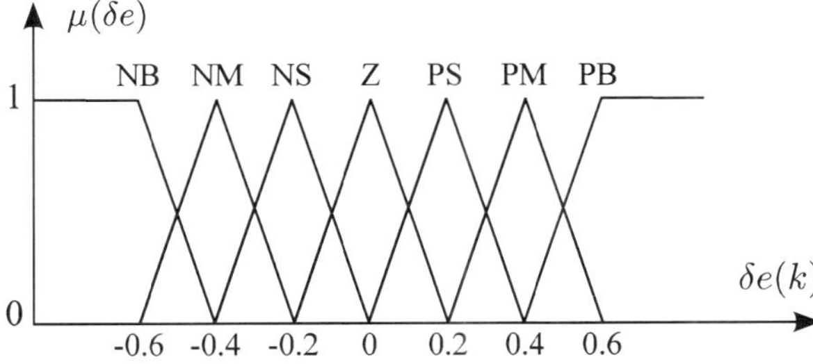

Fig. 10.12   Fuzzy sets of the second input variable δe – increase in the error.

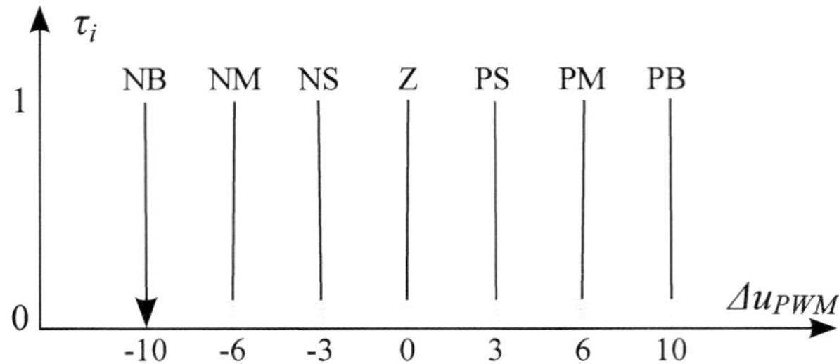

Fig. 10.13   Fuzzy sets of the output variable Δu_{PWM}.

In Figs. 10.11–10.13, the thresholds of fuzzy sets are shown. Linguistic variables assigned to the sets are as follows: NB – negative big, NM – negative medium, NS – negative small, Z – zero, PS – positive small, PM – positive medium, PB – positive big.

$$\mu_L(x, a, b) = \begin{cases} 1 & \text{if} \quad x \le a \\ \dfrac{b - x}{b - a} & \text{if} \quad a < x \le b \\ 0 & \text{if} \quad x > b \end{cases} \tag{10.14}$$

$$\mu_g(x, a, b) = \begin{cases} 0 & \text{if} \quad x \le a \\ \dfrac{x - a}{b - a} & \text{if} \quad a < x \le b \\ 1 & \text{if} \quad x > b \end{cases} \tag{10.15}$$

$$\mu_T(x, a, b) = \begin{cases} 0 & \text{if} \quad x \le a \\ \dfrac{x - a}{b - a} & \text{if} \quad a < x \le b \\ \dfrac{c - x}{c - b} & \text{if} \quad b < x \le c \\ 1 & \text{if} \quad x > c \end{cases} \tag{10.16}$$

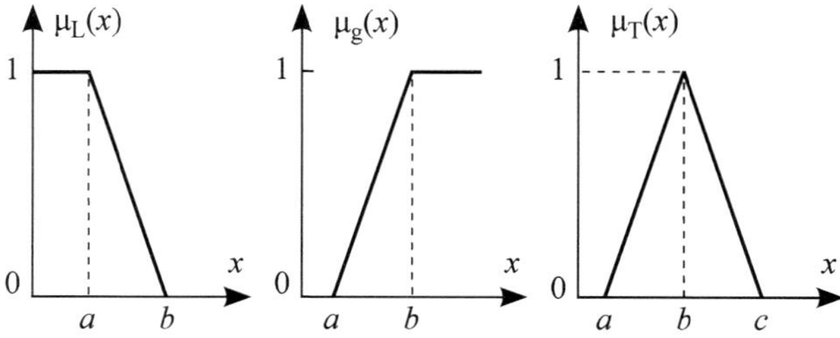

Fig. 10.14   Fuzzy sets of the membership function.

Our RB consists of 49 rules ("if-else" conditional expressions) numbered from $i = 1 \ldots 49$, and $j, k, l = 1 \ldots 7$, respectively to the assumed set of linguistic variables (e.g., NB, NM). An exemplary rule can be expressed as below

$$R_i : \text{ IF } (e(n) = A_j) \text{ AND } (\delta e(n) = B_k) \text{ THEN } (\Delta u_i = C_l), \quad i = 1, 2,$$

where: A_j, B_k – input fuzzy sets, C_l – output fuzzy set, Δu_i – output of the i-th fuzzy rule.

Rules R_i bind both input and output variables together in a cause-effect relations. Table 10.3 contains all possible control rules. This arrangement is the so-called *MacVicar-Whelan RB* and is very common in fuzzy logic PI regulators [Cheong and Lai (2007); MacVicar-Whelan (1977)]. It has been preliminarily assumed in this chapter that such an arrangement is optimal. Both the MacVicar-Whelan RB and its modification done by the authors were tested and compared in the subsequent sections.

Table 10.3 Fuzzy logic PI controller's RB.

e \ δe	NB	NM	NS	Z	PS	PM	PB
NB	NB$_{(0)}$	NB$_{(1)}$	NB$_{(2)}$	NB$_{(3)}$	NM$_{(4)}$	NS$_{(5)}$	Z$_{(6)}$
NM	NB$_{(7)}$	NM$_{(8)}$	NM$_{(9)}$	NM$_{(10)}$	NS$_{(11)}$	Z$_{(12)}$	PS$_{(13)}$
NS	NB$_{(14)}$	NM$_{(15)}$	NS$_{(16)}$	NS$_{(17)}$	Z$_{(18)}$	PS$_{(19)}$	PM$_{(20)}$
Z	NB$_{(21)}$	NM$_{(22)}$	NS$_{(23)}$	Z$_{(24)}$	PS$_{(25)}$	PM$_{(26)}$	PB$_{(27)}$
PS	NM$_{(28)}$	NS$_{(29)}$	Z$_{(30)}$	PS$_{(31)}$	PS$_{(32)}$	PM$_{(33)}$	PB$_{(34)}$
PM	NS$_{(35)}$	Z$_{(36)}$	PS$_{(37)}$	PM$_{(38)}$	PM$_{(39)}$	PM$_{(40)}$	PB$_{(41)}$
PB	Z$_{(42)}$	PS$_{(43)}$	PM$_{(44)}$	PB$_{(45)}$	PB$_{(46)}$	PB$_{(47)}$	PB$_{(48)}$

Figure 10.15 presents the inference scheme of the used controller for two exemplary rules. In the first step, inputs of an error and increase in the error are fuzzified. The algorithm determines to which fuzzy sets they belong to and calculates the value of membership function for each set as well. Next, the firing levels τ_i for each i-th rule are calculated using the t-norm (min() function). To receive the crisp output $\Delta u_{\text{PWM}}(n)$ in Eq. (10.17) from the regulator, the weighted average defuzzification method is used:

$$\Delta u_{\text{PWM}}(n) = \frac{\sum_1^i \tau_i \Delta u_i}{\sum_1^i \tau_i}, \tag{10.17}$$

where: τ_i – firing level of i-th fuzzy rule, Δu_i – outputs of the fuzzy rules, i – number of the particular fuzzy rule.

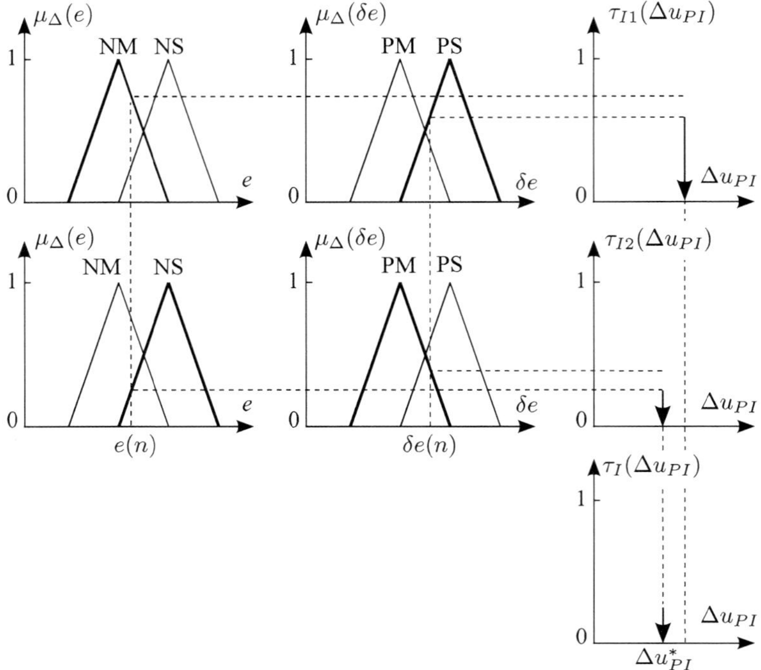

Fig. 10.15   The inference scheme.

10.3.3   *Modification of the Rule Base*

To increase the performance of the fuzzy logic PI algorithm, a modification to the RB was made (see Table 10.4).

Table 10.4   Modified RB for the fuzzy logic PI controller.

e \ δe	NB	NM	NS	Z	PS	PM	PB
NB	$NB_{(0)}$	$NB_{(1)}$	$NB_{(2)}$	$NB_{(3)}$	$NB_{(4)}$	$NM_{(5)}$	$NM_{(6)}$
NM	$NB_{(7)}$	$NB_{(8)}$	$NB_{(9)}$	$NM_{(10)}$	$NM_{(11)}$	$NM_{(12)}$	$NS_{(13)}$
NS	$NB_{(14)}$	$NB_{(15)}$	$NM_{(16)}$	$NM_{(17)}$	$NS_{(18)}$	$Z_{(19)}$	$Z_{(20)}$
Z	$NM_{(21)}$	$NM_{(22)}$	$Z_{(23)}$	$Z_{(24)}$	$Z_{(25)}$	$PS_{(26)}$	$PM_{(27)}$
PS	$Z_{(28)}$	$Z_{(29)}$	$PS_{(30)}$	$PS_{(31)}$	$PM_{(32)}$	$PM_{(33)}$	$PB_{(34)}$
PM	$PM_{(35)}$	$PM_{(36)}$	$PB_{(37)}$	$PB_{(38)}$	$PB_{(39)}$	$PB_{(40)}$	$PB_{(41)}$
PB	$PB_{(42)}$	$PB_{(43)}$	$PB_{(44)}$	$PB_{(45)}$	$PB_{(46)}$	$PB_{(47)}$	$PB_{(48)}$

The modified RB is not symmetric and the main diagonal consisting of "Z" outputs does not exist. This modification has allowed to separate the region of high absolute values of error from the region in which some oscillations occur. Such a change enables one to use higher gain (more aggressive output from the regulator) when the DC motor gains its rotational velocity.

10.3.4 *Test Results*

The tests were carried out for the setpoint speed of the belt pulley at the values of 5, 10 and 15 rpm. Figures 10.16–10.18 show the best achieved step responses for both types of regulators, including the version of the fuzzy logic PI regulator with modified RB (see Tables 10.5 and 10.6). In Table 10.7, a summary of their performance regarding the settling time and a sum of absolute values of error over time is provided. Features like execution time of the algorithm and the number of tuned variables were compared.

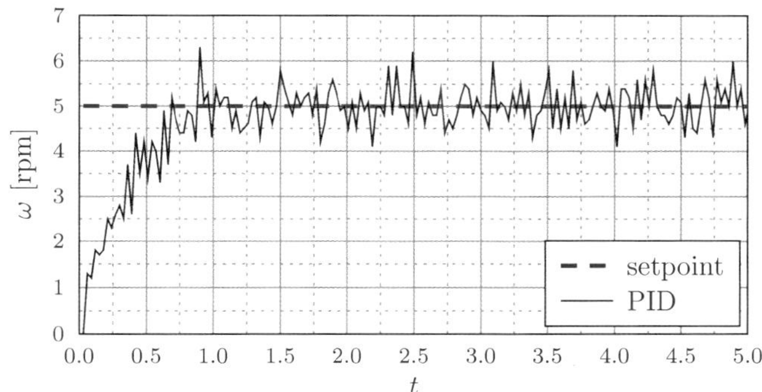

Fig. 10.16   Time history of angular velocity controlled by the tuned PID algorithm ($K_P = 1.2$, $K_I = 2.9$, $K_D = 1.35$, and the setpoint $\omega_{\text{sp}} = 5$ rpm).

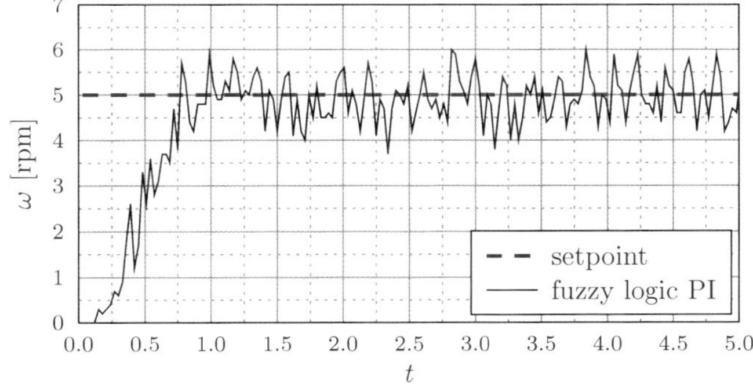

Fig. 10.17   Time history of angular velocity controlled by the tuned fuzzy logic PI algorithm (setpoint $\omega_{\text{sp}} = 5$ rpm).

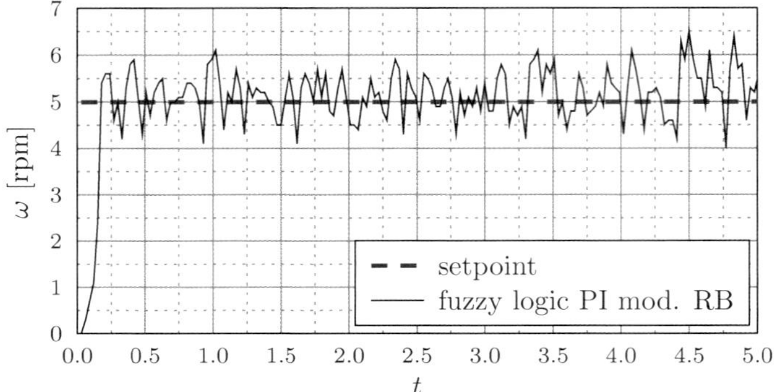

Fig. 10.18 Time history of angular velocity controlled by the tuned fuzzy logic PI algorithm with modified RB (setpoint $\omega_{\mathrm{sp}} = 5$ rpm).

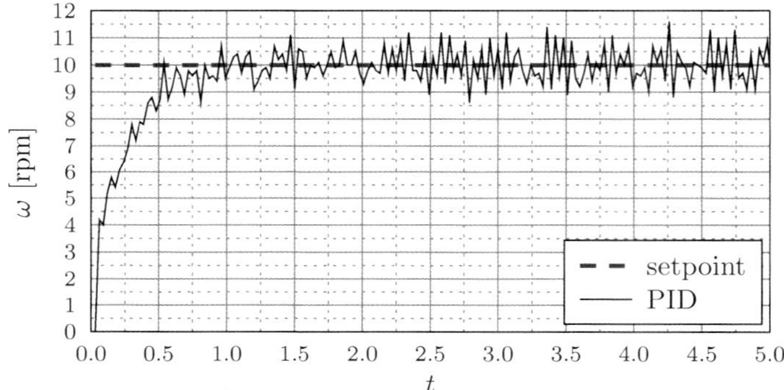

Fig. 10.19 Time history of angular velocity controlled by the tuned PID algorithm ($K_P = 1.2$, $K_I = 2.9$, $K_D = 1.35$, and the setpoint $\omega_{\mathrm{sp}} = 10$ rpm).

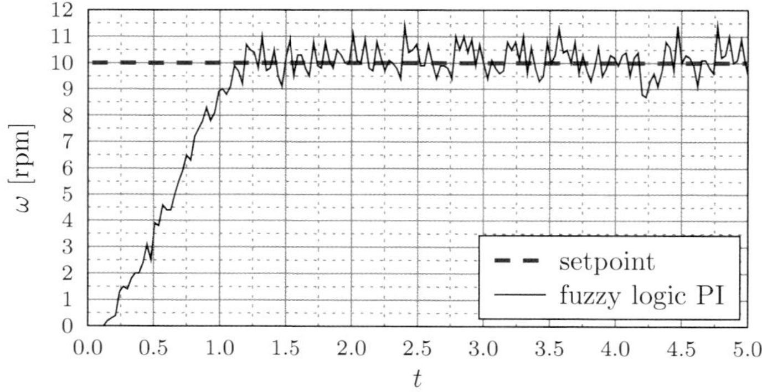

Fig. 10.20 Time history of angular velocity controlled by the tuned fuzzy logic PI algorithm (setpoint $\omega_{\mathrm{sp}} = 10$ rpm).

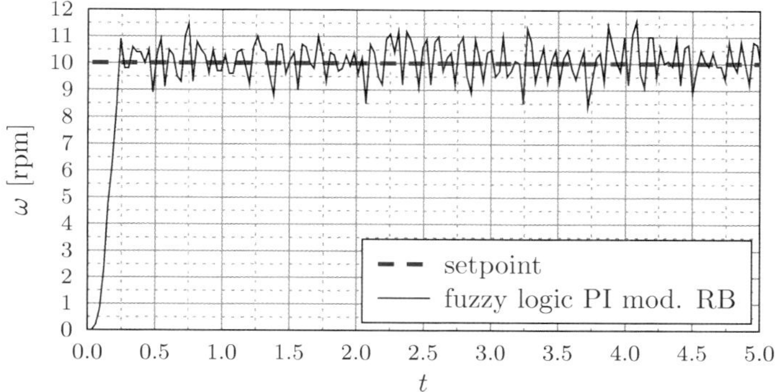

Fig. 10.21 Time history of angular velocity controlled by the tuned fuzzy logic PI algorithm with modified Rule Base (setpoint $\omega_{\mathrm{sp}} = 10$ rpm).

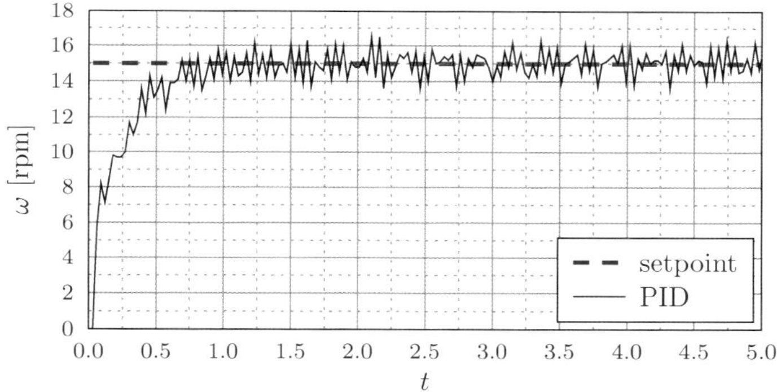

Fig. 10.22 Time history of angular velocity controlled by the tuned PID algorithm ($K_P = 1.2$, $K_I = 2.9$, $K_D = 1.35$, and the setpoint $\omega_{\mathrm{sp}} = 15$ rpm).

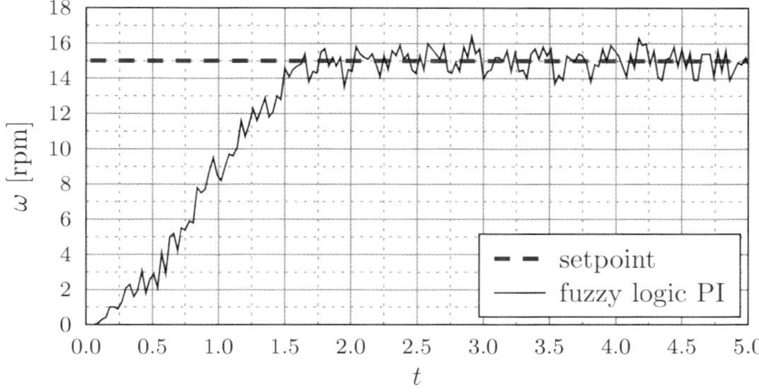

Fig. 10.23 Time history of angular velocity controlled by the tuned fuzzy logic PI algorithm (setpoint $\omega_{\mathrm{sp}} = 15$ rpm).

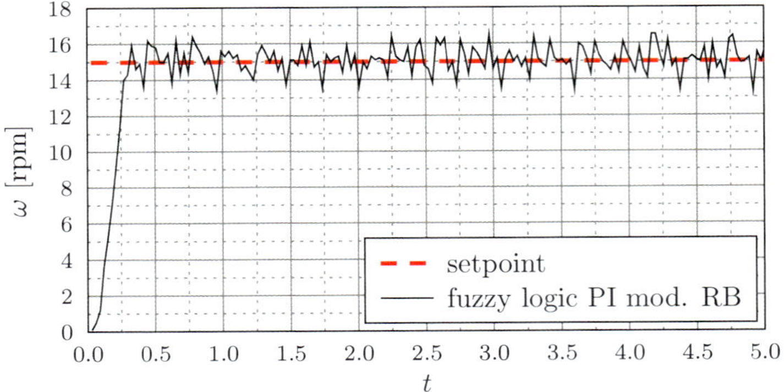

Fig. 10.24   Time history of angular velocity controlled by the tuned fuzzy logic PI algorithm with modified Rule Base (setpoint $\omega_{\mathrm{sp}} = 15\,\mathrm{rpm}$).

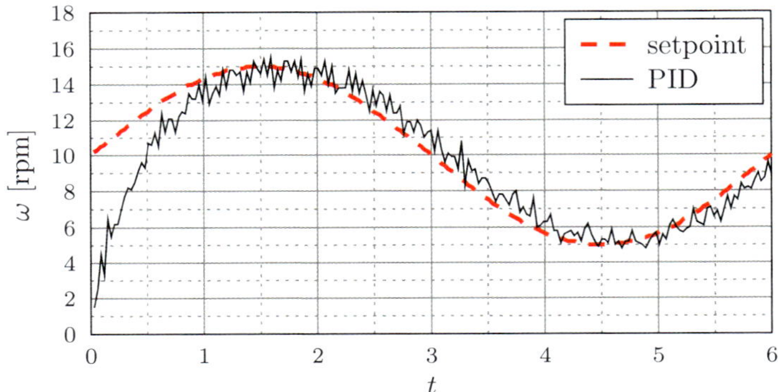

Fig. 10.25   Trajectory tracking by the PID algorithm.

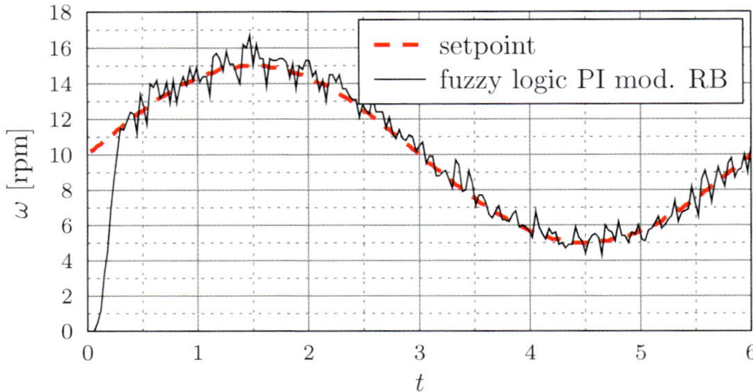

Fig. 10.26   Trajectory tracking by the fuzzy logic PI algorithm with modified RB.

Table 10.5   Fuzzy logic PI controller parameters.

Fig. 10.14		NB	NM	NS	Z	PS	PM	PB
	a	-0.9	-0.9	-0.6	-0.3	0.0	0.3	0.6
e	b	-0.6	-0.6	-0.3	0.0	0.3	0.6	0.9
	c		-0.3	0.0	0.3	0.6	0.9	
	a	-0.6	-0.6	-0.4	-0.2	0.0	0.2	0.4
δe	b	-0.4	-0.4	-0.2	0.0	0.2	0.4	0.6
	c		-0.2	0.0	0.2	0.4	0.6	
u_i		-10	-6	-3	0	3	6	10

Table 10.6   Fuzzy logic PI controller parameters with modified RB.

Fig. 10.14		NB	NM	NS	Z	PS	PM	PB
	a	1.2	-0.8	-0.4	-0.2	0.0	0.5	1.4
e	b	0.6	-0.5	-0.2	0.0	0.4	1.0	2
	c		-0.2	0.0	0.2	0.8	1.5	
	a	0.6	-0.6	-0.4	-0.2	0.0	0.2	0.4
δe	b	0.4	-0.4	-0.2	0.0	0.2	0.4	0.6
	c		-0.2	0.0	0.2	0.4	0.6	
u_i		-12	-6	-3	0	3	8	35

Table 10.7   Performance comparison of the two considered algorithms.

Property	rpm	PID	Fuzzy logic PI	Fuzzy logic PI mod. RB
	5	0.93	0.84	0.21
Settling time [s]	10	0.96	1.12	0.24
	15	0.99	1.65	0.33
	5	101.9	143.0	100.8
Sum of absolute error values over time	10	153.3	277.1	157.6
	15	197.2	516.2	194.1
Lines of code required		105	280	280
Number of tuned variables		3	45	$45 + 49 = 94$
Execution time of algorithms [μs]		73	112	112

As can be observed, only marginal gain in stability of the investigated control system was achieved. Both regulators fail to properly respond to chaotic changes in load parameters. A comparison made basing on the Table 10.7 shows that the basic fuzzy logic PI algorithm does not provide as good control quality as the PID. On the other hand, modification to the RB had a great impact on the settling time of the system. The FLC with tuned RB almost instantly generates adequate output to reach the setpoint speed, while in the case of PID regulator, a decreasing build-up closer to the setpoint can be observed.

In Figs. 10.25 and 10.26, a trajectory tracking trial was presented. An interesting thing to notice is the observable phase displacement (a delay) in output of the PID algorithm. The fuzzy logic controller does not demonstrate such phase shift and achieves the desired trajectory in much shorter time.

It is seen that the modifications to the RB have significantly helped in reducing the settling time, but a negative impact on the stability can be observed comparing the step responses of both algorithms. Another idea to decrease the settling time is to introduce an additional fuzzy rule. The fuzzy set would place itself before the NB set of the error input variable while the rule covering this set would generate a substantial output that would help with gaining the DC motor speed. Applying a fuzzy PID regulator could also help with the sluggish response of the system, but the exponentially growing number of variables to tune could be problematic. Another idea is to develop a hybrid controller. For large values of error of regulation, it could act as a classical PID controller, and for lower values (near the setpoint), the FLC would take control over the system stabilization.

From an engineering point of view, the FLC PI regulator states a bigger challenge than any classical PID controller. While nowadays, the computation time and the volume of code is in most cases less significant, the amount of the tuned variables proves to be a great downside of the fuzzy logic control approach. In Table 10.7, the amount of tuned variables is given by a sum of two values. Number 45 reflects the amount of parameters describing input and output fuzzy sets, and the number 49 is the amount of fuzzy rules in the RB. Initially, the first set of variables was tuned and the RB was assumed to be optimal. In the second approach, both the parameters regarding the fuzzy sets and the RB were tuned. Since the variables of the tuned system have a clear physical interpretations, the tuning process seems to be very intuitive. Unfortunately, in reality, it is very time-consuming and demands a broad knowledge about the controlled object. The study and experimental work also confirm that all parameters should be carefully tuned. In this case, the widely used MacVicar-Whelan RB did not provide satisfactory results.

The fuzzy logic controller described in this chapter provides similar quality of control comparing to the classical solution. Mostly, the settling time has benefited from the use of the fuzzy approach. Despite the interesting concept standing behind it, no substantial improvement in smoothing DC motor's angular velocity in the setpoint region was achieved. Additionally, the very complex tuning process extends the time in which promising results were obtained. To reduce the difficulty of tuning, the fuzzy logic approach requires its own learning algorithm which would pick the right parameters and adopt them actively to the experiment.

Chapter 11

Tracking Control of an Electromechanical System

Stick-slip vibrations appear during relative motion between contacting surfaces of miscellaneous frictional pairs. They depend on the viscous force, Coulomb force or other velocity-dependent forces. These effects appear in almost all mechanical systems, for instance, in positioning systems like servomechanisms, impulse encoders and stepper motors which operate at or about zero velocity of relative motion between shafts and sliding bearings. This chapter extends our study on numerical modeling of a DC (direct current) motor treated as a dynamical system with stick-slip effects which appear in transient motion even while the direction of rotation of its rotor crosses zero velocity speed. These investigations are aimed at some future applications of the control technique serving for explanation of bifurcation phenomena existing in such kind of discontinuous systems. Putting emphasis on nonlinear effects, we apply the well-known but slightly extended sliding surface method allowing for compensation of frictional effects. A limit cycle on a phase plane, as well as time histories of control inputs and system outputs, were obtained using numerical simulations performed in `LabVIEW`.

The natural resistance to the relative motion between nonlubricated surfaces of two contacting bodies is called dry friction. In some dynamical systems modeling nonlinearities caused by dry friction, a controller has to be designed to avoid any steady state tracking errors or vibrations.

An adaptive friction compensation to improve performance of tracking errors without the Stribeck effect has been proposed in [Huang *et al.* (2000)]. A new control strategy for compensation of frictional phenomena including the Stribeck effect, hysteresis, stick-slip limit cycling, pre-sliding displacement and rising static friction has been particularly described and examined. The proposed compensator could be useful for handling significant nonlinearities in motor controls. Similarly, the Lund-Grenoble model of dynamical friction has been used in [Hirschon and Miller (1999)] to control nonlinear effects that the model captures: the Stribeck and Dahl effects, viscous and Coulomb friction [Awrejcewicz and Olejnik (2005a)]. A new Lyapunov-based continuous dynamic controller has been delivered for a more general class of nonlinear systems. It produced better control of a high-speed precision linear tracking table than some tuned PID controllers without direct nonlinear effects compensation. As it has been shown also in [Song *et al.* (1998)], conventional feedback control methods cannot ensure good results in the presence of dry friction (stick-slip) even in the one-degree-of-freedom DC motor system. Because of steady state errors, a traditional PD controller would not achieve satisfactory performance. These errors could be reduced by increasing the P gain, but significant instabilities would be reported while driving the motor with some angular velocities or along the desired rapidly changing time history of its angular position. Very good positioning accuracy have been obtained with the use of a new sliding-mode-based smooth adaptive robust controller designed for dry friction compensation.

A study of control of a mechanism under the influence of low velocity friction has been conducted in [Adams and Payandeh (1996)]. The theoretical and experimental comparative study of linear (PD, PID) and nonlinear (smooth continuous and piece-wise linear discontinuous) compensation algorithms have been proposed. In the case of modeling a two-degree-of-freedom controlled planar manipulator, the nonlinear controllers have proved to have superior performance regarding some P and D gains, compared to any PD controller. Moreover, their tracking performance was also superior to the PID controller, but it provides an oscillatory time dependency of torque. Stability of smooth controllers was much simpler to demonstrate.

Simple active control of the belt-driven oscillator with stick-slip friction in a control system with a feedback loop created by a transducer, frequency filter, phase shifter, amplifier and a shaker attached to the oscillating body has been studied in [Heckl and Abrahams (1996)]. The feedback system allowed for suppression of unstable vibrations at high effectiveness insensitive to errors in phase shift and amplification. Similarly, in [Li *et al.* (2009)], some type of friction-driven oscillator controlled by Lyapunov redesign based on delayed state feedback has been numerically investigated. The authors redesigned a continuous controller on the basis of a delayed state feedback to ensure that the nonsmooth friction-driven system is ultimately bounded. Moreover, by constructing a Lyapunov-Krasovskii function, the sufficient condition of stability for the investigated system was obtained.

Neural networks have the capability of approximating nonlinear functions, therefore they are also demanding when it comes to estimation of frictional behaviors. Much work has been done in this subject [Kim and Lewis (2000); Otten *et al.* (1997)]. Work [Otten *et al.* (1997)], for instance, brings investigations on control of linear motion of motors by means of the learning forward controller that is designed in the discrete state-space. The authors [Driessen and Sadegh (2004)] have also solved the problem of discrete-time iterative learning control for position trajectory tracking of multiple-input, multiple-output systems, including Coulomb friction, bounds on the inputs, static and sliding friction coefficients. On the background of a two-link revolute-joint planar robot arm, some satisfactory learned angular position time histories (at a decrease in position-tracking error) have been shown. In accordance to linear servo motor control, a novel, very interesting approach for designing a wavelet basis function network learning controller for a linear motor control system was considered in [Lin and Huang (2002)]. The proposed wavelet network-based controller dealt with viscous friction and force ripples that occur in motion control of linear synchronous motors.

The considerations presented in the exemplary articles were motivated by examination of control approaches, but the investigation of stability should be also considered. An interesting reference [Chenafa *et al.* (2005)] presents analysis of global stability of linearizing control of induction motors with a new robust nonlinear observer-based approach. Authors used traditional Park's induction model in a stator fixed reference frame related to the stator given by [Mansouri *et al.* (2004)]. They designed a control algorithm based on feedback linearization [Marino *et al.* (1993)]. After assumption of parameters of the induction motor, a detailed scheme of the nonlinear control with an observer has been done in Simulink. The new robust observer based on a nonlinear control scheme offered advantage of only one tuning parameter. The global stability of the whole system consisting of the motor, the controller and the observer was established by means of the precise Lyapunov function that kept observer's dynamics free. More on the initial strategy on input-output linearization can be found in [Chiason (1997)], but on the global stability of the process-observer-controller system in [Lubineau *et al.* (2000)].

Deeper survey through the cited literature provides many references to theoretical derivations and practical implementations confirming permanent interest in control of nonsmooth systems. Basically, control strategies depend on the aim, the friction law, the system at hand and its field of application.

11.1　Problem Statement

This study concerns on a numerical simulation of compensation of frictional effects present in a real system designed for observations and experimental estimation of friction force characteristics, see [Awrejcewicz and Olejnik (2003)]. The system consists of a DC motor driving a wide transmission belt on which a rigid body, being in frictional contact with the surface of the belt, vibrates. For instance, to find bifurcations of sliding solutions [Awrejcewicz and Olejnik (2005d)], after a relative motion observed between contacting surfaces of the investigated coupling, it is required to precisely implement some desired function of changes of angular velocity of the DC motor that drives the belt. Therefore, rotational velocity of the driving motor should vary in a periodic cycle, exactly tracking the desired time-dependent characteristics (triangular, sinusoidal, etc.). However, such a situation does not occur with regard to the existing nonlinear friction characteristics. From the point of view of the control theory, it states a problem of providing a robust tracking control of rotational velocity of the DC motor. Therefore, some close to ideal generation of regular input signal (excitation of the belt) would be possible after application of some tracking control technique that has been already implemented, for instance, to control robot manipulators [Lewis *et al.* (1993)]. The following control errors could be used as input to the method: $e(t) = \varphi(t) - \varphi_d(t)$, $\varepsilon(t) = \dot{\varphi}_d(t) - \lambda e(t)$, where: φ and $\dot{\varphi}$ are, respectively, the angular position and velocity of motor's rotor driving the base by means of a nonstretchable transmission toothed belt; index d denotes desired values of corresponding variables along with the desired phase trajectory.

Particular investigation will be focused on examination of influence of frictional contacts existing during rotation of a rotor of a DC motor. One can distinguish a phenomenon of stick-slip friction that mostly affects accuracy of positioning. Friction of such type was investigated by authors of the monograph in [Awrejcewicz and Olejnik (2005a)], and may result from the following: Coulomb friction that represents maximum static friction $T_{\mathrm{sm}} \operatorname{sgn} \dot{\varphi}(t)$ at a slip phase and $T_{\mathrm{sm}}(1 - \operatorname{sgn}|\dot{\varphi}(t)|)$ at a stick phase, when an input torque generated by a system driven by motor's rotor could by applied, exponential friction described by the Stribeck curve $T_{\mathrm{Stm}}(1 - \exp(-T_0|\dot{\varphi}(t)|) \operatorname{sgn} \dot{\varphi}(t))$, viscous friction $T_{\mathrm{vm}}\dot{\varphi}(t)$, and position-dependent friction $T_{1\mathrm{m}} \sin(T_2\varphi(t)+T_3) \operatorname{sgn}|\dot{\varphi}(t)|$, as proposed in [Slotine and Li (1987)], where: $\operatorname{sgn} \dot{\varphi}$ denotes the sign of the value of angular velocity, φ is an angular displacement, T_{sm} is the maximum static friction torque, T_{Stm} and $T_0 > 0$ are the parameters of Stribeck curve, T_{vm} is the coefficient of viscous friction, T_{1m}, T_2 and T_3 are constants. The mechanical part of the reduced dynamical system of differential equations used for modeling the dynamics of rotational motion of a DC motor holds:

$$J_m\ddot{\varphi}(t) + \left(\frac{c_b c_m}{R_a} + T_{\mathrm{vm}}\right)\dot{\varphi}(t) - T_{\mathrm{Stm}}\left(1 - e^{-T_0|\dot{\varphi}(t)|}\right)\operatorname{sgn}\dot{\varphi}(t) +$$
$$T_{1\mathrm{m}}\sin(T_2\varphi(t) + T_3)\operatorname{sgn}|\dot{\varphi}(t)| + T_{\mathrm{sm}}(1 - \operatorname{sgn}|\dot{\varphi}(t)| +$$
$$\operatorname{sgn}\dot{\varphi}(t)) = c_m\psi_m(t), \qquad (11.1)$$

and the remaining unknown model parameters are as follows: R_a and ψ_m denote the armature resistance and the armature current, respectively, J_m is the moment of inertia of the rotor, c_b is a constant of the back electromotive force, and c_m is the motor torque constant. One rewrites Eq. (11.1) in a form scaled with respect to c_m as follows

$$J\ddot{\varphi}(t) + B\dot{\varphi}(t) + \tau(t) = \psi(t), \qquad (11.2)$$

where the function $\tau(t) = T_v\dot{\varphi}(t) - T_{\mathrm{St}}(1 - \exp(-T_0|\dot{\varphi}(t)|))\operatorname{sgn}\dot{\varphi}(t) + T_1\sin(T_2\varphi(t)+T_3)\operatorname{sgn}|\dot{\varphi}(t)| + T_s(1 - \operatorname{sgn}|\dot{\varphi}(t)| + \operatorname{sgn}\dot{\varphi}(t))$ states for the scaled friction force, and $J, B, T_v, T_s, T_1, T_{\mathrm{St}}$ equal $\frac{J_m}{c_m}, \frac{c_b}{R_a}, \frac{T_{\mathrm{vm}}}{c_m}, \frac{T_{\mathrm{sm}}}{c_m}, \frac{T_{1m}}{c_m}, \frac{T_{\mathrm{Stm}}}{c_m}$, respectively.

It is not possible to exactly describe friction and to correctly assume all values of parameters. A tracking control, which is a point of the study, should also correct any inaccuracies caused by an imprecise system modeling.

11.2   Control Strategy

The objective of control is to design an adaptive controller that would allow to change angular velocity of rotation of motor's rotor according to some desired function $\varphi_d(t)$. Let us begin from the so-called sliding surface method [Slotine and Li (1987)]. On its background, the control error $e(t) = \varphi(t) - \varphi_d(t)$, an auxiliary variable $\varepsilon(t) = \dot{\varphi}_d(t) - \lambda e(t)$ and the definition of the sliding surface $r(t) = \dot{\varphi}(t) - \varepsilon(t) = 0$, where variables with index d denote corresponding desired values, $\lambda > 0$ is for more general multidimensional case a positive definite main diagonal matrix. Having all variables of the sliding surface method introduced, let us propose a control law derived from Eq. (11.2) in the form:

$$\psi(t) = \hat{J}\dot{\varepsilon}(t) + \hat{D}\varepsilon(t) - \hat{T}_{\mathrm{St}}\left(1 - e^{-\hat{T}_0|\dot{\varphi}(t)|}\right)\operatorname{sgn}\dot{\varphi}(t) +$$

$$\hat{T}_s\operatorname{sgn}\dot{\varphi}(t) + \hat{T}_s\left(1 - \operatorname{sgn}|\dot{\varphi}(t)|\right)u_s(t) - u_b(t)\,, \qquad (11.3)$$

where $u_b(t)$ is a condition of bounding function, $u_s(t) = 1 - \operatorname{sgn}|r(t)|$ at term describing sticking phase is a function introduced with respect to the definition of the sliding surface $r(t) = 0$, $\hat{D} = \hat{B} + \hat{T}_v$, and the circumflex ^ above symbols marks estimates of corresponding parameters.

Equation (11.3) will be used for adaptation of unknown estimates in a scheme in which $\psi(t)$ is put to Eq. (11.1) to compensate for linear and nonlinear forces included in it. Such an adaptive feed-forward control loop is good to compensate linear friction forces like Coulomb and viscous ones [Song *et al.* (1998)]. Nonlinear friction forces, like the Stribeck effect and the angular position-dependent friction force, cannot be controlled in the loop, but some adaptation law based on a robust compensator to learn an upper bounding function has to be used [Lewis *et al.* (1993)]. The following bounding function is assumed

$$u_b(t) = k_D r(t) + \hat{\rho} k_T \tanh(r(t)(a + bt))\,, \qquad (11.4)$$

where: a, b and k_D are positive constants, and $k_T > 1$. If parameter $\hat{\rho}$ is an estimate of the upper bound of the nonlinear residual terms, then $u_b(t)|_{r(t)=\lambda e}$ behaves as a proportional gain robustly compensating nonlinear friction forces. It inputs to the control law (11.3) a torque greater than the maximum static friction allowing for compensation.

11.2.1 *Estimation of Linear and Nonlinear Parameters*

In the sliding surface method, the adaptive law validating all unknowns at each step of integration is based on a simple first-order differential equation. Therefore, the following adaptive law [Slotine and Li (1987)] validating estimates of the system parameters at linear terms takes the form:

$$\dot{\hat{J}}(t) = -\delta_1\dot{\varepsilon}(t)r(t)\,, \qquad (11.5)$$

$$\dot{\hat{D}}(t) = -\delta_2\varepsilon(t)r(t)\,, \qquad (11.6)$$

$$\dot{\hat{T}}_s(t) = \begin{cases} -\,\delta_3\operatorname{sgn}\left(\dot{\varphi}(t)\right)r(t), & \text{in a slip,} \\ \delta_3\left(1 - \operatorname{sgn}|\dot{\varphi}(t)|\right)|r(t)|, & \text{in a stick,} \end{cases} \qquad (11.7)$$

where $\delta_{1,\ldots,3}$ are positive constants, $r(t) = \dot{\varphi}(t) - \varepsilon(t)$.

Putting $\psi(t)$ from Eq. (11.3) to (11.2), including $\varepsilon(t) = \dot{\varphi}(t) - r(t)$, $\dot{\varepsilon} = \ddot{\varphi}(t) - \dot{r}(t)$ with $\tilde{T}_{St} = \hat{T}_{St} - T_{St}$ and $\tilde{T}_0 = \hat{T}_0 - T_0$ measuring differences between estimates and their corresponding real values, one gets

$$J\dot{r}(t) + Dr(t) = \hat{J}\dot{\varepsilon}(t) + \hat{D}\varepsilon(t) + \hat{T}_s \operatorname{sgn}\dot{\varphi}(t) + \omega(t) - u_b(t) , \qquad (11.8)$$

where

$$\omega(t) = \left(\hat{T}_{St} e^{-\hat{T}_0|\dot{\varphi}(t)|} - T_{St} e^{-\hat{T}_0|\dot{\varphi}(t)|} e^{-\tilde{T}_0|\dot{\varphi}(t)|} - \tilde{T}_{St} - T_p \right) \operatorname{sgn}\dot{\varphi}(t) . \qquad (11.9)$$

Expanding $\exp(\tilde{T}_0|\dot{\varphi}(t)|) = 1 + \tilde{T}_0|\dot{\varphi}(t)| + \tilde{T}_0|\dot{\varphi}(t)|^2/2 + \tilde{R}$ in a Taylor series about $|\dot{\varphi}(t)| = 0$ and using only the first three terms of the expansion with a reminder $\tilde{R} \le \exp(\tilde{T}_0|\dot{\varphi}(t)|)T_0^3|\dot{\varphi}(t)|^3/6$:

$$\omega(t) = \left[\hat{T}_{St} e^{-\hat{T}_0|\dot{\varphi}(t)|} - T_{St} e^{-\hat{T}_0|\dot{\varphi}(t)|} \left(1 + \tilde{T}_0|\dot{\varphi}(t)| + \tilde{T}_0 \frac{|\dot{\varphi}(t)|^2}{2} + \right. \right.$$
$$\left. \left. \tilde{T}_0^3 \frac{|\dot{\varphi}(t)|^3}{6} e^{\tilde{T}_0|\dot{\varphi}(t)|} \right) - \tilde{T}_{St} - T_p \right] \operatorname{sgn}\dot{\varphi}(t) = \rho \operatorname{sgn}\dot{\varphi}(t) , \quad (11.10)$$

where variable $\rho = -\gamma_1 + \gamma_2 \exp(-\hat{T}_0|\dot{\varphi}(t)|) - \gamma_3|\dot{\varphi}(t)|\exp(-\hat{T}_0|\dot{\varphi}(t)|) - \gamma_4|\dot{\varphi}(t)|^2 \exp(-\hat{T}_0|\dot{\varphi}(t)|)$, $\gamma_1 = \max_{|\dot{\varphi}(t)|\in[0,\infty]}\{|\dot{\varphi}(t)|^3 \exp(-T_0|\dot{\varphi}(t)|)\tilde{T}_0^3 T_{St}/6\} + \tilde{T}_{St} + T_p$, $\gamma_2 = \tilde{T}_{St}$, $\gamma_3 = \tilde{T}_0 T_{St}$, $\gamma_4 = \tilde{T}_0^2 T_{St}/2$. Constants $\gamma_{1,\ldots,4}$ depend on estimates or on their difference from real values.

At this point, let us come back to Eq. (11.4) containing an unknown estimate $\hat{\rho}$. In Eq. (11.9), to get cancellation of reminders not dependent on $r(t), \dot{r}(t), \varepsilon(t)$ and $\dot{\varepsilon}(t)$, $\omega(t) - \hat{\rho}(t)\omega_r(t) \to \infty$, where $\omega_r(t) = k_T \tanh(r(t)(a+bt))$ as introduced in [Cai and Song (1994)]. Therefore, $\rho \operatorname{sgn}\dot{\varphi}(t) \to \hat{\rho}k_T \tanh(r(t)\ (a+bt))$, and if $u_b(t)$ is the upper bounding function of $\omega(t)$, then

$$\hat{\rho}(t) = -\hat{\gamma}_1 + \hat{\gamma}_2 e^{-\hat{T}_0|\dot{\varphi}(t)|} - \hat{\gamma}_3|\dot{\varphi}(t)|e^{-\hat{T}_0|\dot{\varphi}(t)|} - \hat{\gamma}_4|\dot{\varphi}(t)|^2 e^{-\hat{T}_0|\dot{\varphi}(t)|} \qquad (11.11)$$

states the estimate of ρ variable. Similarly to construction of Eq. (11.8), we can calculate the remaining estimate $\hat{\rho}$ by solving the following system of equations:

$$\dot{\hat{\gamma}}_1(t) = \delta_4|r(t)| , \quad \dot{\hat{\gamma}}_2(t) = -\delta_5|r(t)|e^{-\hat{T}_0|\dot{\varphi}(t)|} ,$$
$$\dot{\hat{\gamma}}_3(t) = -\delta_6|r(t)||\dot{\varphi}(t)|e^{-\hat{T}_0|\dot{\varphi}(t)|} , \quad \dot{\hat{\gamma}}_4(t) = -\delta_7|r(t)||\dot{\varphi}(t)|^2 e^{-\hat{T}_0|\dot{\varphi}(t)|} , \qquad (11.12)$$

where $\gamma_{1,\ldots,4}$ are positive constants.

11.2.2   *Voltage Input for Control of Rotational Velocity*

In a real application, one would need to source the motor not with the electric current but with voltage of known function. In this situation, the following full electromechanical system has to be introduced into the analysis:

$$J\ddot{\varphi}_f(t) + B\dot{\varphi}_f(t) + \tau_f(t) = \psi_f(t) , \qquad (11.13)$$
$$L_a\dot{\psi}_f(t) + R_a\psi_f(t) + c_b\dot{\varphi}_f(t) = v_f(t) , \qquad (11.14)$$

where index f is used to denote a full three-dimensional dynamical system, L_a is the armature inductance, $v_f(t)$ is a time-dependent function of voltage required to realize some desired task of control.

One assumes that Eqs. (11.1) and (11.2) mathematically describe the dynamics of the motor of which electrical and mechanical parameters will be taken according to the existing direct current commutation motor PZTK 60-46 J suitable for use in cross-feed drives of numerically controlled machines.

In a full electromechanical system, voltage control requires to regard to Eq. (11.13). If the current-input control of the DC motor works correctly, then the best solution is to maximally reduce the influence of the second equation. In the full system, it provides these unwanted disturbances influencing the optimal current input. The most obvious would be to apply to Eq. (11.14) the voltage input $v_f(t)$ calculated on the basis of $\psi(t)$ which is estimated after solution of only the reduced mechanical system (11.2) given by control law (11.3). Therefore, voltage input necessary to cancel the dynamical disturbances of the complete three-dimensional electromechanical system produced by (11.14) is expected in the form

$$v_f(t) = L_a\dot{\psi}(t) + R_a\psi(t) + c_b\dot{\varphi}(t) + d(t) \tag{11.15}$$

with a limitation that $\psi(t)$ ensures proper tracking current-input control of the reduced model in Eq. (11.2) and $\dot{\varphi}$ states for angular velocity resulting from that control. Function $d(t)$ is a compensator of dynamical differences between state variables of Eqs. (11.14) and (11.15).

After substitution of v_f given by Eq. (11.15) to (11.14), all dynamical terms in Eq. (11.13) have their counterparts canceling them, but some occurring differences are expected to be compensated by $d(t)$ which, if disregarded, makes the substitution working incorrectly and some significant oscillations about zero value are observed. To increase effectiveness of the control strategy, it is proposed to apply a two-dimensional proportional control with a feedback from the object of control described by a full dynamical system of the modeled motor. Therefore, applying

$$d(t) = k_1(\varphi_d(t) - \varphi_f(t)) + k_2(\dot{\varphi}_d(t) - \dot{\varphi}_f(t)) \tag{11.16}$$

to Eq. (11.15) to be used in Eq. (11.13), the following equation of dynamical equilibrium is found:

$$L_a\left(\dot{\psi}_f(t) - \dot{\psi}(t)\right) + R_a\left(\psi_f(t) - \psi(t)\right) + c_b\left(\dot{\varphi}_f(t) - \dot{\varphi}(t)\right) =$$

$$k_1\left(\varphi_d(t) - \varphi_f(t)\right) + k_2\left(\dot{\varphi}_d(t) - \dot{\varphi}_f(t)\right), \tag{11.17}$$

where to get the demanding cancellation of Eq. (11.14), tuning factors k_1 and k_2 should ensure equality of both sides of Eq. (11.17), but at each time instant, solution $\psi_f(t)$ have to be updated in Eq. (11.13), φ_d and $\dot{\varphi}_d$ are the desired coordinates of the phase trajectory of the rotor motion. Having this condition met, solution $\varphi_f(t)$ to Eq. (11.13) should track the optimal solution $\varphi(t)$ of Eq. (11.2). In tracking control, the time history of $v_f(t)$ can be saved and used as input to drive the complete electromechanical dynamical system of the DC motor along with either the desired phase trajectory, angular velocity or angular position of its rotor.

11.3 Numerical Simulation

Efficiency of the two-stage control method is checked with numerical simulations performed for a model of the DC motor PZTK 60-46 J with stick-slip friction occurring in contact zones located between rotor's shaft and bearings. Rotational velocity of the DC motor is required to follow the desired trajectory $\varphi_d(t)$ drawn with a dashed line in Fig. 11.2.

Time history of the desired velocity is formed in the scheme: it increases from 0 to 0.2 rad/s in 0.2 s, then it is held at this value for 0.6 s, it is decreased to 0 in 0.2 s, and without a delay, it changes its value (in the second half of the period) to negative, achieving symmetrically the same thresholds and times of presence as for positive values. After 2 s the cycle is repeated (see the dashed line in Fig. 11.2).

Table 11.1 System and tuning parameters for the numerical simulation.

	Notation	Value	Unit
Motor torque constant	c_m	0.5	N·m/A
Constant of the back electromotive force	c_b	0.011	V/rpm
Armature resistance	R	1.1	Ω
Moment of inertia of the rotor	J_m	2	kg·m^2
Armature inductance	L_a	10^{-3}	H
Coefficient of viscous friction	T_v	8	N·m·s/rad
Max. static friction torque on the Stribeck curve	T_{Stm}	0.5	N·m
Max. static friction torque	T_{sm}	1.5	N·m
Position-dependent friction torque	T_{1m}	0.35	N·m
Constant of the Stribeck curve	T_0	10	–
Constants of position-dependent friction	$[T_2, T_3]$	$[1, 0.5]$	–
Constants of adaptation laws	$\delta_{i=1\dots7}$	$9 \cdot i$	–
Tuning factors: P gain	k_1	$0.21 \cdot 10^3$	–
Tuning factors: D gain	k_2	$1.40 \cdot 10^3$	–

Initial values of parameter estimates at linear terms: $\hat{J}^{(0)} = 1$, $\hat{D}^{(0)} = 1$, $\hat{T}_s^{(0)} = 0.2$, $\hat{T}_{\mathrm{St}}^{(0)} = 1$, $\hat{T}_0^{(0)} = 1$. The initial value of the parameter estimate at nonlinear terms $\hat{\rho}^{(0)} = 0$, and initial values of state variables $\varphi^{(0)} = \dot{\varphi}^{(0)} = 0$, $\psi_f^{(0)} = 0$.

It is important to observe that at the beginning of simulation, exact values of some parameters are not known, but are given by initial values of their counterpart estimates (see convergence in Fig. 11.3). Besides the uncertainty of parameters, there exists some influence of discontinuous terms of frictional torques described earlier.

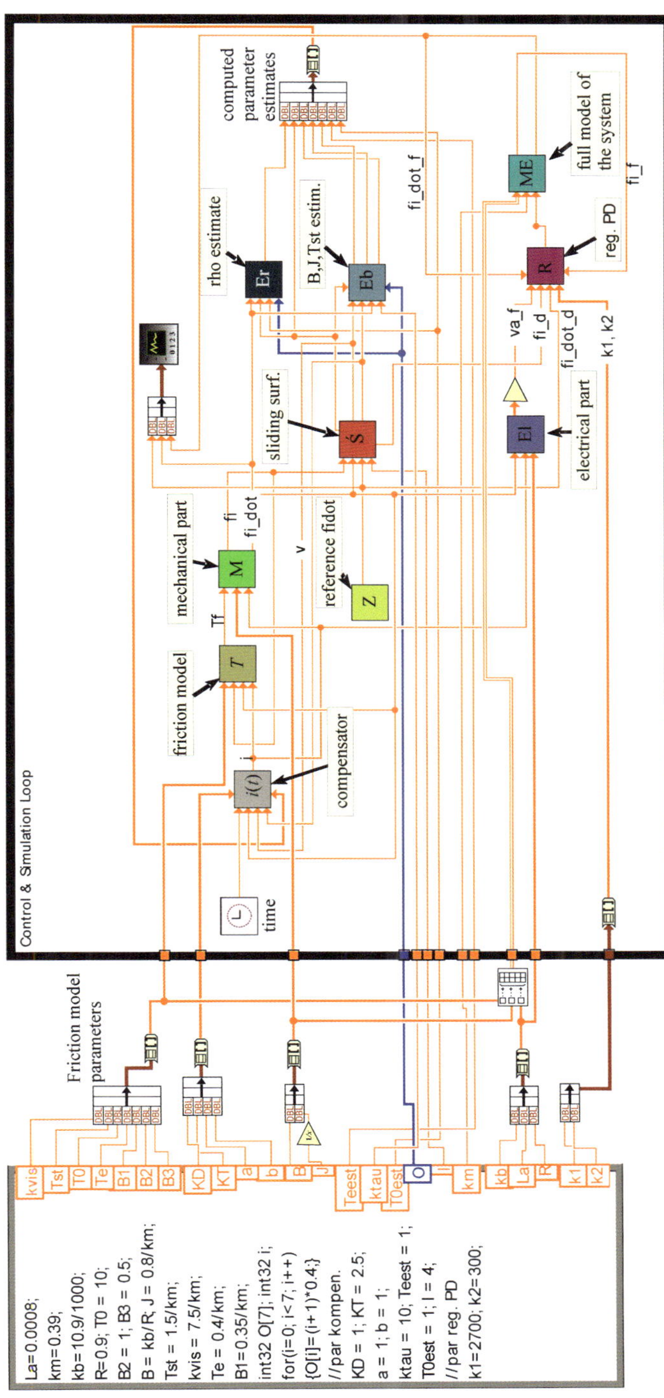

Fig. 11.1   Block diagram of the control system created in LabVIEW.

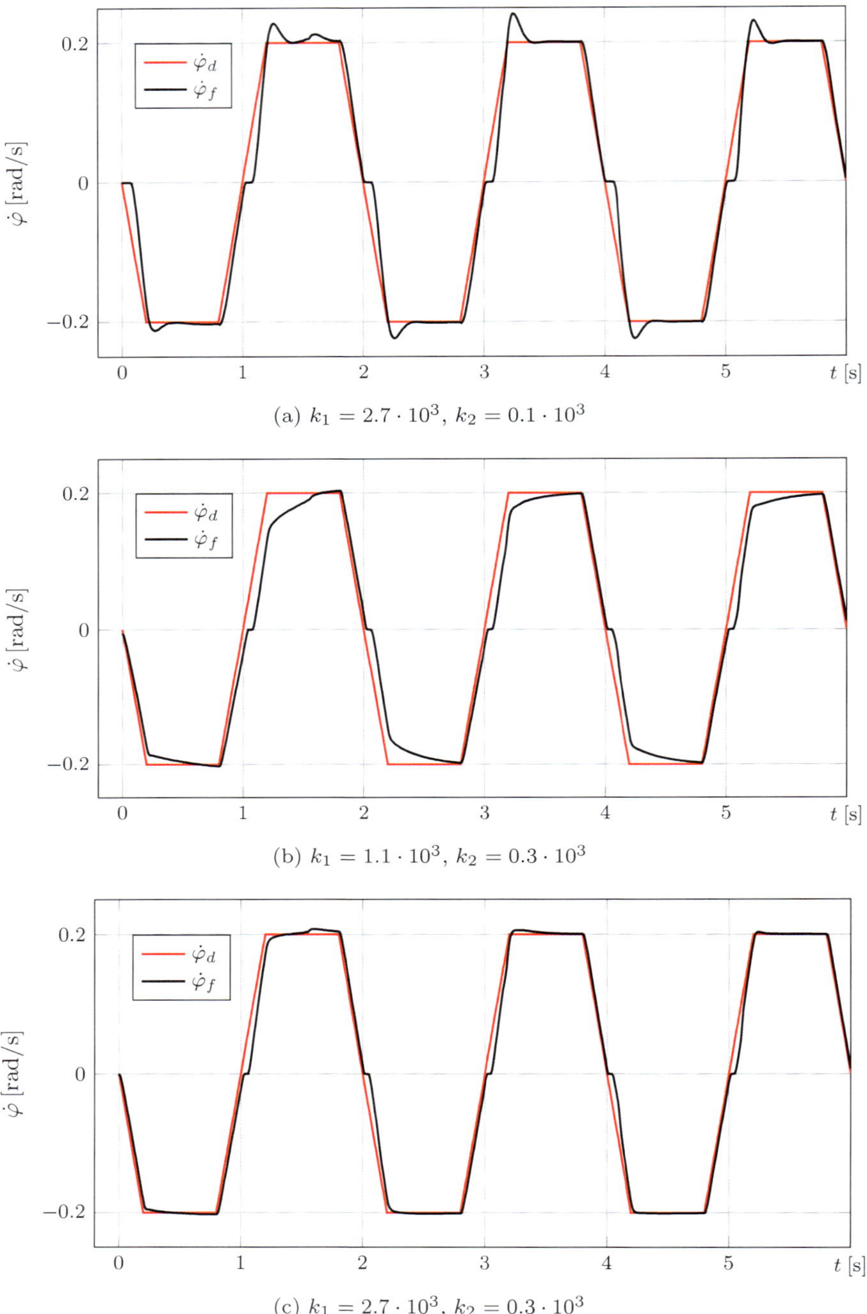

(a) $k_1 = 2.7 \cdot 10^3$, $k_2 = 0.1 \cdot 10^3$

(b) $k_1 = 1.1 \cdot 10^3$, $k_2 = 0.3 \cdot 10^3$

(c) $k_1 = 2.7 \cdot 10^3$, $k_2 = 0.3 \cdot 10^3$

Fig. 11.2 Desired time history of angular velocity $\dot{\varphi}_d(t)$ (dashed line) and the corresponding response $\dot{\varphi}(t)$ (solid line) of the analyzed voltage-controlled simulation model of the DC motor defined by the assumed set of model parameters, PD tuning variables: k_1 and k_2, and initial values of state variables.

Looking at Fig. 11.2, it can be noticed that the system response is inaccurate at first occurrence of the threshold of constant angular velocity (0.2 rad/s). Such transient behavior results from the model and tuning parameters that are not correctly estimated at the corresponding time. The response changes over time to produce an acceptable overlapping of both trajectories at the beginning of the second period (at 2 s). At subsequent ±0.2 rad/s thresholds, the system step response is well damped, smoothly fitting edges of the desired shape. Figures 11.2b, d and f bring much clearer comparison of three solutions: oscillatory, over-dumped and the most accurate, which could be also subjected to some small improvement to get faster convergence to the steady state velocity.

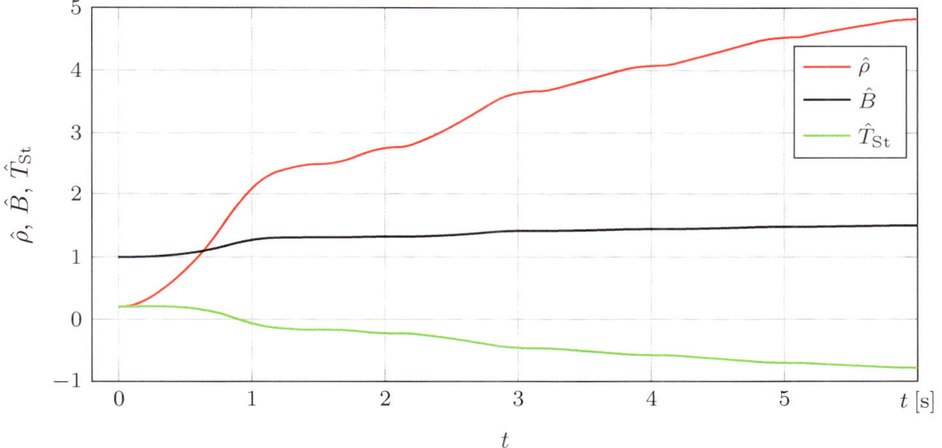

Fig. 11.3   Convergence of selected estimates.

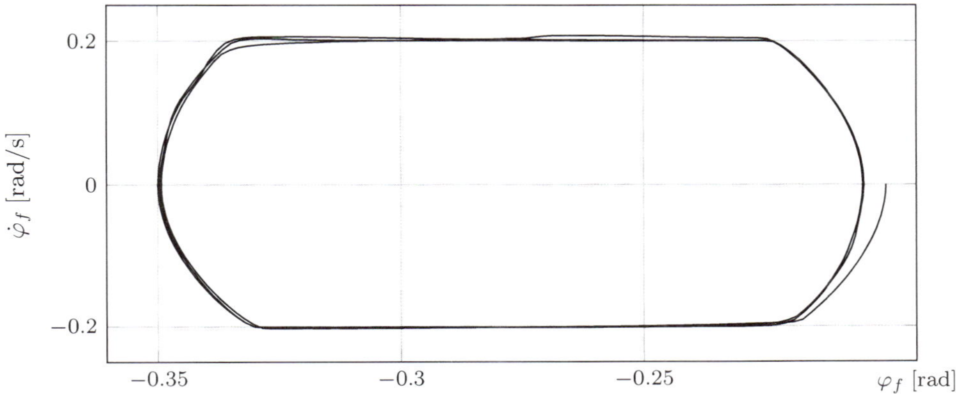

Fig. 11.4   Projection of phase trajectory of the controlled system on the plane $\dot{\varphi}(\varphi)$.

The phase trajectory presented in Fig. 11.4 gives another view on the desired trajectory. It should take a shape of a closed curve bounded between $\dot{\varphi} = \pm 0.2$. To achieve the demanding effect of control, voltage input should be applied accordingly to the time history shown in Fig. 11.5. Amplitude of the demanding voltage control input changes impulsively after crossing $\dot{\varphi} = 0$, for $t = 1, 2, \ldots, n$ [s].

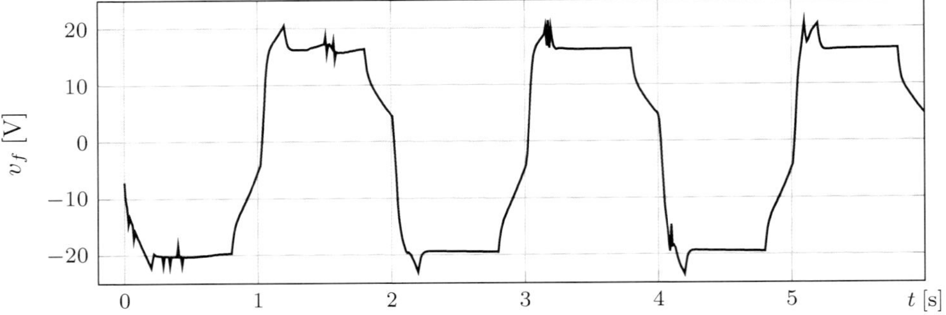

Fig. 11.5   Voltage input $v_f(t)$ applied to the optimally controlled DC motor.

The proposed strategy of voltage tracking control ensures robust adaptation, works correctly, and can be applied to solve of other control objectives related to shaping of time histories of responses of some group belonging to discontinuous dynamical systems. After many trial-and-error attempts of tuning, parameters k_1 and k_2 of the second stage of conducted control have been estimated. They significantly affect local step response (appearing while going on the thresholds of constant angular velocity). Moreover, on the basis of sliding surface based smooth adaptive robust controller for compensation of frictional effects, there was a useful and easily applicable extension of this method for a numerical tracking control of DC motors by means of voltage input proposed. A kind of drawback or an inconvenience in application of the elaborated control strategy is the requirement of estimation of the upper bounding function for the nonlinear stick-slip friction in order to guarantee the closed-loop stability.

Chapter 12

Numerical Modeling of a Shock Response

In this chapter, a lumped mass mechanical model of a thorax subjected to a blast pressure wave is taken into consideration. A thorax spring-dashpot model developed by Lobdell is implemented in numerical modeling of the dynamics of the multibody system. The five-degree-of-freedom mechanical model of a chest adjacent to an elastic backrest is subjected to an impulse loading generated by the blast pressure wave released by an explosion. The so-called coupling of the pressure wave to the thorax is reconsidered. With respect to the evident existence of inherent time delays of displacements, the system of coupled bodies is described by a time delay differential equations that are derived from the large-scale systems approach. Numerical solutions present interesting dynamical behavior of the bio-inspired system, resulting from inherent time delays and a time of arrival of the blast pressure wave. It is pointed out that the inherent state time delays change the dynamical response of the multibody system. Proper time of deployment of the foam-based armor plate reduces relative compression of the thorax, which is to be protected by a bullet-proof waistcoat.

Intrinsic delays in states of physical quantities characterize many dynamical systems in physics, material engineering, ballistics, biology and chemistry [Shi *et al.* (2013); Ananth and Kushari (2013); Martin *et al.* (2013); Hammetter and Zok (2013); Courtemanche *et al.* (1993)]. Natural or artificial control systems have delays occurred from the sensing of a variable and the initiation of appropriate response. Mathematics of systems with time delays poses basic mathematical challenges. These challenges can be described mathematically by delay differential equations, which belong to the class of functional differential equations [Azbelev *et al.* (2007)]. A delay influences dynamical responses of investigated systems and is strongly visible in behavior of multibody systems. The time delay terms of differential equations produce an infinite number of roots of the characteristic equation, making the corresponding dynamical behavior difficult to analyze. One solves such problems indirectly by applying some approximations, but a limitation in accuracy can occur and lead to the instability of systems [Yi *et al.* (2010)]. More effective methods based on an analytic approach aimed at obtaining the complete solution of systems represented by the delay differential equations based on the concept of

Lambert W function was developed in [Asl and Ulsoy (2003)].

From the other side, numerical methods can be used to aid any mathematical modeling.

Nowadays, numerical methods are popular in biodynamics and, in particular, in dynamical analysis of primary blast lung injuries caused by the pressure waves released by explosions [Fitek *et al.* (2011)]. Rapid motion of the human chest wall creates a pressure wave in the lung material [D'yachenko and Manyuhina (2006)]. A concept to protect the chest is to use a layer foam material behind a massive armor plate worn over the chest. Coupling of the blast wave and the thorax causes that soft tissues, placed insie the thorax (e.g. lungs) can sustain large stress and strain rate [Grimal *et al.* (2002)]. To investigate the mechanical responses of internal organs, a complex modeling is required. Homogeneous and linear elasticity material properties are assigned to each part of the model, whereas the human cartilages and bones may have different material properties. Such conditions are taken into consideration by Lobdell's model, which has been reconsidered in this work to perform numerical solutions of a large-scale time delay system.

In this chapter, the problem of modeling of a multibody biomechanical system of a thorax is considered. On the basis of the theory of large-scale continuous-time systems, an uncertain model of the thorax has been rewritten in the representation allowing us to define its parametric uncertainties and complex interactions between its subsystems. The parametric uncertainties make the discontinuous system more difficult to solve, what is compensated by more accurate numerical solution and evident possibility of inclusion of time delays in the shock responses.

The fact is that if we consider an exponential decaying response taking about 0.2 ms, then the time delay superposed on each body of the system plays an important role.

A large-scale dynamical system can be characterized by a large number of state variables, system parametric uncertainties, and a complex interaction between subsystems [Park (2002); Siljak (1978)]. In view of reliability and practical implementation, time delays have to be incorporated into the numerical modeling of the large-scale physical systems due to the real transport of mass, propagation of vibrations and computation times.

12.1 Variation of an Air-Blast Overpressure Wave

Explosions in the air create intense shock waves capable of transferring large transient pressures and impulses to the objects they intercept [Wadley *et al.* (2010)]. The free-field pressure time response from an explosion in the air is described by the known Friedlander's waveform

$$p(x,t) = p_0 e^{(x-v_0 t)/(v_0 t_i)}, \tag{12.1}$$

where $p(x, t)$ is the pressure at the point x and time t, p_0 is the blast overpressure (maximum amplitude), t_i is the wave decaying time and v_0 is the sound speed in the air. A blast wave propagates outwards from an explosion. It consists of a shock front, which precedes a phase of positive pressure and can be followed by a negative pressure phase, which is not taken into the analysis.

In the numerical experiment carried out in this chapter, one assumes that the wave arrived in time t_{arr} after the explosion at the buffer mass m_1 placed at $x = 0$ in front of the investigated multibody system (see Fig. 12.1). Therefore, the simplified time-dependent characteristics of the pressure wave [Fitek *et al.* (2011)] is given

$$p(t) = p_0 e^{-t/t_i}. \tag{12.2}$$

In numerical computations, it will be more useful to represent the pressure wave as the effective impact force $u_b(t)$ per an effective area a of the chest subject to the air shock. Temporal variation of the blast wave which reaches the armor plate is estimated using the following conditions:

(C1) The most harmful positive phase of the blast wave profile is assumed [Goel *et al.* (2012)].
(C2) Maximum amplitude $p_0 = 1\,\mathrm{MPa}$ due to a 10 kg TNT explosion at 2.2 m standoff.
(C3) Wave decay time $t_i = 1.66$ ms causes a decrease in p_0 to zero within $t_r = 1.2$ ms, so that the impulse $I = 0.4\,\mathrm{kPa \cdot s}$ impinges onto the armor buffer plate.
(C4) The effective area $a = 1.825 \cdot 10^{-2}\,\mathrm{m}^2$ of the PUR-foam plate and the chest.
(C5) Maximum amplitude of the effective impact force $u_0 = 18.25\,\mathrm{kN}$ (see Fig. 12.2).
(C6) Impact force $u(t)$ decreases after time t_i to u_0/e.
(C7) Arrival time of blast wave after detonation $t_{\mathrm{arr}} = 1$ ms.
(C8) Mass of the buffer plate $m_1 = 0.365$ kg for the foam density $\rho_f = 20\,\mathrm{kg/m}^2$.
(C9) Reaction force of the armor plate worn over the chest to the blast wave during buffer deployment depends on the foam unloading properties (see $\sigma_u(\varepsilon)$ in Table 12.1).
(C10) For an active mitigation concept, the buffer deployment time depends on $\sigma_u(\varepsilon)$ as well.
(C11) Maximum incident overpressure cannot exceed 13 kN, which is determined from the lung damage threshold on Bowen's curve [Con (1991)] and for duration of positive incident overpressure equal to 1 ms.

If a sensor capable of detecting the electromagnetic emission [Orson *et al.* (2003)] created at the instant of detonation reaches a time delay t_a between detonation and the arrival of the blast wave, then a buffer is deployed by using a high-speed actuator such as a propellant. The force exerted on the buffer as it deploys must assure that the reaction pressure exerted on the protected chest does not exceed 0.3 MPa. For detonations of high explosives, the peak overpressure can cause injury to the thorax.

In our current study based on the work [Olejnik and Awrejcewicz (2015)], the foam deploys with lower peak overpressure determined by $\sigma_u(\varepsilon)$ (see Table 12.1), and the time response impact dynamics is assessed for a model problem consisting of 10 kg of a high explosion TNT at 2.2 m standoff. The relevant ConWep [Con (1991)] computations of peak pressure p_0 over the atmospheric pressure, impulse I and arrival time t_{arr} as a function of range are found in [Wadley *et al.* (2010)].

Finally, we can write a formula for the blast load with u_0 peak pressure and t_i decay coefficient

$$u_b(t) = u_0 e^{-t/t_i}. \tag{12.3}$$

The blast load characteristics will be used to model the impulse loading.

12.2   The Foam-Based Armor With a Buffer Plate

Foam materials have the ability to deform at low stress level while absorbing mechanical energy. Foam is used in impact protection to absorb the kinetic energy of an impact and to reduce the maximum stress on the protected object [Fitek *et al.* (2011)]. The foam material reduces peak acceleration while increasing the duration of the impact.

Pressure waves released by explosions cause the so-called primary blast injuries. They significantly affect the air-containing organs of human body [Horrocks and Brett (2000)], and in particular, lungs. A lung injury caused by the pressure wave occurs after a rapid motion of the chest wall, that creates a following pressure wave in the lung structure. This is referred to as coupling of the blast wave to the thorax. A concept to protect lungs against primary blast injury is to use a layer of foam material behind a massive armor plate worn over the chest [Fitek *et al.* (2011)].

Capability of energy absorbtion is the basic feature of foams. They are deformed and absorb the impact energy [Avalle *et al.* (2001)] while keeping the stress acting on the armor plate loaded with the blast wave. One of the common features of energy absorbing foam materials is that there is a discernible plateau in their compression stress-strain curves. It means that the foam materials can absorb energy by deformation, but keep the stress almost constant [Avalle *et al.* (2001); Han *et al.* (1998)].

The typical shape of the stress-strain curve for solid foams has been presented in [Goog (2011)] and, including hysteresis, in [Del Piero and Pampolini (2010)]. The model proposed by Goog encompasses a few parameters dependent on relative density, because Young's modulus and plateau stress usually increase and densification strain decreases with increasing foam density.

The foam model assumed in this chapter is a little modification of Goog's model that is completed with a hysteresis. It consist of three systems settled in parallel to each other (see in Fig. 12.1):

(G1) Maxwell's arm, containing a spring (linear stiffness k_{16}) and a dashpot (viscosity c_{62}) in series. The first component describes the first and second region of foam compression and deformation while the plateau stress is constant.

(G2) Linear spring k_p. The second stiffness represents the shape of the plateau. It is integrated to describe the increasing or decreasing part of the plateau.

(G3) Nonlinear spring k_D. The third stiffness is responsible for densification part. It is given by the formula

$$k_D(\varepsilon) = \gamma(1 - e^\varepsilon)^n, \quad \varepsilon = (x_1 - x_2)/h_f, \tag{12.4}$$

where γ and n are the model parameters, ε is the strain and h_f is a thickness of the foam.

Each element of the spring-dashpot model will produce the following components of reaction force:

$$F_p = a\sigma_p = ak_p\varepsilon = ak_p(x_1 - x_2)/h_f, \tag{12.5a}$$

$$F_D = a\sigma_D = ak_D\varepsilon = a\gamma(1 - e^\varepsilon)^n(x_1 - x_2)/h_f, \tag{12.5b}$$

$$F_c = a\sigma_c = ac_{62}\dot{\varepsilon}_c = ac_{62}(\dot{x}_2 - \dot{x}_6)/h_f, \tag{12.5c}$$

$$F_k = a\sigma_k = ak_{16}\varepsilon_k = ak_{16}(x_1 - x_6)/h_f, \tag{12.5d}$$

where k_p, γ, c_{62}, k_{16} are stresses [MPa].

Table 12.1   Foam model parameters for estimation of $\sigma(\varepsilon)$ curve with hysteresis.

σ	k_{16}	c_{62}	k_p	γ	n
$\sigma_l(\varepsilon)$	16.46	0.465	0.085	0.176	6
$\sigma_u(\varepsilon)$	3.408	0.111	0.085	0.037	6
$\sigma_d(\varepsilon)$	0	0	0.085	0	0

Polymeric open-cell foams exhibit complex nonlinear behavior [Del Piero and Pampolini (2010)]. As it has been already mentioned, the stress-strain curve for uniaxial compression shows three well distinguishable regimes (G1-G3). The same three regimes are present during unloading of the foam, but the response exhibits a hysteresis loop. For the numerical experiment, loading $\sigma(\varepsilon)|_{\dot{\varepsilon}>0}$ and unloading $\sigma(\varepsilon)|_{\dot{\varepsilon}<0}$ curves are estimated, respectively, for the model parameters of foam densities $\rho_l = 50\,\text{kg/m}^3$ and $\rho_u = 40\,\text{kg/m}^3$ (divided by two) listed in Table 12.1 (compare with [Goog (2011)]).

The reader should note that the study is not oriented on the precise modeling of the foam behavior, but it tends to stress the importance of time delays in modeling of impulse responses of multibody systems using only the simplified description. It might be significant in the case of any physical object of some inertia that is subject to rapid excitation or deformation.

12.3 Physical Model of the System

The problem of time varying position- and velocity-dependent system parameters is reflected in the modeling by many factors, i.e., Maxwell's elements. Skeletal deflection of the thorax was determined by the difference between displacements of its anterior and posterior walls. The parameter discontinuity, observed in connections between the directly coupled lumped masses of the mechanical idealization, appears in four cases (U1-U5).

(U1) If a relative displacement $x_{21}(t)-x_{31}(t) < 0$ measured between the front and back walls of the thorax exceeds $d = 3.8\,\text{cm}$, then a bilinear spring stiffness k_{23} doubles its value that satisfies Kroell's corridors at large deflection. Stiffness of contents of the thorax changes due to the nonlinear material behavior of the rib cage.

(U2) If a velocity of relative displacement of front and back wall of the thorax becomes negative, i.e., $x_{22}(t) - x_{32}(t) < 0$, then a parameter c_{23} of viscous damping doubles its value. A different damping coefficient, responsible for elongation and compression, is assumed in order to satisfy the descending part of Kroell's corridors.

(U3) If a foam material is compressed (in loading state) or relaxed (in unloading state), then two different phenomenological solid foam models, which are nonlinear according to (G1-G3), are assumed (see $\sigma_l(\varepsilon)$ and $\sigma_u(\varepsilon)$ in Table 12.1).

(U4) If the foam is fully deployed and there is no compressing force acting on it, its modeling is changed from the full nonlinear viscoelastic foam model (G1-G3) to a reduced one, which is modeled by a spring connection of elasticity k_p between the proof mass m_1 and the front wall's mass m_2 of the thorax (see $\sigma_d(\varepsilon)$ in Table 12.1).

(U5) If the foam achieves a state of full deployment, then the proof mass m_1 (the armor plate) starts to pull the posterior wall of the chest via the waistcoat. In consequence, the spring-dashpot model is activated, so k_{13} and c_{13} become different from zero. It is because a deployable foam-based armor plate is assumed in the experiment to be integrated on an outer side of the bullet-proof waistcoat worn over the chest.

In general, discontinuity sources (U1-U5) in system's stiffness and damping produce coexisting time varying uncertainties. Modeling of the investigated system dynamics is quite complicated, but the large-scale system's representation does make it easier.

An idealized biomechanical model of a thorax being supported from behind has been depicted in Fig. 12.1a. Extended models of a sitting human are analyzed in [Piersol and Paez (2010)]. The concept of the thorax model originates from Lobdell's approach, which was developed by General Motors to study the response of the human thorax in automobile crashes [Lobdell *et al.* (1973)]. An application

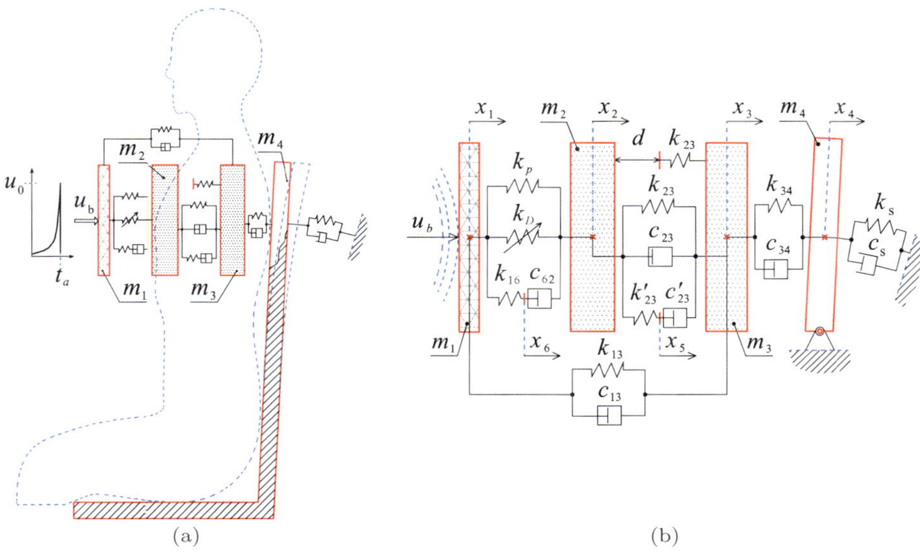

Fig. 12.1 A biomechanical model of a thorax (a) subjected to the frontal blast overpressure wave (m_1 – proof mass of the armor plate, m_2, m_3 – masses of posterior and anterior walls of the chest, respectively, m_4 – mass of the backrest). Detailed view (b) of the lumped mass mechanical model of a thorax subjected to the frontal blast pressure wave (k_{13} and c_{13} – spring-dashpot model of a bullet-proof waistcoat).

of the model has been presented in [Fitek *et al.* (2011)], where the model consisted of a configuration of springs and dashpot elements. An injury, which results from the pressure wave released by an explosion, is referred to as primary blast injury [Wadley *et al.* (2010)]. Primary blast injuries most significantly affect the air-containing organs of the body [Horrocks and Brett (2000)].

Detailed view of the biomechanical model of the thorax subjected to frontal blast pressure wave has been shown in Fig. 12.1b. From the left side: the effective foam armor's mass m_1 under the pressure impact, the front body of the mass m_2 in the anterior surface of the chest wall, the rear body of mass m_3 in the posterior surface of the chest wall, m_4 – mass of the support.

Lobdell's model was developed through measuring the thoracic response of a human subjected to a shock loading. The use of the model has been extended by researchers to the field of protection against air-blasts to predict the thoracic response to an air-blast wave [Chan *et al.* (2010)].

The model and some associated concepts of limiting injuries caused by the aforementioned air blasts are reconsidered here in one system, and also supplemented. The following concepts are incorporated:

(a) a support attached to the thorax from behind,
(b) basic approximation of an air pressure wave given in Friedlander's form [Dewey (2010)],

(c) a phenomenological model for solid foams introduced in [Goog (2011)],

(d) a deployable waistcoat with integrated (initially compressed) foam-based armor plate.

An idealized air-blast pressure wave reaches an effective area of the foam-based armor plate represented by the body of the mass m_1 which is elastically attached to the second body of the mass m_2 in the assumed foam model (front wall of the thorax, see Fig. 12.1b).

12.3.1 *Formulation of the Large-Scale Problem*

Let us take into consideration a class of uncertain continuous-time system composed of N coupled subsystems as follows:

$$\frac{d\bar{x}_i(t)}{dt} = \big(A_i + \Delta A_{ii}(t)\big)\bar{x}_i(t) + \sum\nolimits_{j\neq i}^{N}\big(A_{ij} + \Delta A_{ij}(t)\big)\bar{x}_j(t'_j)$$
$$+ \big(B_i + \Delta B_i(t)\big)\bar{u}(t), \tag{12.6a}$$

$$\bar{y}_i(t) = C_i\bar{x}_i(t) + D_i\bar{u}(t), \quad \text{for} \quad i = 1,\dots,N, \tag{12.6b}$$

where $\bar{x}_i(t) \in \mathbb{R}^{n_i\times 2}$, $\bar{u}(t) \in \mathbb{R}^{m_i}$, and $\bar{y}_i(t) \in \mathbb{R}^{l_i}$ denote, respectively: vectors of system states, control inputs, and system outputs, $t'_j = t - \tau_j$.

The dynamical system (12.6) of i coupled subsystems is described by the internal behavior time independent state matrices $A_i^{n_i\times n_i}$, while the control inputs matrix $B_i^{n_i\times m_i}$, the system output matrix $C_i^{l_i\times n_i}$ and the control inputs transition matrix $D_i^{l_i\times m_i}$ represent connections between the external world and the system, τ_j is the time delay of j-th coupled system. Control inputs do not directly influence the system outputs in the investigations, therefore the matrix $D_i^{l_i\times m_i}$ is zero. A controlled case of the analyzed multi-body system has been taken into consideration by the authors in Chap. 13 and [Olejnik and Awrejcewicz (2011, 2010)].

For the purpose of solution of the analyzed problem, in Eq. (12.6), time-dependent matrices $\Delta A_{ii}^{n_i\times n_i}$ $\Delta A_{ji}^{n_i\times n_i}$ and $\Delta B_i^{n_i\times m_i}$ are introduced that define, respectively: the system state and control input uncertainties. It allows for a more or less precise inclusion of the system parameters disturbances (U1-U5) given in a form of known time-dependent function or in a quite different form of a function dynamically dependent on the internal system state (time- or state-varying properties are allowed to be modeled as well). Matrices $\Delta A_{ji}^{n_i\times n_i}$ represent all possibilities of connections between interconnected subsystems forming the entire system.

Equations (12.6) can be expanded to the following forms:

$$\dot{\bar{x}}_1(t) = (A_{11} + \Delta A_{11})\bar{x}_1(t) + (A_{12} + \Delta A_{12})\bar{x}_2(t_2') + \Delta A_{13}\bar{x}_3(t_3')$$
$$+ \Delta A_{16}\bar{x}_6(t_6') + B_1\bar{u}(t), \tag{12.7a}$$

$$\dot{\bar{x}}_2(t) = (A_{22} + \Delta A_{22})\bar{x}_2(t) + (A_{21} + \Delta A_{21})\bar{x}_1(t_1') + (A_{23} + \Delta A_{23})\bar{x}_3(t_3')$$
$$+ A_{25}\bar{x}_5(t_5') + \Delta A_{26}\bar{x}_6(t_6') \tag{12.7b}$$

$$\dot{\bar{x}}_3(t) = (A_{33} + \Delta A_{33})\bar{x}_3(t) + (A_{32} + \Delta A_{32})\bar{x}_2(t_2') + \Delta A_{31}\bar{x}_1(t_3')$$
$$+ A_{34}\bar{x}_4(t_4') + A_{35}\bar{x}_5(t_5') + B_3\bar{u}(t), \tag{12.7c}$$

$$\dot{\bar{x}}_4(t) = A_{44}\bar{x}_4(t) + A_{43}\bar{x}_3(t_3'), \tag{12.7d}$$

$$\dot{\bar{x}}_5(t) = A_{55}\bar{x}_5(t) + A_{52}\bar{x}_2(t_2') + A_{53}\bar{x}_3(t_3'), \tag{12.7e}$$

$$\dot{\bar{x}}_6(t) = \Delta A_{66}\bar{x}_6(t) + \Delta A_{61}\bar{x}_1(t_1') + A_{62}\bar{x}_2(t_2'), \tag{12.7f}$$

$$y(t) = \begin{bmatrix} 0 & 1 & -1 & 0 \\ 0 & 0 & 0 & 0 \end{bmatrix} \bar{x}_i, \tag{12.7g}$$

where: $\bar{x}_i = \begin{bmatrix} x_{i1}, x_{i2} \end{bmatrix}^T$, $\dot{\bar{x}}_i = \begin{bmatrix} \dot{x}_{i1}, \dot{x}_{i2} \end{bmatrix}^T$, $\bar{u} = \begin{bmatrix} u_b, u_s \end{bmatrix}^T$, an impact force of the air-blast pressure wave $u_b(t)$ is given by Eq. (12.3), and a reaction force of the support

$$u_s(t) = -k_s x_{41}(t) - c_s x_{42}(t). \tag{12.8}$$

In Eq. (12.7g), a difference $x_{21}(t) - x_{31}(t)$ in displacements of bodies denoted by m_2 and m_3 will be the observed system output. In all subequations of Eq. (12.7), zero matrices are neglected.

Note that displacements x_5 and x_6 (see in Fig. 12.1b) of massless points in Maxwell's elements are expressed in Eq. (12.7e) by x_{51} and in Eq. (12.7f) by x_{61}, while corresponding velocities are obtained directly from a two-point method for approximating the derivative of displacements

$$\dot{x}_{i2} = \frac{x_{i1}(t+h) - x_{i1}(t-h)}{2h}, \quad i = 5, 6.$$

Nonzero state-space matrices are as follows:

$$A_{11} = \begin{bmatrix} 0 & 1 \\ \dfrac{-k_p}{m_1} & 0 \end{bmatrix}, \; A_{12} = \begin{bmatrix} 0 & 0 \\ \dfrac{k_p}{m_1} & 0 \end{bmatrix}, \; A_{22} = \begin{bmatrix} 0 & 1 \\ \dfrac{-k_{23}-k_{23}'-k_p}{m_2} & \dfrac{-c_{23}}{m_2} \end{bmatrix}, \; A_{21} = \begin{bmatrix} 0 & 0 \\ \dfrac{k_p}{m_2} & 0 \end{bmatrix},$$

$$A_{23} = \begin{bmatrix} 0 & 0 \\ \dfrac{k_{23}}{m_2} & \dfrac{c_{23}}{m_2} \end{bmatrix}, \; A_{25} = \begin{bmatrix} 0 & 0 \\ \dfrac{k_{23}'}{m_2} & 0 \end{bmatrix}, \; A_{33} = \begin{bmatrix} 0 & 1 \\ \dfrac{-k_{23}-k_{34}}{m_3} & \dfrac{-c_{23}-c_{23}'-c_{34}}{m_3} \end{bmatrix}, \; A_{32} = \begin{bmatrix} 0 & 0 \\ \dfrac{k_{23}}{m_3} & \dfrac{c_{23}}{m_3} \end{bmatrix}$$

$$A_{34} = \begin{bmatrix} 0 & 0 \\ \dfrac{k_{34}}{m_3} & \dfrac{c_{34}}{m_3} \end{bmatrix}, \; A_{35} = \begin{bmatrix} 0 & 0 \\ 0 & \dfrac{c_{23}'}{m_3} \end{bmatrix}, \; A_{44} = \begin{bmatrix} 0 & 1 \\ \dfrac{-k_{34}}{m_4} & \dfrac{-c_{34}}{m_4} \end{bmatrix}, \; A_{43} = \begin{bmatrix} 0 & 0 \\ \dfrac{k_{34}}{m_4} & \dfrac{c_{34}}{m_4} \end{bmatrix},$$

$$
A_{55} = \begin{bmatrix} -\frac{k'_{23}}{c'_{23}} & 0 \\ 0 & 0 \end{bmatrix}, \, A_{52} = \begin{bmatrix} \frac{k'_{23}}{c'_{23}} & 0 \\ 0 & 0 \end{bmatrix}, \, A_{53} = \begin{bmatrix} 0 & 1 \\ 0 & 0 \end{bmatrix}, \, A_{62} = \begin{bmatrix} 0 & 1 \\ 0 & 0 \end{bmatrix}, \, B_1 = \begin{bmatrix} 0 & 0 \\ \frac{1}{m_1} & 0 \end{bmatrix},
$$

$$
B_4 = \begin{bmatrix} 0 & 0 \\ 0 & \frac{1}{m_4} \end{bmatrix}, \, \Delta A_{11} = \begin{bmatrix} 0 & 0 \\ -\frac{k_{\mathrm{D}}(t)-k_{16}(t)-k_{13}(t)}{m_1} & -\frac{c_{13}(t)}{m_1} \end{bmatrix}, \, \Delta A_{12} = \begin{bmatrix} 0 & 0 \\ \frac{k_{\mathrm{D}}(t)}{m_1} & 0 \end{bmatrix},
$$

$$
\Delta A_{16} = \begin{bmatrix} 0 & 0 \\ \frac{k_{16}(t)}{m_1} & 0 \end{bmatrix}, \, \Delta A_{13} = \begin{bmatrix} 0 & 0 \\ -\frac{k_{13}(t)}{m_1} & \frac{c_{13}(t)}{m_1} \end{bmatrix}, \, \Delta A_{21} = \begin{bmatrix} 0 & 0 \\ \frac{k_{\mathrm{D}}(t)}{m_2} & 0 \end{bmatrix},
$$

$$
\Delta A_{22} = \begin{bmatrix} 0 & 0 \\ \frac{-k_{23}(t)-k_{\mathrm{D}}(t)}{m_2} & \frac{-c_{23}(t)-c_{62}(t)}{m_2} \end{bmatrix}, \, \Delta A_{23} = \begin{bmatrix} 0 & 0 \\ \frac{k_{23}(t)}{m_2} & \frac{c_{23}(t)}{m_2} \end{bmatrix}, \, \Delta A_{26} = \begin{bmatrix} 0 & 0 \\ 0 & \frac{c_{62}(t)}{m_2} \end{bmatrix},
$$

$$
\Delta A_{33} = \begin{bmatrix} 0 & 0 \\ \frac{-k_{23}(t)-k_{13}(t)}{m_3} & \frac{-c_{23}(t)-c_{13}(t)}{m_3} \end{bmatrix}, \, \Delta A_{31} = \begin{bmatrix} 0 & 0 \\ \frac{k_{13}(t)}{m_3} & \frac{c_{13}(t)}{m_3} \end{bmatrix},
$$

$$
\Delta A_{32} = \begin{bmatrix} 0 & 0 \\ \frac{k_{23}(t)}{m_3} & \frac{c_{23}(t)}{m_3} \end{bmatrix}, \, \Delta A_{66} = \begin{bmatrix} \frac{-k_{16}(t)}{c_{62}(t)} & 0 \\ 0 & 0 \end{bmatrix}, \, \Delta A_{61} = \begin{bmatrix} \frac{k_{16}(t)}{c_{62}(t)} & 0 \\ 0 & 0 \end{bmatrix}.
$$

12.3.2 Uncertainties and the Switching Matrices

The problem definition uncovers some new features of the investigated bio-inspired system shown in Fig. 12.1b. The problem of time-varying parameters that introduces some uncertainties to the model have been numerically solved by definition of multivalued matrices switched in accordance to cases (U1-U5). Stiffness of the chest interior with organs increases at condition (U1) about two times to appropriately approximate the real chest compression. It obviously means that stiffness of the rheological coupling increases discontinuously with regard to a greater than d compression of the thorax x_r, and that damping ability of the coupling varies in time as the thorax undergoes suitable compression or relaxation. Such discontinuity in subsystem's stiffness and damping produces parameter uncertainties dependent on state variable.

Uncertainties (U1-U2).

In equations (12.7b) and (12.7c), one can encounter four $\Delta A_{ij}(t)$ parameter uncertainties of the analyzed dynamical system:

$$
\Delta A_{ij}(t) = \begin{bmatrix} 0 & 0 \\ \frac{\sigma(i,j)k_{23}(t)}{m_i} & \frac{\sigma(i,j)c_{23}(t)}{m_i} \end{bmatrix} = D_i F(t) E_{ij}, \tag{12.9a}
$$

$$
\sigma(i,j) = -\mathrm{sgn}\big((-1)^{i+j}\big), \tag{12.9b}
$$

where $i,j \in (2,3)$.

Distribution of entries in $\Delta A_{ij}(t)$ does not change over their all possibilities. Therefore, the unknown time-varying real-valued matrix $F(t)$ is assumed to be defined in the same way, but $F^T(t)F(t) \leq I$ must hold for $t \in \mathbb{R}^+$, where I is the identity matrix. The introduced decomposition DFE includes some known constant real-valued matrices D_i and E_{ij} of appropriate dimensions that need to be estimated to use them, for instance, in a solution to LMI problems [Park (2002); Mukaidani *et al.* (2004)].

Dependently on $x_r(t)$ and $v_r(t)$, the following forms of the uncertainty matrix $\Delta A_{22}(t) = \Delta A_{22}^{(k)}$ are possible in Eq. (12.7b):

$$
\Delta A_{22}^{(k)} = \begin{cases} \begin{bmatrix} 0 & 0 \\ 0 & 0 \end{bmatrix} & \text{if } s_1 \quad \text{then } k = 1\,, \\[2ex] \begin{bmatrix} 0 & 0 \\ \frac{-k_{23}}{m_2} & 0 \end{bmatrix} & \text{if } s_2 \quad \text{then } k = 2\,, \\[2ex] \begin{bmatrix} 0 & 0 \\ 0 & \frac{-c_{23}}{m_2} \end{bmatrix} & \text{if } s_3 \quad \text{then } k = 3\,, \\[2ex] \begin{bmatrix} 0 & 0 \\ \frac{-k_{23}}{m_2} & \frac{-c_{23}}{m_2} \end{bmatrix} & \text{if } s_4 \quad \text{then } k = 4\,, \end{cases}
\tag{12.10}
$$

where the remaining $\Delta A_{ij}(t)$ (for $i, j = 2, 3$) are to be defined in a similar way, $s_k = \{x_r(t), v_r(t) : x_r(t) < d \wedge v_r(t) > 0; \ x_r(t) \geq d \wedge v_r(t) > 0; \ x_r(t) < d \wedge v_r(t) \leq 0; \ x_r(t) \geq d \wedge v_r(t) \leq 0\}$ defines rheological properties of the biomechanical system. Switching conditions s_i will select only one of k possibilities $\Delta A_{ij}^{(k)}$ for $k = 1, \ldots, 4$, dependently on values of pairs $(x_r(t), v_r(t))$ creating the discontinuous time history of uncertainties $\Delta A_{ij}(t)$ of the state matrices of coupled two-degree-of-freedom neighboring subsystems.

To find a better description of the existing switching nature, it is now required to choose D_i and E_{ij} matrices of a decomposition. For instance, an exemplary decomposition of $\Delta A_{22}^{(k)}$, for $k = 3$ could be made accordingly to the scheme presented in [Olejnik and Awrejcewicz (2013)].

Perturbation matrix $F(t)$ will depend on k cases that have been delivered in the case statement (12.10). Therefore, with regard to $F^T(t)F(t) \leq I$ and using Eq. (12.9), the following formula reads

$$
\Delta A_{i,j}^{(k)} = D_i F^{(k)} E_{ij} = \begin{bmatrix} 0 & 0 \\ 0 & \delta_{4i} \end{bmatrix} F^{(k)} \begin{bmatrix} \frac{\sigma(i,j)\beta_1}{\gamma} & 0 \\ 0 & \frac{\sigma(i,j)\beta_4}{\gamma} \end{bmatrix},
\tag{12.11}
$$

where: $i, j \in (2, 3)$, $k = 1, \ldots, 4$, $\sigma(i, j)$ is given in Eq. (12.9b), and $F(t)$ will be switched accordingly to:

$$
F^{(k)} = \begin{cases} \begin{bmatrix} 0 & 0 \\ 0 & 0 \end{bmatrix} & \text{if } s_1 \quad \text{then } k = 1\,, \\[2ex] \begin{bmatrix} 0 & \pm\gamma \\ \pm\gamma & 0 \end{bmatrix} & \text{if } s_2 \quad \text{then } k = 2\,, \\[2ex] \begin{bmatrix} \pm\gamma & 0 \\ 0 & \pm\gamma \end{bmatrix} & \text{if } s_3 \quad \text{then } k = 3\,, \\[2ex] \begin{bmatrix} 0 & \pm\gamma \\ \pm\gamma & \pm\gamma \end{bmatrix} & \text{if } s_4 \quad \text{then } k = 4\,. \end{cases}
\tag{12.12}
$$

The case statement (12.12) captures switching properties of $\Delta A_{ij}(t)$ described in comments to Eq. (12.10). As it was expected, $F(t)$ is defined in the same way for four uncertainties of the model. Matrices D_i and E_{ij} are constant and their entries depend on the subsystem that they are related to.

To check correctness of the previously made estimations, let the following parameters be assigned for the uncertainties: $\delta_{42} = 1/m_2$, $\delta_{43} = 1/m_3$, $\beta_1 = k_{23}$, $\beta_4 = c_{23}$, $\gamma \neq 0$. For example, $\Delta A_{2,2}^{(4)} = D_2 \bar{F}^{(4)} E_{22} = [[0,0],[0,1/m_2]] \cdot [[0,\gamma],[\gamma,\gamma]] \cdot [[-k_{23}/\gamma, 0],[0, -c_{23}/\gamma]] = [[0,0],[-k_{23}/m_2, -c_{23}/m_2]]$ what is in agreement with the fourth case of Eq. (12.10). Parameter $\gamma = 0.618033$ was estimated in a semi-analytical way presented in Sec. 12.4.

Uncertainties (U3-U4).

While a foam material is compressed (in loading state) or relaxed (in unloading state), two different phenomenological solid foam models are distinguished $\sigma_l(\varepsilon)$ and $\sigma_u(\varepsilon)$, which are nonlinear according to (G1-G3) (see in Table 12.1). If the foam is fully deployed and there is no compressing force acting on it, then its modeling is changed from the full nonlinear viscoelastic foam model (G1-G3) to a reduced one $\sigma_d(\varepsilon)$ which states a spring connection of elasticity k_p between the proof mass m_1 and the front wall mass m_2 of the chest (see in Table 12.1). The mentioned models are cast by the case statement:

$$\sigma = \begin{cases} \sigma_l(\varepsilon) & \text{if} \quad \varepsilon > 0 \quad \text{and} \quad \dot{\varepsilon} > 0\,, \\ \sigma_u(\varepsilon), & \text{if} \quad \varepsilon > 0 \quad \text{and} \quad \dot{\varepsilon} \leq 0\,, \\ \sigma_d(\varepsilon), & \text{if} \quad \varepsilon \leq 0\,. \end{cases} \tag{12.13}$$

Uncertainty (U5).

While the foam achieves a state of full deployment, the proof mass m_1 (the armor plate) starts to pull the posterior part of the thorax of the mass m_3 via the waistcoat. The spring-dashpot model is activated, k_{13} and c_{13} become different from zero. It really holds because in the experiment, a deployable foam-based armor plate is integrated on an outer side of the bullet-proof waistcoat which is worn over the chest. From this point of view, a force between bodies 1 and 3 is activated:

$$F_{13}(t) = \begin{cases} -k_{13}(x_{11} - x_{31}) - c_{13}(x_{12} - x_{32}), & \text{if} \quad \varepsilon \leq 0\,, \\ 0, & \text{if} \quad \varepsilon > 0\,, \end{cases} \tag{12.14}$$

Numerical integration of 12 first-order differential equations is a bit conditioned, but it better approximates complex dynamics of the biomechanical model of the chest subjected to an impulsive loading. It is equipped with additional bodies like the armor plate and the backrest of a seat, which have an influence on behavior of the investigated multibody system.

12.4   Semi-Analytical Estimation of the Optimal Parameter

Conducting our research according to [Olejnik and Awrejcewicz (2013)], a semi-analytical estimation of γ parameter, which has been introduced in Eq. (12.11), is given below.

Matrix $F(t)$ in Eq. (12.12) will depend on k cases that have been delivered in the case statement (12.7b). Therefore, with regard to $F^T(t)F(t) \leq I$, let us assume $F^{(3)T}F^{(3)} = \gamma I$ holds for $k = 3$ and $\gamma \leq 1$, then:

$$F^{(3)T}F^{(3)} = \begin{bmatrix} f_1 & f_3 \\ f_2 & f_4 \end{bmatrix} \begin{bmatrix} f_1 & f_2 \\ f_3 & f_4 \end{bmatrix} = \begin{bmatrix} \gamma & 0 \\ 0 & \gamma \end{bmatrix},$$

$$\begin{cases} f_1^2 + f_3^2 = \gamma, \\ f_2^2 + f_4^2 = \gamma, \\ f_1 f_2 + f_3 f_4 = 0 . \end{cases} \tag{12.15}$$

In the next step, expansion of Eq. (12.9) holds

$$\Delta A_{22}^{(3)} = \begin{bmatrix} d_1 & d_2 \\ d_3 & d_4 \end{bmatrix} \begin{bmatrix} f_1 & f_2 \\ f_3 & f_4 \end{bmatrix} \begin{bmatrix} e_1 & e_2 \\ e_3 & e_4 \end{bmatrix} = \begin{bmatrix} p_1 e_1 + p_2 e_3 & p_1 e_2 + p_2 e_4 \\ p_3 e_1 + p_4 e_3 & p_3 e_2 + p_4 e_4 \end{bmatrix}, \tag{12.16}$$

where: $p_1 = d_1 f_1 + d_2 f_3$, $p_2 = d_1 f_2 + d_2 f_4$, $p_3 = d_3 f_1 + d_4 f_3$, $p_4 = d_3 f_2 + d_4 f_4$.

Comparison of Eq. (12.16) with $\Delta A_{22}^{(3)}$ in Eq. (12.10) yields:

$$p_1 e_1 + p_2 e_3 = 0 , \tag{12.17a}$$

$$p_1 e_2 + p_2 e_4 = 0 , \tag{12.17b}$$

$$p_3 e_1 + p_4 e_3 = 0 , \tag{12.17c}$$

$$p_3 e_2 + p_4 e_4 = \beta_4 , \tag{12.17d}$$

where β_4 represents a nonzero entry of the decomposed matrix $\Delta A_{22}^{(3)}$, while other entries are equal to zero. Let us reduce the number of equations (12.17).

Putting e_1 from (12.17a) to (12.17c) and e_2 from (12.17b) to (12.17d), one finds $e_3 \pi = 0$, which will be satisfied if $e_3 = 0$ or $\pi = 0$, but with regard to $e_4 \pi = \beta_4$, (where $\beta_4 \neq 0$ and $\pi = p_4 - p_3 p_2 / p_1$) π and e_4 cannot be zero, so $e_3 = 0$ must be set. After that assumption one gets $e_1 = 0$ and, in a consequence, Eqs. (12.17a) and (12.17c) vanish. Two equations remaining in (12.17) can be rewritten:

$$d_1 \phi_1 + d_2 \phi_2 = 0 , \tag{12.18a}$$

$$d_3 \phi_1 + d_4 \phi_2 = \beta_4 , \tag{12.18b}$$

where:

$$\phi_1 = f_1 e_2 + f_2 e_4 , \tag{12.19a}$$

$$\phi_2 = f_3 e_2 + f_4 e_4 . \tag{12.19b}$$

The two cases can be distinguished: (i) if $\phi_1 = 0$, then from Eq. (12.18b) $\phi_2 \neq 0$, so $d_2 = 0$ to satisfy Eq. (12.18a); (ii) if $\phi_2 = 0$, then from (12.18b) $\phi_1 \neq 0$, so $d_1 = 0$ to satisfy (12.18a). We choose the first case, then $d_4 = \beta_4 / \phi_2$ for $\phi_2 \neq 0$.

According to Eq. (12.18), $\phi_2 = \beta_4 d_1 / (d_1 d_4 - d_2 d_3)$. Selecting $d_3 = 0$, $\phi_2 = \beta_4 / d_4$ is confirmed while $d_4 \neq 0$, and $\phi_1 = d_2 = 0$ in Eq. (12.18a), so let $d_1 = 0$ be arbitrarily set.

Having the above derived, let $f_3 = 0$ in Eq. (12.15). Then, $f_1 = f_4 = \pm\gamma$ and $f_2 = 0$. Now, one writes from Eq. (12.18b) that $d_4 f_4 e_4 = \beta_4$ at $\phi_1 = 0$. Putting $d_4 = \delta_4 \neq 0$, one gets $e_4 = \pm\beta_4/\gamma$. Finally, $e_2 = 0$ to satisfy Eq. (12.19a), and the decomposition (12.16) finds the following continuation

$$\Delta A_{22}^{(3)} = \begin{bmatrix} 0 & 0 \\ 0 & \delta_4 \end{bmatrix} \begin{bmatrix} \pm\gamma & 0 \\ 0 & \pm\gamma \end{bmatrix} \begin{bmatrix} 0 & 0 \\ 0 & \pm\dfrac{\beta_4}{\gamma} \end{bmatrix}. \tag{12.20}$$

It is possible to obtain for $k = 2$ and $k = 4$ the remaining cases of the uncertainty matrix $\Delta A_{22}^{(k)}$ (see Eq. (12.10)) in a similar way, as shown below:

$$\Delta A_{22}^{(2)} = \begin{bmatrix} 0 & 0 \\ 0 & \delta_4 \end{bmatrix} \begin{bmatrix} 0 & \pm\gamma \\ \pm\gamma & 0 \end{bmatrix} \begin{bmatrix} \pm\dfrac{\beta_1}{\gamma} & 0 \\ 0 & 0 \end{bmatrix}, \tag{12.21a}$$

$$\Delta A_{22}^{(4)} = \begin{bmatrix} 0 & 0 \\ 0 & \delta_4 \end{bmatrix} \begin{bmatrix} 0 & \pm\gamma \\ \pm\gamma & \pm\gamma \end{bmatrix} \begin{bmatrix} \pm\dfrac{\beta_1}{\gamma} & 0 \\ 0 & \pm\dfrac{\beta_4}{\gamma} \end{bmatrix}. \tag{12.21b}$$

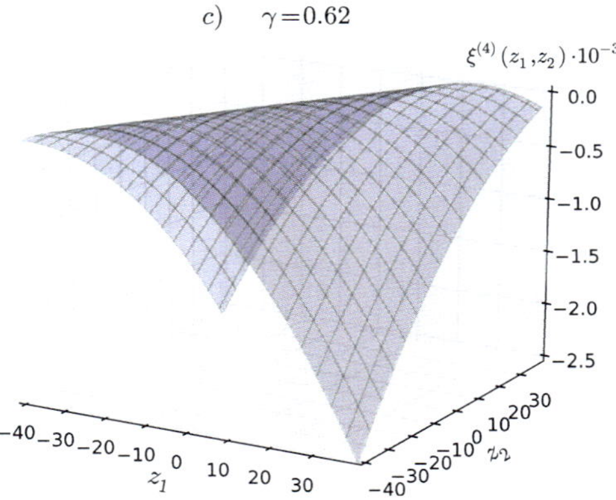

Fig. 12.2   Optimal γ parameter-dependent $\xi^{(4)}(z_1, z_2)$ plot.

The last condition taken into consideration is as follows: $F^{(k)T} F^{(k)} \leq I \iff \xi^{(k)} = F^{(k)T} F^{(k)} - I \leq 0$ for $k = 1, \ldots, 4$. The $n \times n$ real symmetric matrix $\xi^{(k)}$ is negative semi-definite if $z^T \xi^{(k)} z \leq 0$ for all nonzero vectors $z \in R^n$, where z^T denotes the transpose of z. One could check that $\xi^{(k)}$ is symmetric and in particular: $\xi^{(1)} = 0$, $\xi^{(2)} = \xi^{(3)} = (\gamma^2 - 1)(z_1^2 + z_2^2)$, $\xi^{(4)} = \gamma^2(z_1^2 + 2z_1 z_2 + 2z_2^2) - z_1^2 - z_2^2$, and then, for $\xi^{(k)} \leq 0$ the following bounds are determined $\{\gamma : -1 \leq \gamma \leq 1, |\gamma| \leq (z_1^2 + z_2^2)^{1/2}(z_1^2 + 2z_1 z_2 + 2z_2^2)^{-1/2}\}$. It is seen that only the fourth condition on $\xi^{(4)}$ is sensitive on the selection of γ parameter. For the case of presence of two entries in the bottom row of E_{ij}, one would require to accomplish the task: maximize γ subject to $\xi^{(4)}(z_1, z_2) \leq 0$, $z_1 \neq 0$, $z_2 \neq 0$. The maximum value of parameter

$\gamma^* = 0.618033$ was estimated numerically and the corresponding $\xi^{(4)}(z_1, z_2)$ surface plot is shown in Fig. 12.2.

The γ estimate satisfies $-\gamma^* \leq \gamma \leq \gamma^*$, but to achieve the similar decomposition, the presented derivation could follow another way. To sum up, the results are useful in numerical integration of a class of discontinuous-state large-scale dynamical systems. The switching nature of the system parameters defined by a set of matrices $F^{(k)}$ must be assumed if an exact numerical modeling of the investigated multibody system has to be achieved.

12.5   Numerical Experiments

Two sources of variability are applied in the numerical simulation to investigate responses of the analyzed multibody system:

1. Various arrival times of impact force $u_b(t)$ at foam-based armor plate.
2. Various inherent time delays τ_i of each subsystem of the thorax.

Parameters of the simulation: $m_1 = 0.365$, $m_2 = 0.45$, $m_3 = 27$, $m_4 = 10$ [kg]; $k_{13} = 10^5$, $k_{23} = 0.263 \cdot 10^5$, $k'_{23} = 0.132 \cdot 10^5$, $k_{34} = 0.05 \cdot 10^5$, $k_s = 0.1 \cdot 10^5$ [N/m]; $c_{13} = 0.02 \cdot 10^3$, $c_{23} = 0.52 \cdot 10^3$, $c'_{23} = 0.18 \cdot 10^3$, $c_{34} = 0.1 \cdot 10^3$, $c_s = 0.1 \cdot 10^3$ [N/m]; foam thickness $h_f = 0.1$ [m]; number of iterations $N = 5 \cdot 10^4$; step time of numerical integration $h = 2 \cdot 10^{-6}$; time delays: $t'_{1i} = t'_{2i} = t'_{3i} = ih$ $(i = 1, \ldots, 5)$; all initial conditions of state variables are zero except for $x_{11}(0) = 0.7 h_f$ (compressed foam thickness equals 3 cm); the remaining conditions are provided in (C1-C11), (G1-G3), (U1-U5) and Table 12.1.

Stress F_{ch} in the thorax model is calculated according to the formula

$$F_{\mathrm{ch}} = k_{23}(x_{21} - x_{31}) + c_{23}(x_{22} - x_{32}), \tag{12.22}$$

where $k_{23}, \bar{x}_2, \bar{x}_3$ are time-dependent variables.

Deployment of the foam initiates all the numerical experiments presented below.

Influence of air-blast wave time arrivals.

Figures 12.3 exhibit significant differences in the dynamical behavior of the system. The time histories were computed for a few time delays of arrival (ranging from 2.5 ms to 0) of the air-blast pressure wave that reaches the foam-based armor plate. It is seen that for t_a equal c.a. 0.5 ms, the relative deformation is c.a. 4 cm. If the foam deployment delay is greater, then the strain in the thorax rises comparing to the nondelayed counterpart. The results prove that even small time delay affects the dynamics, but in this particular case, both the deformation (Fig. 12.5) and maximum stress (Fig. 12.6) decrease.

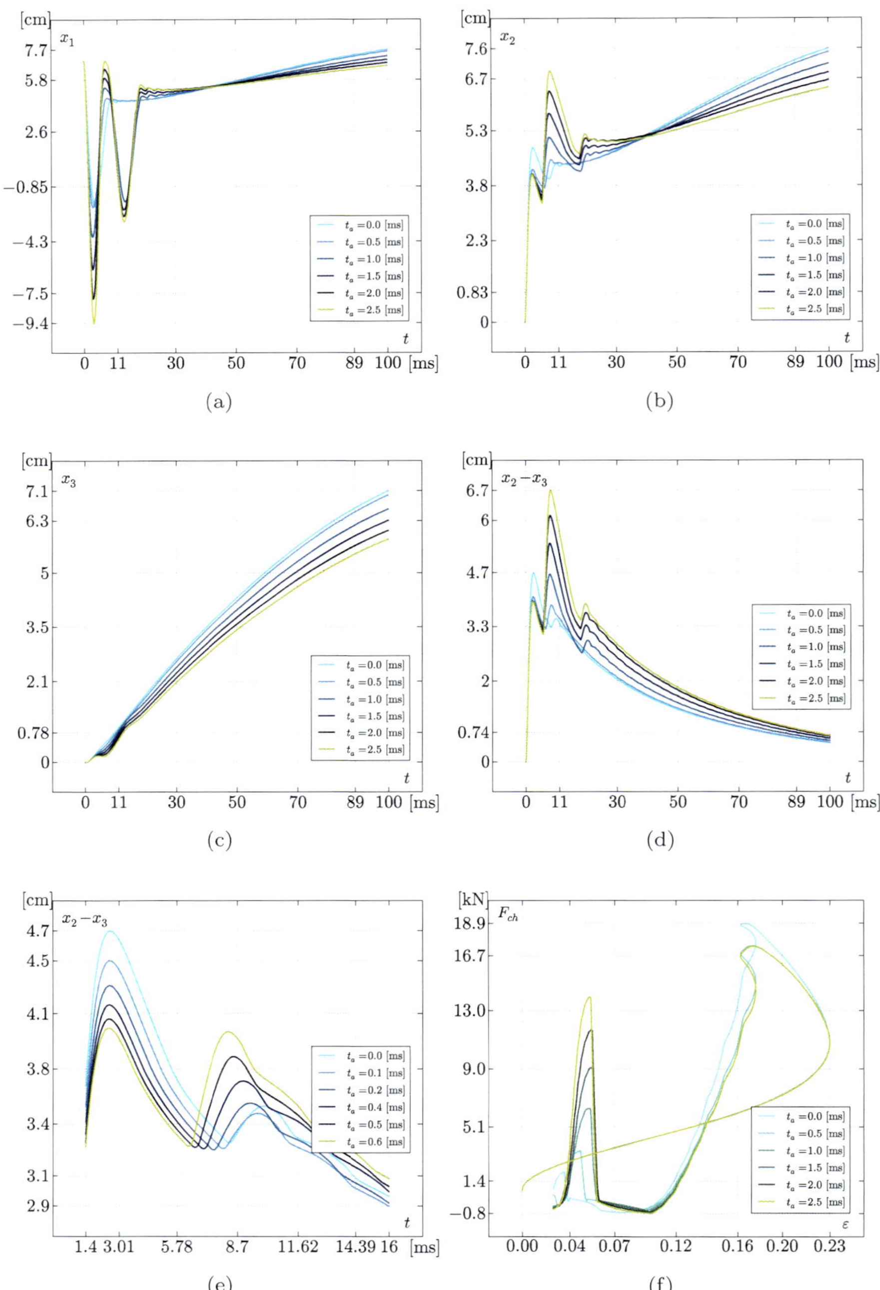

Fig. 12.3　Time histories of displacements x_1, x_2, x_3 (a, b, c), time histories of a relative displacement $x_{21} - x_{31}$ (d, e) and the corresponding stress-strain curve F_{ch} (f) as a function of blast wave time arrivals t_a.

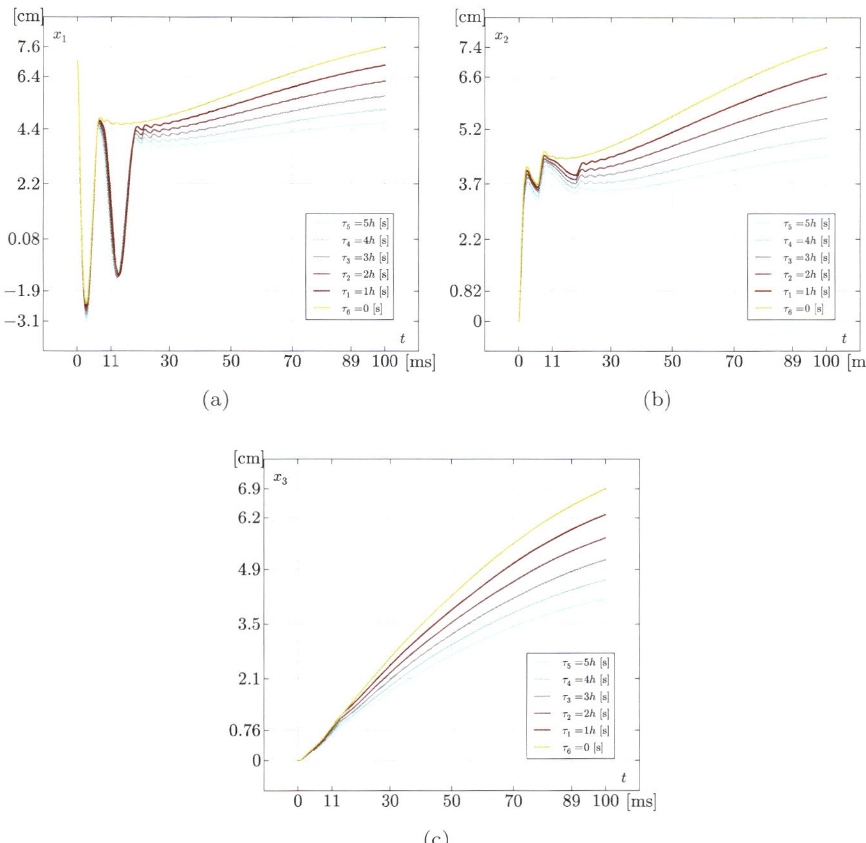

Fig. 12.4 Time histories of displacements $x_1(t)$, $x_2(t)$ and $x_3(t)$ (a, b, c) as a function of inherent time delay τ_i.

Influence of inherent time delays.

The time histories were computed at very small time delays (ranging from $5h$ to $h = 2 \cdot 10^{-6}$) and compared with the nondelayed counterpart. The results prove that even small time delay affects the dynamics as well as a side effect appears in a form of small amplitude vibrations of higher frequency.

Figure 12.6 presents a comparison of stress-strain characteristics of the chest model deformation estimated for nondelayed (for τ_6) and delayed displacements of bodies m_i, $i = 1, 2, 3$. One can observe that even small time delays of about a few time steps of numerical integration play significant role in the blast pressure wave response. Inherent time delays are important in modeling of the chest. Condition (C11) will be satisfied if the time delay $\tau = 5h$ (see Fig. 12.6).

Importance of time delays in numerical solution of the analyzed multibody system is confirmed. Modeling of a response of the thorax energized by air-blast over-

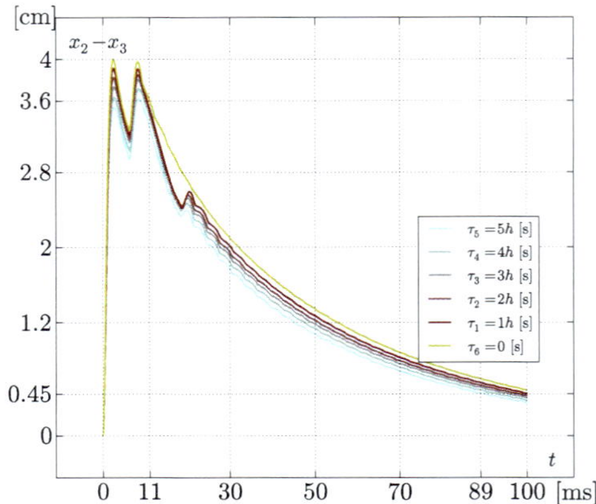

Fig. 12.5   Time history of the relative displacement $x_{21}(t) - x_{31}(t)$ as a function of τ_i.

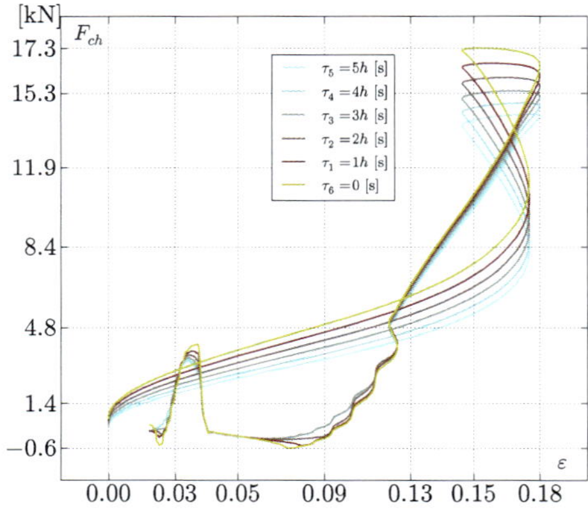

Fig. 12.6   Stress-strain curve as a function of τ_i.

pressure has gained new quality. Some attempts related to the study were mentioned in the literature overview, but significance of time delays in such dynamical systems have been sufficiently emphasized in this chapter of the monograph. The investigated system has received a new useful representation by application of the large-scale systems approach.

Interesting dynamical behavior of the bio-inspired system is solved numerically. It is pointed out that the inherent state time delays change the dynamical response of the multibody system. Proper time of deployment (initiated about 0.6 ms before an impact of the blast wave) of the foam-based armor plate reduces (at some conditions of the experiment) relative compression of the thorax.

Chapter 13

Control of a Multibody System Response to a Suddenly Applied Force

Active control can be adopted in optimization of fast impulsive response systems occurring in body interacting biomechanics. Such kinds of control of mechanical or biological structures are not new, but can be still explored and successively used. This chapter focuses on application of one controlling force to minimize a relative compression of a human chest which has been caused by some impacting action of an elastic external force. A virtual actuator controlling deformation in the analyzed rheological dynamical system of three-degree-of-freedom acts between the back side of the human thorax and the back rest. Reduction of internal displacements in the thorax has been estimated solving the linear quadratic regulator (LQR) optimization problem. Time histories of the controlled and uncontrolled system responses, evaluation of response's shape after changing coefficients of the control method as well as dependency of the objective function estimation on the proportional gain vector are presented and discussed.

13.1 Introduction

It is obvious that today's modern high technology cannot exists without modeling and computer simulations. These tools bring an important information about specific properties of the investigated model, which could be checked and validated in the final design of many products. Usually, it is one of the earliest stages of products design, its meaning and observations are seriously taken into consideration. The most of computer simulations in the field of biomechanics of a human, animals or flora are shared between some finite elements methods [Plank *et al.* (1994)] and classical rigid and multibody mechanics. Active control of building structures was a motivation for the approach shown in the paper, therefore a few representative examples of the second branch can be found in [Jalihal and Utku (1998)].

Investigations on tension or deformation of the parts of our surrounding nature, the shape of which is fully described, lie in the scope of interest of the first field of science. This chapter follows the second way to illustrate the possibility of global action on the analyzed structure, where the dynamics of the bodies represented by point masses is reconstructed. It gives some valuable advantages like, for ex-

305

ample, less complexity of the mathematical description, low costs of testing of the prototype before setting it to the construction, etc. There appear also, as usually, some drawbacks resulting, for example, from omission of internal structure deformations of the investigated bodies, and others. Disregarding of wear phenomena or an influence of temperature fields may serve as instances of these both concepts.

Nowadays, numerical methods are popular in biodynamics and, in particular, in dynamical analysis of communication accidents [Noureddine *et al.* (1996)]. Generally, frontal impacts are considered to be the most common vehicle collision causing numerous injuries [Pietrabissa *et al.* (2002)]. When a human body is exposed to an impact load, soft tissues of the internal organs can sustain large stress and strain rate. To investigate the mechanical responses of the internal organs, sometimes complex modeling of the organs is required. Homogeneous and linear elasticity material properties are assigned to each part of the model, whereas the human cartilages and bones may have different material properties. In order to have a more realistic representation, more complex tissue material properties should be applied [Harrigan and Hamilton (1994)].

Model development often converges to some advance and some complex analyzes, including passive or active control of some weak points. This work focuses on the linear quadratic regulator (LQR) method that is known to have very good robustness properties. These properties are independent of choosing the weighting matrices in the objective function estimation (the cost functional), so if a control system belongs to the class, these robustness properties are definitely assured. It has been confirmed by some applications of the method in [Alavinsab *et al.* (2006); Bernussou *et al.* (1989); Uchida *et al.* (1988)].

Observe that in the existing literature, the use of linear quadratic regulation of impulsive load acting on an elastic and damped biomechanical model of human organs is not reported, hence our considerations are the first attempt to check such a possibility. Methodology concerns a dynamical modeling of a human chest as an elastic connection of its main components, considered separately as a mechanical rigid body having a point-focused mass. Such an assumption does some averaging in the behavior of the thorax, but is very useful for the examination of its controlled realization.

13.2   Dynamical Modeling of the Analyzed Problem

Equilibrium of forces in the gravitational field leads to a system of three second-order differential equations and a one of the first-order (because of the massless point of coupling at x_4 (see Fig. 13.1 b), valid for the rheological properties of the inner part of the thorax). One could said that we deal with a three and a half degree-of-freedom mechanical model.

The system of equations has been written (for further numerical purpose) in the

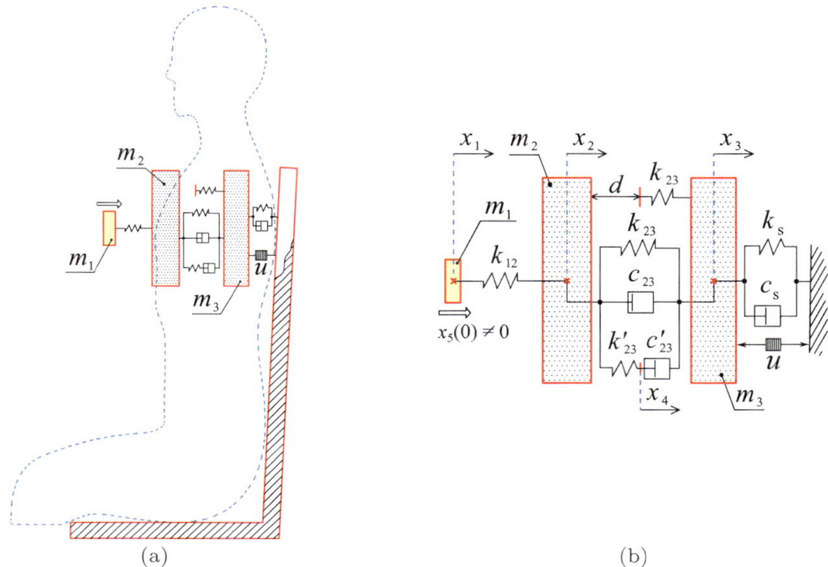

Fig. 13.1 A simplified scheme of a sitting human body (a) and the marked out investigated region of the chest. Particular redraw of the third degree of freedom of the dynamical system (b) with the mass m_1 elastically impacting the chest in the posterior surface.

form of seven first order differential equations:

$$\dot{x}_i = x_{i+4}, \quad (i = 1, 2, 3), \tag{13.1}$$

$$\dot{x}_4 = x_7 + \frac{\bar{k}_{23}}{\bar{c}_{23}}(x_2 - x_4), \tag{13.2}$$

$$\dot{x}_5 = \frac{1}{m_1}\big(k_{12}(x_2 - x_1)\big), \tag{13.3}$$

$$\dot{x}_6 = \frac{1}{m_2}\big(k_{12}(x_1 - x_2) - k_{23}(x_2 - x_3) - \bar{k}_{23}(x_2 - x_4) - c_{23}(x_6 - x_7)\big), \tag{13.4}$$

$$\dot{x}_7 = \frac{1}{m_3}\big(k_{23}x_2 - (k_{23} + k_s)x_3 + c_{23}x_6 - (c_{23} + \bar{c}_{23} + c_s)x_7 + \bar{c}_{23}x_8 - u\big) \tag{13.5}$$

where:

- $m_1, \ldots, m_3$ denote the separated point masses of the model;
- $\bar{k} = [k_{12}, k_{23}, k'_{23}, k_s]$, $\bar{c} = [c_{23}, c'_{23}, c_s]$ are the vectors of system stiffness and damping, respectively;
- $\bar{x}_d = [x_1, \ldots, x_4]$ is the vector of system displacements in each direction;
- $\bar{x}_v = [x_5, \ldots, x_8]$ is the vector of system velocities, but with regard to the introduced rheological description of the model, the massless point reduces the dimension of the system to 7. The object of control (the plant) is described by an odd-dimension system state vector $\bar{x} = [\bar{x}_d, x_5, x_6, x_7]$.

Rheological properties of mechanical rigid bodies (here, masses focused in a point) connection model are introduced twice: firstly, when a relative displacement

$x_r = x_2 - x_3$ (the controlled reference distance) between the front and back sides of the thorax ranges over $d = 3.8$ cm, then k_{23} doubles its value, and secondly, when relative velocity $v_r = x_6 - x_7$ becomes negative, then c_{23} doubles its value as well. It means that stiffness of the rheological coupling increases discontinuously with regard to a bigger than d compression of chest x_r and that damping ability of the coupling varies in time as the thorax remains under compression or depression. Such a discontinuity in system's stiffness and damping vectors is very interesting and will need a special attention during estimation of controlling force.

13.3　Control Methodology

In this section, a numerical investigation of possible modification of standard LQR optimization of proportional system control is presented. A qualitative change in the system dynamics comes from discontinuous changes of system parameters. Such a behavior results from the assumed biomechanical system of a human thorax some material properties of which (also in reality) introduce the discontinuities of stiffness k_{23} and damping c_{23} to its dynamics. The way of their evaluation has been explained above, in Sec. 13.2.

For this particular case, the analyzed problem has been mathematically described in [Olejnik and Awrejcewicz (2010)]. The state-space representation of the dynamical system takes the form:

$$\frac{d\bar{x}}{dt} = A\bar{x} + B\bar{u} =$$

$$= \begin{bmatrix} 0 & 0 & 0 & 0 & 1 & 0 & 0 \\ 0 & 0 & 0 & 0 & 0 & 1 & 0 \\ 0 & 0 & 0 & 0 & 0 & 0 & 1 \\ 0 & \frac{\bar{k}_{23}}{\bar{c}_{23}} & 0 & \frac{-\bar{k}_{23}}{\bar{c}_{23}} & 0 & 0 & 1 \\ \frac{-k_{12}}{m_1} & \frac{k_{12}}{m_1} & 0 & 0 & 0 & 0 & 0 \\ \frac{k_{12}}{m_2} & \frac{-a_{62}}{m_2} & \frac{k_{23}}{m_2} & \frac{\bar{k}_{23}}{m_2} & 0 & \frac{-c_{23}}{m_2} & \frac{c_{23}}{m_2} \\ 0 & \frac{-a_{72}}{m_3} & \frac{-a_{73}}{m_3} & \frac{-\bar{k}_{23}}{m_3} & 0 & \frac{c_{23}}{m_3} & \frac{-a_{77}}{m_3} \end{bmatrix} \bar{x} + \begin{bmatrix} 0 \\ 0 \\ 0 \\ 0 \\ 0 \\ 0 \\ \frac{-1}{m_3} \end{bmatrix} \bar{u}, \qquad (13.6)$$

$$\bar{y} = C\bar{x} + D\bar{u} =$$
$$= \begin{bmatrix} 0 & 1 & -1 & 0 & 0 & 0 & 0 \end{bmatrix} \bar{x}, \qquad (13.7)$$
$$\bar{x}(t_0) = [0, 0, 0, 0, x_5(0) \neq 0, 0], \qquad (13.8)$$

where: the state vector $\bar{x} = [x_1 \ldots x_7]^T$, control input vector $\bar{u} = u$, state-space representation matrices: A – system matrix, B – input matrix, C – output matrix, D – input transforming matrix and constants: $a_{62} = k_{12} + a_{72}$, $a_{72} = k_{23} + \bar{k}_{23}$, $a_{73} = k_{23} + k_s$, $a_{77} = c_{23} + c_s$. Nonzero initial velocity $x_5(0)$ of the impacting mass (the impacting mass is a part of the whole system) states the external excitation.

Our task focuses on searching for the control force $u(t)$ that, at some weighting matrices, would satisfactorily minimize the objective function J in time $t \in [t_0; t_f]$:

$$J(t_0, t_f) = \frac{1}{2}\int_{t_0}^{t_f} \begin{bmatrix} \bar{x}(t) \\ \bar{u}(t) \end{bmatrix}^T \begin{bmatrix} Q & 0 \\ 0 & R \end{bmatrix} \begin{bmatrix} \bar{x}(t) \\ \bar{u}(t) \end{bmatrix} dt$$

$$= \frac{1}{2}\int_{t_0}^{t_f} \begin{bmatrix} x_1 \\ \vdots \\ x_7 \\ u \end{bmatrix}^T \begin{bmatrix} q_1 & & \\ & \ddots & \\ & & q_7 \\ & & & r \end{bmatrix} \begin{bmatrix} x_1 \\ \vdots \\ x_7 \\ u \end{bmatrix} dt$$

$$= \frac{1}{2}\int_{t_0}^{t_f} \left(\sum_{i=1}^{n=7} (q_i x_i^2(t)) + r u^2(t) \right) dt , \tag{13.9}$$

where weighting matrices of the LQR control method are as follows: quality matrix Q has nonzero elements only on its main diagonal, reaction matrix $R = r$ is reduced to a single constant. Computation of integral J will be done by means of the numerical trapezoidal integration. One needs to note that $\bar{x}(t)$ represents a state-space vector of dynamical changes of the controlled system. If $u(t)$ is present in Eq. 13.5, then k_s and c_s are omitted.

The control law for the minimal realization of J in Eq. 13.9 follows

$$u(t) = -r B^T K_x \bar{x}_f(t). \tag{13.10}$$

Equation 13.10 introduces a new matrix K_x of dimension (7×7), called Riccati's matrix. This matrix is symmetrical along the main diagonal, so we get 28 unknown elements ($\xi_{21} = \xi_{12}$, $\xi_{31} = \xi_{13}$ and so on) that need to be estimated. Observe that the sought control law $u(t)$ is governed by a proportional relation to the state vector solution $\bar{x}_f(t)$ of the free system. It is confirmed here that the best method of estimation of the K_x matrix, and thereby of estimation of $u(t)$, is the utilization of a proper convergent numerical procedure. This procedure solves the following matrix equation

$$\left(\dot{K}_x + K_x A + A^T K_x - K_x B \frac{1}{r} B^T K_x + Q \right) \bar{x}_f(t) = 0 , \tag{13.11}$$

Solving Eq. (13.11) with respect to all provided forms of matrices brings 28 first order differential equations for each independent element of the symmetric matrix K_x:

$$\dot{\xi}_{11} = (2k_{12}(\xi_{15}e_7 - \xi_{16}e_6) + \xi_{17}^2 e_1)/e_5 - q_1,$$

$$\dot{\xi}_{12} = (c'_{23}k_{12}(\xi_{25}e_7 - \xi_{26}e_6) + \xi_{27}\xi_{17}e_9 - e_4(\xi_{14}e_{10} + \xi_{15}c'_{23}e_8))/(c'_{23}e_5)$$
$$- (\xi_{16}e_{11}e_{14} + \xi_{17}e_9e_{13})/(c'_{23}e_5),$$

$$\dot{\xi}_{13} = (k_{12}(\xi_{35}e_7 - \xi_{36}e_6) + \xi_{37}\xi_{17}e_1 - m_1 m_3 r(\xi_{16}k_{23}(t)m_3 - \xi_{17}m_2 e_{15}))/e_5,$$

$$\dot{\xi}_{14} = \xi_{45}k_{12}e_7 m_1(\xi_{46}k_{12}e_3 + \xi_{47}\xi_{17}m_2)/e_5$$
$$- (m_1 k'_{23}e_4(\xi_{14}e_2 + c'_{23}(\xi_{16}m_3 - \xi_{17}m_2)))/(c'_{23}e_5),$$

$$\dot{\xi}_{15} = (\xi_{55}k_{12}e_7 - \xi_{56}k_{12}e_6 + \xi_{57}\xi_{17}e_1)/e_5 - \xi_{11},$$

$$\begin{aligned}
\dot{\xi}_{16} =&(\xi_{56}k_{12}e_7 - m_1(\xi_{66}k_{12}e_3 + \xi_{67}\xi_{17}m_2))/e_5 - \xi_{16}c_{23}(t)e_4m_1m_3/e_5 \\
&- \xi_{17}c_{23}(t)e_4m_1m_2/e_5 - \xi_{12},
\end{aligned}$$

$$\begin{aligned}
\dot{\xi}_{17} =&(\xi_{57}k_{12}e_7 - m_1(\xi_{67}k_{12}e_3 + \xi_{77}\xi_{17}m_2))/e_5 \\
&+ m_1e_4(\xi_{14}e_2 + \xi_{16}c_{23}(t)m_3 - \xi_{17}m_2e_{12}))/e_5 - \xi_{13},
\end{aligned}$$

$$\begin{aligned}
\dot{\xi}_{22} =& -(2\xi_{25}k_{12}e_7 - m_1(2\xi_{26}e_3e_{14} + m_2\xi_{27}^2 - 2\xi_{27}e_4e_{13}))/e_5 \\
&- 2m_1\xi_{24}k'_{23}e_3/(c'_{23}e_5) - q_1,
\end{aligned}$$

$$\begin{aligned}
\dot{\xi}_{23} =&\xi_{26}k_{23}(t)e_6/e_5 + (\xi_{27}e_9(\xi_{37} - e_4e_{15}) + e_4(\xi_{34}e_{10} + \xi_{35}c'_{23}e_8))/(c'_{23}e_5) \\
&- e_4(\xi_{36}e_{11}e_{14} - \xi_{37}e_9e_{13})/(c'_{23}e_5),
\end{aligned}$$

$$\begin{aligned}
\dot{\xi}_{24} =& -(\xi_{26}c'_{23}k'_{23}e_6 - \xi_{27}e_9(\xi_{47} + k'_{23}e_4) + e_4(\xi_{44}e_{10} + \xi_{45}c'_{23}e_8))/(c'_{23}e_5) \\
&- m_1e_4(\xi_{46}m_3e_{14} - \xi_{47}m_2e_{13})/e_5 + \xi_{24}k'_{23}e_2/(c'_{23}e_5),
\end{aligned}$$

$$\begin{aligned}
\dot{\xi}_{25} =& -(\xi_{45}e_{10} + \xi_{55}c'_{23}e_8 - \xi_{56}e_{11}e_{14} + \xi_{57}e_9e_{13})/(c'_{23}m_1m_2m_3) \\
&- \xi_{12} + \xi_{27}\xi_{57}/e_3,
\end{aligned}$$

$$\begin{aligned}
\dot{\xi}_{26} =&\xi_{26}e_6c_{23}(t)/e_5 + (\xi_{27}e_9(\xi_{67} - c_{23}(t)e_4) - e_4\xi_{46}e_{10})/(c'_{23}e_5) \\
&- e_4(\xi_{56}e_8 - m_1(\xi_{66}m_3e_{14} - \xi_{67}m_2e_{13} - \xi_{22}e_2))/e_5,
\end{aligned}$$

$$\begin{aligned}
\dot{\xi}_{27} =& -\xi_{26}e_6c_{23}(t)/e_5 - (\xi_{27}e_9(\xi_{77} + e_4e_{12}) + e_4\xi_{47}e_{10})/(c'_{23}e_5) \\
&+ e_4(\xi_{57}e_8 - m_1(\xi_{67}m_3e_{14} - \xi_{77}m_2e_{13} - e_2(\xi_{23} + \xi_{24})))/e_5,
\end{aligned}$$

$$\dot{\xi}_{33} = -(2\xi_{36}k_{23}(t)e_3 - \xi_{37}m_2(\xi_{37} + 2e_4e_{15}))/e_7 - q_2,$$

$$\begin{aligned}
\dot{\xi}_{34} =&\xi_{34}k'_{23}/c'_{23} - (\xi_{36}k'_{23}e_3 - \xi_{37}m_2(\xi_{47} + k'_{23}e_4) + e_4\xi_{46}k_{23}(t)m_3)/e_7 \\
&- e_4\xi_{47}m_2e_{15}/e_7,
\end{aligned}$$

$$\dot{\xi}_{35} = \xi_{37}\xi_{57}/e_3 - (\xi_{56}k_{23}(t)m_3 - \xi_{57}m_2e_{15})/e_2 - \xi_{13},$$

$$\begin{aligned}
\dot{\xi}_{36} =&(\xi_{36}c_{23}(t)e_3 + \xi_{37}m_2(\xi_{67} - c_{23}(t)e_4)/e_7 - e_4\xi_{66}k_{23}(t)m_3/e_7 \\
&- e_4m_2(\xi_{67}e_{15} - \xi_{23}m_3)/e_7,
\end{aligned}$$

$$\begin{aligned}
\dot{\xi}_{37} =& -\xi_{33} - \xi_{34} - (\xi_{36}c_{23}(t)e_3 - \xi_{37}m_2(\xi_{77} + e_4e_{12}))/e_7 \\
&- e_4(\xi_{67}k_{23}(t)m_3 - \xi_{77}m_2e_{15})/e_7,
\end{aligned}$$

$$\dot{\xi}_{44} = (2\xi_{44}k'_{23}e_7 - c'_{23}(2\xi_{46}k'_{23}e_3 - \xi_{47}m_2(\xi_{47} + 2k'_{23}e_4)))/(c'_{23}e_7) - q_3,$$

$$\dot{\xi}_{45} = \xi_{45}k'_{23}/c'_{23} + \xi_{47}\xi_{57}m_2/e_7 - e_4(\xi_{56}k'_{23}m_3 - m_2(\xi_{57}k'_{23} - \xi_{14}m_3))/e_7,$$

$$\dot{\xi}_{46} = \xi_{46}e_3c_{23}(t)/e_7 + k'_{23}m_2/(c'_{23}e_7) + \xi_{47}m_2(\xi_{67} - c_{23}(t)e_4)/e_7 \\ - e_4(\xi_{66}k'_{23}m_3 - m_2(\xi_{67}k'_{23} - \xi_{24}m_3))/e_7,$$

$$\dot{\xi}_{47} = -\xi_{34} - \xi_{44} - (\xi_{46}c_{23}(t)e_3 + \xi_{47}m_2(\xi_{77} + e_4e_{12}))/e_7 \\ + e_4\xi_{47}m_2k'_{23}m_3/(c'_{23}e_7) - k'_{23}e_4(\xi_{67}m_3 - \xi_{77}m_2))/e_7,$$

$$\dot{\xi}_{55} = \xi_{57}^2/e_3 - 2\xi_{15} - q_4,$$

$$\dot{\xi}_{56} = -\xi_{25} - (\xi_{56}c_{23}(t)e_3 + m_2(\xi_{57}(\xi_{67} - c_{23}(t)e_4) - \xi_{16}e_3))/e_7,$$

$$\dot{\xi}_{57} = -\xi_{35} - \xi_{45} - (\xi_{56}c_{23}(t)e_3 - m_2(\xi_{57}(\xi_{77} + e_4e_{12}) - \xi_{17}e_3))/e_7,$$

$$\dot{\xi}_{66} = -2\xi_{26} + (2\xi_{66}c_{23}(t)e_3 + \xi_{67}m_2(\xi_{67} - 2c_{23}(t)e_4))/e_7 - q_5,$$

$$\dot{\xi}_{67} = -\xi_{27} - \xi_{36} - \xi_{46} - \xi_{66}c_{23}(t)e_3/e_7 \\ + (\xi_{67}(\xi_{77}m_2 + e_4(m_2(c_{23}(t) + c_s) + m_3)) - \xi_{77}c_{23}(t)e_2r)/e_7,$$

$$\dot{\xi}_{77} = -2(\xi_{37} + \xi_{47}) - (2\xi_{67}c_{23}(t)e_3 - \xi_{77}m_2(\xi_{77} + 2e_4e_{12}))/e_7 - q_6, \qquad (13.12)$$

where: $e_1 = m_1m_2$, $e_2 = m_2m_3$, $e_3 = m_3^2r$, $e_4 = m_3r$, $e_5 = e_1m_3^2r$, $e_6 = m_1m_3^2r$, $e_7 = m_2m_3^2r$, $e_8 = k_{12}e_2$, $e_9 = c'_{23}e_1$, $e_{10} = k'_{23}e_1m_3$, $e_{11} = c'_{23}m_1m_3$, $e_{12} = c_{23}(t) + c_s$, $e_{13} = k_{23}(t) + k'_{23}$, $e_{14} = k_{12} + e_{13}$, $e_{15} = k_{23}(t) + k_s$. Solving equations (13.12) for zero initial conditions, all the coefficients ξ_{ij} for $i,j \in (1,\ldots,7)$ of the matrix K_x are estimated during evaluation of the numerical control algorithm.

Equation (13.10) can be used now for calculation of the control law:

$$u(t) = \bar{f}_x \cdot \bar{x}_f(t) = \left[\frac{\xi_{i,7}}{rm_3}\right] \cdot \bar{x}_f(t) = \sum_{i=1}^{7} \frac{\xi_{i,7}x_{f,i}}{rm_3}, \qquad (13.13)$$

where $\bar{f}_x = -rB^TK_x$ is the proportional gain of the control feedback loop. Because of the specification of the analyzed rheological dynamical system, $\bar{f}_x$ will switch between two states, as it has been described below.

Before the scalar product of a time-dependent vector of state and the vector of constant gain $\bar{x} \cdot \bar{f}_{x,i}$ is calculated, one needs to consider the rheology of the bio-inspired model and to select the proper vector of proportional gain. Because of two 2-state switching blocks, there are four combinations like: $[k_{23}, c_{23}]$, $[k_{23}, 2c_{23}]$, $[2k_{23}, c_{23}]$ and $[2k_{23}, 2c_{23}]$. As the numerical solution to the free system dynamics confirms, x_r is always greater than d. Therefore, a Riccati matrix equation associated with the controlled system have to be solved only twice (this time, the two last combinations are unnecessary), and during numerical solution to the control problem, as relative velocity v_r varies in time, the first element of the damping vector $\bar{c}$ will switch between two values c_{23} and $2c_{23}$, generating some unexpected disturbances of gain.

13.4   Numerical Simulation

The following set of system parameters is assumed in the numerical model: $m_1 = 1.6$, $m_2 = 0.45$, $m_3 = 27$ kg, $d = 3.8$ cm, $k_{12} = 281$, $k_{23} = 26.3$, $\bar{k}_{23} = 13.2$, $k_s = 10$ $\cdot 10^3$ N/m, $c_{23} = 1.23$, $\bar{c}_{23} = 0.18$, $c_s = 0.11$ $\cdot 10^3$ N·s/m and initial conditions: $\bar{x}_f(t_0 = 0) = [0, 0, 0, 0, 13.9, 0, 0, 0]$.

The proportional control law is realized by a virtual actuating impulsive mechanism characterized by a rapid force response having a shape of $u(t)$, the time dependency visible in Fig. 13.2. Actual relative displacement x_c of the controlled system supported with the actuating mechanism shown in Fig. 13.1 can be simultaneously observed.

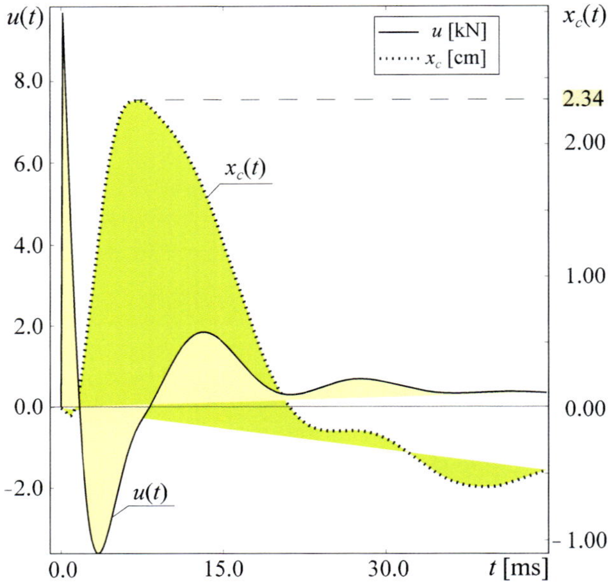

Fig. 13.2   Time history of the control force $u(t)$ applied on the anterior surface of the thorax model on the background of the controlled system relative displacement $x_c(t)$. For a better picture, time range t_f of the time history is shortened to 40 ms.

Before $x_c(t)$ reaches its peak value at 2.34 cm, the control force $u(t)$ exposes very rapid changes of amplitude from about 10 to -4 kN, what is the crucial stage of the response taking about 10 ms. In practice, this time interval is probably longer, and making it longer here could be possible after introduction of some time delays (see Chap. 12) either in response of the mechanism or in a reaction of the shock-excited body. Nevertheless, one confirms that the method of active control can be used for finding a shape of force characteristics which has to be realized by the actuator.

The final result of control obtained with attention to $J_{min} = 1.39 \cdot 10^7$ is presented in Fig. 13.3. The set of parameters of the control procedure listed in its area has been estimated on the basis of observation of minimal amplitude of $x_c(t)$.

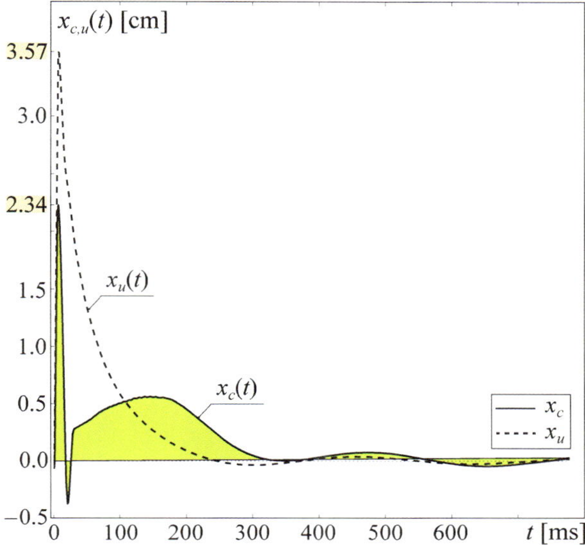

Fig. 13.3　Relative displacement characteristics $x_u(t)$ and $x_c(t)$ of masses m_2 and m_3 of the thorax, respectively before (dashed line) and after (solid line) application of the control algorithm.

The linear quadratic regulator method can be applied with a success in control of very fast, parameter-discontinuous dynamical systems. It requires some additional coding in numerical procedures and switching function's block located in the controlled system feedback loop.

In the presented example of interesting biomechanical application, reduction of maximal amplitude of the output signal was obtained. Dependently on requirements, the output response of the controlled system can be optimized with respect to a minimum value of the performance index of the LQR method or even with respect to the minimal amplitude.

Thinking about practical implementation of the estimated shape of the controlling force to compensate a shock-induced compression, one would need to design a very fast reaction forcing actuator that could be located between the back side of the human thorax and the back rest.

Bibliography

(1991). *Conventional Weapons Effects ("ConWep", 1991), is a computer programme, based on the content of technical manual TM 5-855-1*, Structural Laboratory, Waterways Experimental Station for the Department of US Army Corps of Engineers.

Adams, J. and Payandeh, S. (1996). Methods for low-velocity friction compensation: Theory and experimental study, *J. Robotic Systems* **13**, 6, pp. 391–404.

Alavinsab, A., Moharami, H., and Khajepour, A. (2006). Active control of structures using energy-based LQR method, *Comput.-Aided Civ. Infrastruct. Eng.* **21**, pp. 605–611.

Ananth, S. M. and Kushari, A. (2013). Quasi-steady prediction of coupled bending-torsion flutter under classic surge, *J. Appl. Mech.* **80**, 5, pp. 051010–051010, doi: 10.1115/1.4023617.

Arrofiq, M. and Saad, N. (2008). A PLC-based self-tuning PI-fuzzy controller for linear and non-linear drives control, in *IEEE International Conference on Power and Energy* (Johor Bahru), pp. 701–706.

Asl, F. M. and Ulsoy, A. G. (2003). Analysis of a system of linear delay differential equations, *J. Dyn. Syst. Meas. Contr.* **125**, 2, pp. 215–223, doi:10.1115/1.1568121.

Avalle, M., Belingardi, G., and Montanini, R. (2001). Characterization of polymeric structural foams under compressive impact loading by means of energy-absorption diagram, *Int. J. Impact Eng.* **25**, pp. 455–472.

Awrejcewicz, J. (1991). *Bifurcation and Chaos in Coupled Oscillators* (World Scientific, Singapore).

Awrejcewicz, J. (2012a). *Classical Mechanics. Dynamics* (Springer, Berlin).

Awrejcewicz, J. (2012b). *Classical Mechanics. Kinematics and Statics* (Springer, Berlin).

Awrejcewicz, J. (2014). *Ordinary Differential Equations and Mechanical Systems* (Springer).

Awrejcewicz, J. and Koruba, Z. (2012). *Classical Mechanics. Applied Mechanics and Mechatronics* (Springer, Berlin).

Awrejcewicz, J. and Krysko, V. (2008). *Chaos in Structural Mechanics* (Springer, Berlin).

Awrejcewicz, J. and Olejnik, P. (2002). Numerical analysis of self-excited by friction chaotic oscillations in two-degrees-of-freedom system using exact Hénon method, *Machine Dynamics Problems* **26**, 4, p. 920.

Awrejcewicz, J. and Olejnik, P. (2003). Stick-slip dynamics of a two degree-of-freedom system, *Int. J. Bif. and Chaos* **13**, 4, pp. 843–861.

Awrejcewicz, J. and Olejnik, P. (2005a). Analysis of dynamic systems with various friction laws, *Appl. Mech. Rev.* **58**, 6, pp. 389–411.

Awrejcewicz, J. and Olejnik, P. (2005b). Friction pair modeling by 2-DOF system: numerical and experimental investigations, *International Journal of Bifurcations and Chaos* **15**, 6, pp. 1931–1944.

Awrejcewicz, J. and Olejnik, P. (2005c). Friction pair modeling by 2-DOF system: Numerical and experimental investigations, *Int. J. Bifurcation Chaos* **15**, 6, pp. 1931–1944.

Awrejcewicz, J. and Olejnik, P. (2005d). Sliding solutions of a simple two degrees-of-freedom dynamical system with friction, in *Proceedings of 5-th EUROMECH, Nonlinear Dynamics Conference ENOC*, Vol. 01-196 (Eindhoven, The Netherlands), pp. 277–282.

Awrejcewicz, J. and Olejnik, P. (2007). Occurrence of stick-slip phenomenon, *J. Theoret. Appl. Mech.* **45**, 1, pp. 33–40.

Azbelev, N., Maksimov, V., and Rakhmatullina, L. (2007). *Introduction to the Theory of Functional Differential Equations: Methods and Applications* (Hindawi Publishing Corporation).

Bai, Y. and Wang, D. (2006). Fundamentals of fuzzy logic control – fuzzy sets, fuzzy rules and defuzzifications, in Y. Bai, H. Zhuang, and D. Wang (eds.), *Advanced Fuzzy Logic Technologies in Industrial Applications* (Springer, London), pp. 17–36.

Bašić, M., Vukadinović, D., and Polić, M. (2013). Fuzzy logic vs. classical PI voltage controller for a self-excited induction generator, in N. Trisovic and D. Rasteiro (eds.), *Mathematical Applications in Science and Mechanics* (WSEAS Press, Dubrovnik), pp. 189–194.

Bernussou, J., Peres, P., and Geromel, J. (1989). A linear programming oriented procedure for quadratic stabilization of uncertain systems, *Syst. Control Lett.* **13**, pp. 65–72.

Bialkowski, B. (2014). *Modeling and control of electrohydraulic servomechanisms in LabVIEW*, Master Thesis under the supervision of dr. Pawel Olejnik, Lodz University of Technology, Department of Automation, Biomechanics and Mechatronics.

Borzi, A. and Schulz, V. (2012). *Computational Optimization of Systems Governed by Partial Differential Equations* (The Society for Industrial and Applied Mathematics, SIAM, Philadelphia, USA).

Boyd, S., Kim, S.-J., Vandenberghe, L., and Hassibi, A. (2007). A tutorial on geometric programming, *Optim Eng* **2007**, 8, pp. 67–127.

Brand, M., Neaves, S., and Smith, E. (2015). Lodestone. museum of electricity and magnetism, .

Brandstadter, J. and Taylor, J. (1974). Fire control system, US Patent 3,844,196.

Budynas, R. and Nisbett, J. (2015). *Shigley's Mechanical Engineering Design*, 10th edn., Series in Mechanical Engineering (McGraw-Hill Education).

Butenko, S. and Pardalos, P. (2014). *Numerical Methods and Optimization: An Introduction*, Chapman and Hall/CRC Numerical Analysis and Scientific Computing Series (Chapman and Hall/CRC).

Cai, L. and Song, G. (1994). Jointstick-slip friction compensation of robot manipulators by using smooth robust controllers, *J. Robotic Syst.* **11**, 6, pp. 451–470.

Carlston, J. (189). Lodestone compass: Chines or olmec primacy?: Multidisciplinary analysis of an olmec hematite artifact from san lorenzo, veracruz, mexico, *Science* **189**, pp. 753–760.

Ceramics, P. (2008). Complete material set pic181, Tech. rep., PI Ceramics.

Cetinkunt, S. (2007). *Mechatronics* (John Wiley & Sons, Inc.).

Chaber, W. and Chaber, R. (1891). *Chamber's Encyclopedia. A Dictionary of Universal Knowledge: Magnesia and Magnetism* (William and Robert Chambers, Ltd., London and Edinburgh).

Chan, K. and Chu, X. (2009). Design of a fuzzy PI controller to guarantee proportional delay differentiation on web servers, in *Algorithmic Aspects in Information and Management* (Springer-Verlag), pp. 389–398.

Chan, P., MacFadden, L., Dang, X., Ho, K., Maffeo, M., Carboni, M., and DeCristofano,

B. (2010). Study of material effects on blast lung injury using normalized work method, in *Personal Armour Systems Symposium* (Quebec City, Canada), pp. pp. 1–10.

Chang, S., Rogacheva, N., and CC, C. (1995). Analysis of methods for determining electromechanical coupling coefficients of piezoelectric elemenets. *IEEE Trans. Ultrason. Ferroelectr. Freq. Control* **42**, 4, pp. 630–640.

Chenafa, M., Mansouri, A., Bouhenna, A., Etien, E., Belaidi, A., and Denai, M. (2005). Global stability of linearizing control with a new robust nonlinear observer of the induction motor, *Int. J. Appl. Math. Comput. Sci.* **15**, 2, pp. 235–243.

Cheong, F. and Lai, R. (2007). Simplifying the automatic design of a fuzzy logic controller using evolutionary programming, *Soft Computing* **11**, 9, pp. 839–846.

Chiason, J. (1997). A new approach to dynamic feedback linearization control of an induction motor, *IEEE Trans. Automat. Contr.* **43**, 3, pp. 391–397.

Courtemanche, M., Glass, L., and Keener, J. P. (1993). Instabilities of a propagating pulse in a ring of excitable media, *Phys. Rev. Lett.* **70**, 14, pp. 2182–2185.

Craig, R. and Kurdila, A. (2006). *Fundamentals of Structural Dynamics* (Wiley, New Jersey).

Del Piero, G. and Pampolini, G. (2010). On the rate-dependent properties of open-cell polyurethane foams, *Technische Mechanik* **30**, 1-3, pp. 74–84.

Delfour, M. (2012). *Introduction to Optimization and Semidifferential Calculus* (The Society for Industrial and Applied Mathematics, SIAM, Philadelphia, USA).

Dewey, J. (2010). The shape of the blast wave: studies of the friedlander equation, in *21st International Symposium on Military Aspects of Blast and Shock* (Israel).

Di Paola, A. and Cicirelli, G. (eds.) (2010). *Mechatronic Systems Applications* (InTech).

Driessen, B. J. and Sadegh, N. (2004). Convergence theory for multi-input discrete-time iterative learning control with Coulomb friction, continuous outputs, and input bounds, *Int. J. Adapt. Control Signal Process.* **18**, pp. 457–471.

D'yachenko, A. and Manyuhina, O. (2006). Modeling of weak blast wave propagation in the lung, *J. Biomech.* **39**, pp. 2113–2122.

Edminister, J. and Hahvi, M. (2011). *Electromagnetics* (McGraw Hill, New York).

Ewins, D. (2000). *Modal Testing. Theory Practice and Application,* 2nd edn. (Research Studies Press).

Fitek, J., Carboni, M., DeCristofano, B., and Maffeo, M. (2011). A simple model of a plate and foam armor system for primary blast lung injury protection, in *RTO Human Factors and Medicine Panel (HFM) Symposium* (Halifax, Canada), pp. (32–1)–(32–24).

Fleisch, D. (2008). *A Student's Guide to Maxwell's Equations* (Cambridge University Press, Cambridge).

Fowler, D. and O'Connor, S. (1976). Aerodynamic surface control feel augmentation system, US Patent 3,960,348.

Gao, S. and Meng, G. (2011). Research of the spindle overhang and bearing span on the system milling stability, *Archive of Applied Mechanics* **81**, 10, pp. 1473–1486, doi:10.1007/s00419-010-0498-4.

Goel, M., Matsagar, V., Gupta, A., and Marburg, S. (2012). An abridged review of blast wave parameters, *Defence Sci. J.* **62**, 5, pp. 300–306.

Goog, V. (2011). New phenomenological model for solid foams, in J. Muriń, V. Kompiš, and V. Kutiš (eds.), *Computational Modelling and Advanced Simulations* (Springer), pp. 67–82.

Grimal, Q., S., N., and A., W. (2002). A study of transient elastic wave propagation in a bimaterial modeling the thorax, *Int. J. Solids Struct.* **39**, pp. 5345–5369.

Guillon, M. (1966). *Theory and Calculation of Hydraulic Systems* (WNT, Warsaw), in Polish.

Hammetter, C. I. and Zok, F. W. (2013). Compressive response of pyramidal lattices embedded in foams, *J. Appl. Mech.* **81**, 1, pp. 011006–011006, doi:10.1115/1.4024408.

Han, F., Zhu, Z., and Gao, J. (1998). Compressive deformation and energy absobing characteristic of foamed aluminum, *Metallurgical Mater. Trans. A* **29A**, pp. 2497–2511.

Harrigan, T. and Hamilton, J. (1994). Necessary and sufficient conditions for global stability and uniqueness in finite element simulations of adaptive bone, *Int. J. Solids Struct.* **31**, 1, pp. 97–107.

Heckl, M. A. and Abrahams, I. D. (1996). Active control of friction-driven oscillations, *J. Sound and Vibrations* **193**, 1, pp. 417–426.

Hirschon, R. M. and Miller, G. (1999). Control of nonlinear systems with friction, *IEEE Trans. Control Syst. Technology* **7**, 5, pp. 588–595.

Holtrop, J. (1983). Electrohydraulic control system, US Patent 4,418,610.

Horrocks, C. and Brett, S. (2000). Blast injury, *T. Anaesth. Cr. Care* **11**, pp. 113–119.

Huang, S. N., Tan, K. K., and Lee, T. H. (2000). Adaptive friction compensation using neural network approximations, *IEEE Trans. Syst., Man. Cybern. – Part C: Applications and Reviews* **30**, 4, pp. 551–557.

Ikeda, T. (1990). *Fundamentals of Piezoelectrocity* (Oxford University Press).

Jager, R. (1995). *Fuzzy Logic in Control*, Ph.D. thesis, TU Delft.

Jalihal, P. and Utku, S. (1998). Active control in passively base isolated buildings subjected to low power excitations, *Comput. Struct.* **68**, 2–3, pp. 211–224.

Jang, J. (1991). Fuzzy modelling using generalized neural networks and Kalman filter algorithm, in *AAAI'91: Proceedings of the Ninth National Conference on Artificial Intelligence*, Vol. 2 (Anaheim), pp. 762–767.

Jee, S. and Korem, Y. (2004). Adaptive fuzzy logic controller for feed drives of a CNC machine tool, *Mechatronics* **14**, pp. 299–326.

Kamlah, M. (2001). Ferroelectric and ferroelastic piezoceramics – modeling of electromechanical hysteresis phenomena, *Continuum Mech. Termodyn.* **13**, pp. 219–268.

Kanno, Y. (2011). *Nonsmooth Mechanics and Convex Optimization* (CRC Press).

Karnopp, D., Margolis, D., and RC, R. (2012). *System Dynamics: Modeling, Simulation, and Control of Mechatronic Systems*, 5th edn. (John Wiley & Sons, Inc.).

Kelley, C. (1999). *Iterative Methods for Optimization* (The Society for Industrial and Applied Mathematics, SIAM, Philadelphia, USA).

Kim, Y. H. and Lewis, F. L. (2000). Optimal design of cmac neural network controller for robot manipulators, *IEEE Trans. Syst., Man. Cybern.– Part C: Applications and Reviews* **30**, 1, pp. 22–31.

Kunikowski, W. (2014). *Fuzzy Logic in Angular Velocity Control Algorithms of DC Motors*, Master's thesis, Lodz University of Technology, Department of Automation, Biomechanics and Mechatronics, Lodz.

Kunikowski, W., Czerwiński, E., Olejnik, P., and Awrejcewicz, J. (2015). An overview of ATmega AVR microcontrollers used in scientific research and industrial applications, *Pomiary Automatyka Robotyka* **19**, 1, pp. 15–20.

Kunikowski, W., Olejnik, P., and Awrejcewicz, J. (2013). Efficiency of a PLC-based PI controller in stabilization of a rotational motion affected by the chaotic disturbances, in J. Awrejcewicz, M. Kaźmierczak, P. Olejnik, and J. Mrozowski (eds.), *Dynamical Systems – Applications* (Publishing House of Lodz University of Technology, Lodz), pp. 173–184.

Leondes, C. (ed.) (1981). *Structural Dynamic Systems Computational Techniques and Op-*

timization (Gordon and Breach Publishers International Series In Engineering, Technology and Applied Science).

Lerner, R. and Trigg, G. (eds.) (2005). *Encyclopedia of Physics*, 3rd edn. (Wiley-VCH).

Lewandowski, D. (1971). *Static and Dynamic Properties of a Hydrostatic Bearing*, Ph.D. thesis, Lodz University of Technology.

Lewandowski, D. (1996). *Aerostatic Bearing of High-Speed Spindle Systems* (Lodz University of Technology Press, Lodz).

Lewandowski, D. (2005). Pneumatics in machine tools to window frames made of PCV, *Pneumatyka* **3**, pp. 37–40, in Polish.

Lewandowski, D. and Awrejcewicz, J. (2012). Modeling of a hydro-mechanical system controled by a proportional valve, in *Proceedings of the International Scientific-Technical Conference on Hydraulics and Pneumatics* (Wroclaw, Poland), pp. 121–127.

Lewis, F., Abdallah, C., and Dawson, D. (1993). *Control of Robot Manipulators* (Macmillan Publishing Company, New York).

Li, Z., Wang, Q., and Gao, H. (2009). Friction driven oscillator control by Lyapunov redesign based on delayed state feedback, *Acta Mech. Sin.* **25**, pp. 257–264.

Lin, C. L. and Huang, H. T. (2002). Linear servo motor control using adaptive neural networks, *Proc. Instn. Mech. Engrs., Part I: J. Systems and Control Engineering* **216**, pp. 407–427.

Lobdell, T., Kroell, C., Schneider, D., Hering, W., and Nahum, A. (1973). Impact response of the human torax, in *Symposium of Human Impact Response: Human impact response: measurement and simulation.*, General Motors Research Laboratories (Plenum Press, New York London), pp. 201–245.

Lorenz, N. and Wanka, G. (2012). Scalar and vector optimization with composed objective functions and constraints, in *Chemnitz Scientific Computing Preprints*, Vol. 2012-01 (Technical University of Chemnitz), pp. 1–15.

Lubineau, D., Dion, J., Dugard, L., and Roye, D. (2000). Design of an advanced non linear controller for induction motors and experimental validation on an industrial benchmark, *Eur. Phys. J. Appl. Phys.* **9**, pp. 165–175.

MacVicar-Whelan, P. (1977). Fuzzy sets for man-machine interactions, *Int. J. Man Mach. Stud.* **8**, 6, pp. 687–697.

Mamdani, E. (1974). Application of fuzzy algorithms for the control of a simple dynamic plant, in *Proc IEEE*, pp. 121–159.

Mansouri, A., Chenafa, M., Bouhenna, A., and Etien, E. (2004). Powerful nonlinear observer associated with the field-oriented control of the induction motor, *Int. J. Appl. Math. Comput. Sci.* **14**, 2, pp. 209–220.

Marino, R., Peresada, S., and Valigi, P. (1993). Adaptive input-output linearizing control of induction motors, *IEEE Trans. Automat. Control* **38**, 2, pp. 208–221.

Martin, H. T., Boyer, E., and Kuo, K. K. (2013). Effect of initial temperature on the interior ballistics of a 120-mm mortar system, *J. Appl. Mech.* **80**, 3, pp. 031408–031408 doi:10.1115/1.4023318.

Mason, W. (1950). *Piezoelectric Crystals and Their Application to Ultrasonics* (D. van Nostrand, New York, Toronto, London).

Mason, W. (ed.) (1964). *Physical Acoustics – Principles and Methods*, Vol. 1 (Academic Press, New York).

Maugin, G. (1988). *Mechanics of Electromagnetic Solids* (Elsevier, Amsterdam).

Meitzler, A., Tiersten, H., AW, W., Berlincourt, D., Coquin, G., and Welsh, F. (1988). Ieee standard on piezoelectricity, Tech. rep., Institute of Electrical and Electronics Engineers.

Milič, V., Situm, Z., and Essert, M. (2010). Robust position control synthesis of an electro-hydraulic servo system, *ISA Trans.* **49**, 4, pp. 535–542, doi: http://dx.doi.org/10.1016/j.isatra.2010.06.004.

Mukaidani, H., Tanaka, Y., and Mizukami, K. (2004). Guaranted cost control for large-scale systems under control gain perturbations, *Electr. Eng. Jpn.* **146**, pp. 118–129.

Nakayama, Y. and Boucher, R. (2000). *Introduction to Fluid Mechanics* (Elsevier).

Neogy, S., Das, A., and Bapat, R. (eds.) (2009). *Modeling, Computation and Optimization, Statistical Science and Interdisciplinary Research*, Vol. 6 (World Scientific).

Nersessian, N. (2002). *Reading Natural Philosophy*, chap. Maxwell and the Method of Physical Analogy: Model-Based Reasoning, Generic Abstraction, and Conceptual Change (Open Court, Chicago).

Neto, P., Mendes, N., Pires, J., Norberto, J., and Moreira, A. (2010). CAD-based robot programming: The role of fuzzy-PI force control in unstructured environments, in *IEEE Conference on Automation Science and Engineering* (Toronto), pp. 362–367.

Noureddine, A., Digges, K., and Bedewi, N. (1996). An evaluation of deformation based chest injury criteria using a hybrid iii finite model, *Int. J. Crashworthiness* **1**, pp. 181–189.

Olejnik, P. (2002). *Numerical and experimental analysis of self-excited regular and chaotic vibrations in a two-degrees-of-freedom system with friction*, Ph.D. thesis, Lodz University of Technology, Department of Automation, Biomechanics and Mechatronics, Lodz.

Olejnik, P. (2013). *Numerical Methods of Solution, Analysis and Control of Discontinuous Dynamical Systems*, no. 1151 in Scientific Books (Lodz University of Technology Press, Lodz).

Olejnik, P. and Awrejcewicz, J. (2010). Reduction of deformation in a spring-mass realisation of human chest occurred after action of impact, *J. KONES Powertrain and Transport* **7**, 1, pp. 327–335.

Olejnik, P. and Awrejcewicz, J. (2011). One-dimensional discrete LQR control of compression of the human chest impulsively loaded by fast moving point mass, *Commun. Nonlinear Sci. Numer. Simulat.* **16**, 5, pp. 2225–2229, doi: 10.1016/j.cnsns.2010.04.058.

Olejnik, P. and Awrejcewicz, J. (2013). Modeling of time delays in numerical solution of an uncertain large-scale system, in J. Awrejcewicz, M. Kaźmierczak, P. Olejnik, and J. Mrozowski (eds.), *Dynamical Systems – Applications* (Lodz University of Technology Press, Lodz), pp. 515–526.

Olejnik, P. and Awrejcewicz, J. (2015). Time delays in numerical modeling of frontal thoracic blast pressure wave responses, *Int. J. Dyn. Contr.* **3**, 1, pp. 109–119.

Orson, J., Bagby, W., and Perram, G. (2003). Infrared signatures from bomb detonations, *Infrared Phys. Technol.* **44**, pp. 101–107.

Ostanin, A. (2009). *Optimization Methods in MATLAB. Laboratory* (NAKOM, Poznan), in Polish.

Otten, G., de Vries, T. J. A., van Amerongen, J., Rankers, A. M., and Gaal, E. W. (1997). Linear motor motion control using a learning feedforward controller, *IEEE/ASME Trans. Mechatr.* **2**, 3, pp. 179–187.

Pardalos, P., Tseveendorj, I., and Enkhbat, R. (eds.) (2003). *Optimization and Optimal Control, Series on Computers and Operations Research*, Vol. 1 (World Scientific).

Park, J. H. (2002). Robust decentralized stabilization of uncertain large-scale discrete-time system with delays, *J. Optimz. Theory App.* **113**, 1, pp. 105–119.

Pena, G. and Andrade-Filho Jde, S. (2010). Analogies in medicine: valuable for learning, reasoning, remembering and naming, *Adv Health Sci Educ Theory Pract* **15**, 4, pp.

609–619.

Pert, G. (2013). *Introductory Fluid Mechanics for Physicists and Mathematicians* (Wiley).

Petrella, R., Tursini, M., Peretti, L., and Zigliotto, M. (2007). Speed measurement algorithms for low-resolution incremental encoder equipped drives: A comparative analysis, in *ACEMP '07, International Aegean Conference on Electrical Machines and Power Electronics* (Bodrum), pp. 780–787.

Piefort, V. (2001). *Finite Element Modelling of Piezoelectric Active Structures*, Ph.D. thesis, Universite Libre de Bruxelles, Active Structures Laboratory, Department of Mechanical Engineering and Robotics.

Piersol, A. and Paez, T. (2010). *Harris' Shock and Vibration Handbook*, 6th edn. (McGraw-Hill).

Pietrabissa, R., Quaglini, V., and Villa, T. (2002). Experimental methods in testing of tissues and implants, *Meccanica* **37**, pp. 477–488.

Plank, G., Kleinberger, M., and Eppinger, R. (1994). Finite element modeling and analysis of thorax restrain system interaction, in *Proceedings of the 14th International Technical Conference on the Enhanced Safety of Vehicles* (Munich, Germany).

Prakosa, T., Wibowo, A., and Ilhamsyah, R. (2013). Optimizing static and dynamic stiffness of machine tools spindle shaft, for improving machining product quality, *Journal of KONES* **20**, 4, pp. 363–370.

Rabie, M. (2009). *Fluid Power Engineering* (McGraw-Hill, New York), ISBN 9780071626064.

Rosenberg, E. (1989). Optimal module sizing in VLSI floorplanning by nonlinear programming, *Methods and Models of Operations Research* **33**, 2, pp. 131–143.

Ruano, A. (ed.) (1999). *Intelligent Control Systems Using Computational Intelligence Techniques, Control Engineering Series*, Vol. 70 (IET).

Rutkowski, L., Korytkowski, M., Scherer, R., Tadeusiewicz, R., Zadeh, L., and Zurada, J. (eds.) (2015). *Artificial Intelligence and Soft Computing, 14th International Conference, ICAISC 2015*, Vol. 2 (Springer-Verlag, Zakopane, Poland).

Rydberg, K.-E. (2008). Hydraulic servo systems, Tech. rep., Linkopings Universitet, access: 2014-06-12.

Sadeghieh, A., Sazgar, H., Goodarzi, K., and Lucas, C. (2012). Identification and real-time position control of a servo-hydraulic rotary actuator by means of a neurobiologically motivated algorithm, *ISA Trans.* **51**, 1, pp. 208–219, doi: http://dx.doi.org/10.1016/j.isatra.2011.09.006.

Samal, M., Seshu, P., Parashar, S., von Wagner, U., Hagedorn, P., Dutta, B., and Kushwaha, H. (2006). Nonlinear behaviour of piezoceramics under weak electric fields, *Int. J. Solids Struct.* **43**, pp. 1422–1436.

Scherz, P. and Monk, S. (2013). *Practical Electronics for Inventors*, 3rd edn. (Mc Graw Hill).

Schmid, D. (ed.) (2002). *Mechatronics. Handbook for Students of Medium and Vocational Technical Schools* (REA), in Polish.

Schonecker, M. (2009). *Traveling Wave Ultrasonic Motors Based on the Piezoelectric Shear Effect*, Ph.D. thesis, Technischen Universitat Darmstadt.

Schwartz, M. (1972). *Principles of Electrodynamics* (Dover Publications, New York).

Sherwani, N. (1999). *Algorithms for VLSI design automation* (Kluwer Academic, Dordrecht).

Shi, Y. (2001). *Multiple Criteria and Multiple Constraint Levels Linear Programming.Concepts, Techniques and Applications* (World Scientific).

Shi, Y., Wu, P., Lloyd, D., and Li, D. (2013). Effect of rate sensitivity on necking behavior of a laminated tube under dynamic loading, *J. Appl. Mech.* **81**, 5, pp. 051010–051010, doi:10.1115/1.4025839.

Shu-hua, L. (1954). Origine de la boussole, *Aimant et Boussole* **11**, p. 175.

Siljak, D. (1978). *Large-Scale Dynamic Systems: Stability and Structure* (North Holland, Amsterdam, Holland).

Slotine, J. and Li, W. (1987). On the adaptive control of robot manipulators, *Int. J. Rob. Res.* **6**, 3, pp. 49–59.

Song, G., Cai, L., Wang, Y., and Longman, R. W. (1998). A sliding-mode based smooth adaptive robust controller for friction compensation, *Int. J. Robust Nonlinear Control* **8**, pp. 725–739.

Sreekumar, K. and Ramchandani, A. (2005). Launch stabilisation system for vertical launch of a missile, *Defence Sci. J.* **55**, 3, pp. 223–230.

Sterrett, S. (2006). Models of machines and models of phenomena, *International Studies in the Philosophy of Science* **20**, pp. 69–80.

Sugeno, M. and Kang, G. (1988). Structure identification of fuzzy model, *Fuzzy Sets and Systems* **28**, 1, pp. 15–33, doi:http://dx.doi.org/10.1016/0165-0114(88)90113-3.

Teeter, J., Chow, M., and Brickley, J. (1996). A novel fuzzy friction compensation approach to improve the per-formance of a DC motor control system, *IEEE Trans. Ind. Electron.* **43**, 1, pp. 113–120.

Uchida, K., Shimemura, E., Kubo, T., and Abe, N. (1988). The linear quadratic optimal control approach to feedback control design for systems with delay, *Automatica* **24**, pp. 773–780.

Ulitko, A. (1977). Theory of electromechanical energy conversion in nonuniformly deformable piezoceramics, *Int. Appl. Mech.* **13**, 10, pp. 115–123.

Velagić, J. and Galijasević, A. (2009). Design of fuzzy logic control of permanent magnet DC motor under real constraints and disturbances, in *IEEE Multi-Conference on Systems and Control* (Saint Petersburg), pp. 461–466.

Viswanath, S. and Nagarajan, R. (2002). Helicopter hydraulic system, in *ICAS 2002 Congress*.

Vittoria, C. (2011). *Magnetics, Dielectrics, and Wave Propagation with MATLAB Codes* (CRC Press / Taylor and Francis, Boca Raton).

von Wagner, U. (2003). *Nichtlineare Effekte bei Piezokeramiken unter schwachem elektrischen Feld: Experimentelle Untersuchung und Modellbildung*, Habilitationsschrift (GCA-Verlag, TU Darmstadt, Herdecke).

von Wagner, U. (2004). Nonlinear longitudinal vibrations of non-slender piezoceramic rods, *Int. J. Non Linear Mech.* **39**, 4, pp. 673–688.

von Wagner, U. and Hagedorn, P. (2002). *Piezo-beam systems subjected to weak electric field: Experiments and modeling of nonlinearities*, Vol. 256 (Journal of Sound and Vibration).

von Wagner, U., Hagedorn, P., and Nguyen, M. (2001). Nonlinear behavior of piezo-beam-systems subjected to weak electric field, in *Proceedings of ASME DETC 2001*, VIB 21488 (Pittsburgh).

Wadley, H., Dharmasena, K., He, M., McMeeking, R., Evans, A., Bui-Thanh, T., and Radovitzky, R. (2010). An active concept for limiting injuries caused by air blasts, *Int. J. Impact Eng.* **37**, 3, pp. 317–323.

Weinreb, L. (2005). *The Use of Analogy in Legal Argument* (Cambridge University Press).

Xie, M. (2003). *Fundamentals of Robotics. Linking Perception to Action, Series in Machine Perception and Artificial Intelligence*, Vol. 54 (World Scientific).

Yi, S., Nelson, P., and Ulsoy, A. (2010). *Time-Delay Systems. Analysis and Control Using the Lambert W Functions* (World Scientific Publishing, Singapore).

Younkin, G. (2002). *Industrial Servo Control Systems Fundamentals And Applications*,

2nd edn. (CRC Press, New York).

Zadeh, L. (1989). Knowledge representation in fuzzy logic, *IEEE Trans. Knowl. Data Eng.* **1**, 1, pp. 89–101.

Zhu, Z. H. (2009). Modelling of servo hydraulic control of recovery assist cable system for shipborne helicopter, in *3rd International Conference on Integrity, Reliabilityand Failure.*

Zierep, J. (1978). *Similarity Laws and Principles of Modeling in Mechanics of Fluids* (PWN, Warsaw).

Index